Human Geography of the UK

This new core textbook for introductory courses in human geography provides first- and second-year undergraduates with a comprehensive thematic approach to the changing human geography of the UK at the end of the twentieth century and the beginning of the twenty-first. It brings to life the complex processes shaping the UK and the uneven geographical landscapes they have produced. The concern is not only with broad regional differences, but also with more complex patterns of geographical inequality – within regions and cities, and between town and country. The focus is on the 1990s and the dynamic forces that have moulded the geography of the UK since the tumultuous years of Conservative rule. The upheavals of the 1980s, which so shocked geographers two decades ago, are put into context, and the contemporary condition of the UK is set within a longer period of historical and geographical change. Economic, social, political and cultural forces are given equal attention together with positioning issues like gender and ethnicity, all central to the social and economic transformations of recent years.

Especially designed to be user-friendly for students, and written by a team of authors who are actively engaged in researching the contemporary UK, it explores issues of immediate relevance to students such as crime, unemployment, social exclusion and AIDS. It will interest teachers and students studying the contemporary UK in geography departments in Britain and beyond.

Irene Hardill is Professor of Economic Geography at Nottingham Trent University. **David T. Graham** is Senior Lecturer in Geography at Nottingham Trent University. **Eleonore Kofman** is Professor of Human Geography at Nottingham Trent University.

Human Geography of the UK

An introduction

**IRENE HARDILL,
DAVID T. GRAHAM
AND ELEONORE KOFMAN**

London and New York

First published 2001
by Routledge
11 New Fetter Lane, London EC4P 4EE

Simultaneously published in the USA and Canada
by Routledge
29 West 35th Street, New York, NY 10001

Routledge is an imprint of the Taylor & Francis Group

Typeset in Palatino by
Keystroke, Jacaranda Lodge, Wolverhampton
Printed and bound in Great Britain by
Bell & Bain Ltd, Glasgow

British Library Cataloguing in Publication Data
A catalogue record for this book is available from the
British Library

Library of Congress Cataloging in Publication Data
Hardill, Irene, 1951–
    Human geography of the United Kingdom: an
introduction / Irene Hardill, David T. Graham, and
Eleonore Kofman.
       p. cm.
    Includes bibliographical references and index.
    1. Human geography—Great Britain. 2. Great
Britain—Social conditions.   I. Graham, David T.,
1953–   II. Kofman, Eleonore.   III. Title.

GF551 .H27 2001
941.086—dc21   00–053332

ISBN 0–415–21425–4 (hbk)
ISBN 0–415–21426–2 (pbk)

For our parents

Contents

Figures

Tables

Boxes

Acknowledgements

The authors would like to thank Pete Shirlow of the University of Ulster for his significant contribution of material on various aspects of Northern Ireland.

We are extremely grateful to the following people for advice, guidance, encouragement and help: Mark Baker, Chris Bellamy, John Alexander Burnett, Tony Champion, Andrew Crowhurst, Mike Danson, Linda Dawes, Maureen Docherty, Ranji Devadason, Danny Dorling, Claire Dowsing, Janet Elkington, Pete George, Jeff Healy, Helen Lawton Smith, Ruth Lupton, Linda McDowell, Andrew Mould, Ines Newman, Alistair Owens, Monder Ram, Al Stewart, Moira Taylor, Don van Vliet and Cecilia Wong. We would like to extend special thanks to Olwyn Ince for the tireless, cheerful and efficient manner in which she hunted down information.

We are also grateful to the following for permission to reproduce copyright material: *This England*, the Child Poverty Action Group, *Private Eye* and cartoonists Philip Berkin, John Cooper, Ken Pyne and Robert Thompson, *The Independent* Syndication and Cherry Norton (text), John Voos and Chris Jones (photographs) for article 'Workaholic Britain', 27 July 1999, *The Guardian* and Andrew Testa and Gary Weaser for Reclaim the Streets and the Bluewater Shopping Mall, Kent photos, *The Sunday Telegraph Review* and Mark Pinder and Alexander Bratell for Unwanted Housing in Benwell and Desirable Property in Fulham photos, photographer Darren Regnier for the cover photo, Popperfoto and photographer Michael Crabtree for the Countryside March photo.

Abbreviations

AIDS	Acquired immune deficiency syndrome
BSE	Bovine spongiform encephalopathy
CAP	Common Agricultural Policy (EU)
DETR	Department of the Environment, Transport and the Regions
DfEE	Department for Education and Employment
DoE	Department of the Environment
EEC	European Economic Community
EOC	Equal Opportunities Commission
EU	European Union
GDP	Gross domestic product
GLA	Greater London Authority
GLC	Greater London Council
GOR	Government Office Region
GRO(S)	General Register Office (Scotland)
HIV	Human immunodeficiency virus
ICT	Information and communications technology
MAFF	Ministry of Agriculture, Fisheries and Food
MEP	Member of the European Parliament
MP	Member of Parliament
MSP	Member of the Scottish Parliament
NHS	National Health Service
nvCJD	New variant Creutzfeldt–Jakob disease
ONS	Office for National Statistics
OPCS	Office for Population Censuses and Surveys
PFIs	private finance initiatives
PHLS	Public Health Laboratory Service (England and Wales)
PR	proportional representation
QUANGO	Quasi-autonomous non-governmental organisation
RDAs	Regional Development Agencies
RDC	Rural Development Commission
SCC	Scottish Constitutional Convention
SEU	Social Exclusion Unit
SRB	Single Regeneration Budget
SNP	Scottish National Party
SSR	Standard Statistical Region
TB	Tuberculosis
UA	Unitary Authority

PART  THE HUMAN GEOGRAPHY OF THE UK

Some issues

APPROACHES TO THE HUMAN GEOGRAPHY OF THE UK

- **Introduction**

- **The human geography of the UK: some problems**

- **Approaches to the human geography of the UK**

1.1 INTRODUCTION

A number of authors have recently published works that critically reflect on the United Kingdom, and in particular on what it is to be British (Davies 1999; Grant and Stringer 1995; Marr 2000). Other writers are exploring issues of national **identity** (*see Glossary) for those countries within the 'Union' – England (Barnes 1998; Davey 1999; Jones 1998), Scotland (Devine 1999a; Nairn 2000), Wales (Bowie 1993) and Ireland (Graham 1997). The United Kingdom will be 300 years old in 2007, and in 2001 it will be 200 years since the United Kingdom of Great Britain and Ireland was established when the Irish Parliament ceased and power was transferred to Westminster, an event which will deliberately go unmarked by Tony Blair's government. This is in stark contrast to the recent Millennium celebrations, or the way in which the government led the fiftieth anniversary celebrations marking the arrival of the *Empire Windrush* landing Jamaican immigrants in London in 1948, an event which started the wave of post-war non-white immigration leading to the present 'new' multiculturalism.

In 2001, the United Kingdom is a very different place to that of 1901 or 1801. In writing this book we want you critically to examine what the UK is today, and what it means to be British at this critical period in history. We focus on the geography of the 1990s and the likely changes in the first few years of this new decade and millennium. The shift in the UK's position in the world, its increasing economic and social integration into the European sphere, and the effect of a deregulated economy have produced changes in the geography of the UK. The introduction of market principles and private sector into the state have produced new forms of governance combining in different ways the public and private. New work and social practices, a continuing increase in female economic activity rates and changes in family structures, have made British society more diversified and polarised. Though geographically uneven, new modes of **consumption** have substantially altered everyday life. So too have post-war dimensions of the UK population resulted in new social and cultural forms, especially in the metropolitan centres. Rural areas too are far more economically and socially diversified than many stereotypical images would suggest.

The South East of England continues to attract the largest number of both international migrants and services, while some of the poorer regions, such as the North East of England, remain buffeted by the uncertainties of inward investment. At the same time, the minority countries with nationalist movements (Scotland and Wales) now have devolved power, though to varying degrees, but the status of the English regions has yet to be fully resolved. However, the attempt to resolve the formerly intractable 'Irish Question' has been one of the major developments of the last few years. These are most of the key transformations that have modified the human geography of the UK in the past decade and will continue to do so at the beginning of this new century.

The four of us involved in writing this book are all British citizens, though for one it has been an acquired status. However, being a British citizen means different things to us, partly because of our different histories and relationship to the UK. **Citizenship**, as we shall see, may connote for some little more than legal membership of a state; for others it represents in addition a deep attachment to a national community or a regional identity. We each now reflect on our own identity and positionality, which allows for the incorporation of difference and acknowledges the partiality and situatedness of our knowledges. We also invite you, if you are a British citizen, to reflect what on what 'being British' means to you, and to reflect on the historical ties that bind British citizens together. We are thus confronting each of you to think deeply about the United Kingdom as we are all participants in the historical/political process, though some are more passive than others.

Irene Hardill: When I am asked where I come from I always reply that I am a British citizen but also that I am English. While I do have a strong sense of being English, my Englishness stems from being from Yorkshire. I have a very strong sense of identity, of a rootedness to place, a small mill town in west Yorkshire, even though I have spent more years living outside the county than in it. My strong sense of identity is tied up with my family, and their class position in a small mill town. This position was derived from the fact that my great-grandfather established an iron foundry in 1874. This small town and the county of Yorkshire are the foci of my 'sense of belonging', and I thus have a strong regional identity.

David Graham: Born and raised in Dundee, Scotland I have spent most of my life outside that city and much of my life outside that country – including spells in Canada, the USA, Mexico, Spain, Northern Ireland and England. My sense of identity stems from the civic, rather than ethnic, identity that is the basis of the new post-devolution Scottish identity. It is informed by a working-class upbringing, a first-class education and extensive national and international travel. As someone who has lived in most of the constituent countries of the UK I realise just how complex and heterogeneous this small archipelago is in comparison to other states.

Eleonore Kofman: In the past century each generation in my family has migrated internationally. I came as a postgraduate student to the UK from Australia. However, I am not a patrial; that is, someone with a recent British heritage who has automatic rights of settlement. My relationship to the UK as a citizen is a mixture of legal status coupled with a lengthy residence so that I share the discussions of what it is to live in the UK. I consider myself to be cosmopolitan with a strong attachment to and knowledge of several places. I was brought up in Sydney, have lived for several decades in London, while also feeling at home in Paris where I have spent much time.

Pete Shirlow: When I first went to live in America and then England virtually everyone treated me as Irish. I was called 'Paddy' and endured stereotyped ideas of Irish identity. This was not simply upsetting because of the racism but also because I don't consider myself to be Irish. However, this does not mean that I am simply British. Although I

grew up in a Protestant community and was taught the glories of the British Empire and the perceived benefits of Britishness, much of this identity dimension has been lost in recent years. In many ways I have a mixed Irish and British identity, but not one simply based upon nationalism. Instead I have chosen an identity which celebrates the uniqueness of British and Irish identity in Northern Ireland, but which is post-nationalist. For me, my wife, kids, friends, music and politics make up my identity. An identity based upon real issues as opposed to cultural imaginings and flag waving. At the end of the day you can't eat a flag so what use is it!

In the remaining part of this introductory section of this book we explore a number of themes beginning with a review of approaches to the human geography of the United Kingdom, followed by a chapter exploring the UK in a global context. Chapter 3 examines the processes of socioeconomic change, while Chapter 4 highlights the impact of economic and social changes since the 1970s that are creating a more socially polarised society, although a number of axes of polarisation are examined. Chapter 5 focuses on cultural change and is followed by a chapter on political change.

1.2 THE HUMAN GEOGRAPHY OF THE UK: SOME PROBLEMS

Throughout this book we emphasise that the UK is not a nation-state, rather a multinational state made up of four separate nations which have been profoundly shaped by international migration flows over several centuries. The history and traditions of these four distinct units means that any study of the UK becomes much more complex than a more unitary state like France, or even a federal state like Australia or the USA. Data are rarely collected on a UK-wide basis. For example, the Registrar-General for Scotland (RGS) and Registrar-General for Northern Ireland (RGNI) collect and produce demographic data for their respective countries, while the Office for National Statistics (ONS) does the same for England and Wales. The creation of the Office for National Statistics might have been seen as a godsend to those studying the UK, but it is not entirely clear which nation is involved – England, England and Wales, Great Britain, the UK?

The Census of Population – the biggest and most comprehensive source of socioeconomic and demographic data – in Northern Ireland is entirely separate from that in Great Britain. Up until relatively recently the Census arrangements in England and Wales were entirely separate from those of Scotland. Recently there has been closer collaboration between ONS and RGS. Even the questions asked vary. In Northern Ireland there has long been a question on religion – no such question is asked in Scotland, England or Wales. In Northern Ireland, Scotland and Wales, questions are asked on language (see Chapter 11). The questions on ethnic origin, asked in Scotland, England and Wales, do not apply in Northern Ireland. Housing questions are different in the Scottish Census, and there is a Survey of English Housing and a separate Scottish House Condition Survey. Even in a publication that does give UK-wide coverage, for example *Regional Trends*, the writers are at some pains to point out that crime data are not comparable between Northern Ireland, Scotland, and England and Wales. This is because of the entirely different legal systems in England and Wales and Scotland and the special conditions in Northern Ireland. Similarly with education, there are different systems in Northern Ireland, Scotland, and England and Wales. Further, as noted in Chapter 12 and elsewhere, governmental structures and policies vary between the different nations – the 1967 Abortion Act, for example, does not apply in Northern Ireland.

The complexities of functional and territorial administrative responsibilities in the UK can be illustrated by the Forestry Commission. There are four ministers responsible for forestry in the UK – the agriculture ministers for England, Wales, Scotland and Northern Ireland. Northern Ireland has its own Forestry Service. For Great Britain the Forestry Commission, which has its own civil service, acts as both regulatory body for forestry and the major forester in Britain. The three Great Britain ministers are responsible for the Forestry Commission, but The Scottish Office is the 'lead department', reflecting both the Commission's headquarters location in Edinburgh and the high proportions of forests in Scotland. Internally, the Forestry Commission is divided into the Forestry Authority, which deals with the regulatory and advice side, and Forest Enterprise, which manages the Commission's own forestry activities. The Forest Authority and Forest Enterprise have completely different geographical structures at both first and second levels.

(Hogwood 1996: 6)

The visitor to the UK is often bewildered by these anomalies rarely found in other countries. What does the visitor from the USA, where a dollar is the same design throughout the fifty states, make of a country where there are seven or eight versions of the five-pound note – that of the Bank of England, the Bank of Scotland, the Clydesdale Bank, the Royal Bank of Scotland, the Northern Bank, Ulster Bank and so on? It is for these reasons that many writers concentrate their analyses on specific components of the UK – usually England and Wales, which have been linked together for the longest period. We have attempted to be as inclusive as possible. Nevertheless there are some discussions within this book that concentrate on specific parts of the UK. Where this is done is clearly identified.

1.3 APPROACHES TO THE HUMAN GEOGRAPHY OF THE UK

Since Mackinder's classic study *Britain and the British Seas*, published in 1902, a number of books have explored the changing geography of the UK. Each book is a product of its time, reflecting in part the deterministic legacy of geography, and has addressed the geographical issues and debates of the day. The scope of these books has differed in two key respects: first, in their area of coverage; second, in their geographical approach. With reference to the area of coverage this in some way has been shaped by the political history of the islands that make up the British Isles.

These islands encompass both the Channel Islands and the Isle of Man, which have maintained their own separate (from Westminster) system of government. The largest set of islands is Great Britain, comprising England, Wales and Scotland which have been joined under one monarch since 1603. For the period 1801–1920, Ireland was ruled directly from Westminster. Thus, for books printed before 1920, the British Isles was the United Kingdom of Great Britain and Ireland (and the Channel Islands, the Isles of Scilly and the Isle of Man). Since 1920, Ireland has been divided into Northern Ireland and Eire (the Republic of Ireland). Some texts produced in the period of post-Irish partition explored the British Isles (Demangeon 1927; Stamp and Beaver 1933; Watson and Sissons 1964); others the UK (Gardiner and Matthews 2000; Mohan 1999) and some Great Britain (Mitchell 1962).

The second major difference is in their approach to the study of geography. Some adopted a **regional** approach, others a **systematic** approach, while others have combined the two approaches, as indeed we do in this book. A regional approach is essentially a Vidalienne approach (Vidal de la Blache 1922) which incorporated nature as a dynamic element in human geography. His approach was concerned with the *milieu*, the basic differentiation of the earth's surface; *genres de vie*, the lifestyles of a particular region, reflecting the economic, social, ideological and physiological identities imprinted on landscapes; and *circulation*, the disruptive process by which human contact and progress took place between regions (Unwin 1992).

A regional approach attempts to give a portrait of a region. Until the mid-1970s, regional geography was seen as a central component of mainstream geographical scholarship, combining aspects of physical and human geography. This is reflected in the literature. For example, Mackinder's (1902) classic, *Britain and the British Seas*, was part of a *Regions of the World* series. In the *Provinces of England*, originally published in 1919, Fawcett (1960) proposes a number of natural regions including Wessex, Peakdon, Trent and Bristol. A typical text from the interwar period, *Great Britain: Essays in Regional Geography*, looks at twenty-three regions of Britain (Ogilvie 1928). The series prepared for the British Association (e.g. see Carter 1962; Edwards 1966; Jones 1968; Miller and Tivy 1958), concentrated on regional issues. The *Regions of the British Isles* series, published in the 1960s and edited by W.G. East, included volumes on the *North of England*, *East Midlands and the Peak*, *Wessex*, the *Bristol Region*, among others. A strong regional approach is also evident in school texts, for example *A Geography of Britain* (Tolson and Johnstone 1970). A good example of the regional approach is illustrated by the 1962 text edited by Jean Mitchell, which was a thematic study of the regions of the islands of Great Britain. Each author selected a theme or themes to bring out the essential character of the area as s/he saw it. The book avoided an analysis of regional frontiers, since 'on the ground they rarely exist' (Mitchell 1962: xi). The conceptualisation of a region is intimately bound up with the wider debate about the conceptualisation of space and place (see Chapters 13 and 14). One groundbreaking book, *Regional Variations in Britain*, combined the regional approach with the techniques of the quantitative revolution in geography, though, despite its title, it included Northern Ireland in the analytical framework (Coates and Rawstron 1971). Another regional approach was adopted by the contributors to *Regional Problems, Problem Regions and Public Policy in the United Kingdom* (Damesick and Wood 1987). In this work the contributions do reflect the title. Another important book of the period, *The North–South Divide: Regional Change in Britain in the 1980s*, extended the regional approach to look at the macro-geographical divide (Lewis and Townsend 1989). Although most of the contributions look at the three countries of Great Britain, the ones on education and health concentrate on England and Wales. Recent work on the region, notably by Doreen Massey (1984), embodies a strongly relational approach to thinking about space and place (see also Allen *et al.*

1998), which we embrace in some chapters in this book.

Stamp and Beaver (1933), on the other hand, adopted the systematic approach; this consisted of an examination of the natural environment and resources of the British Isles as well as the use made of them. They also examined the consequences of past exploitation and the capital accumulated. The systematic surveys of the sectors of the economy reflect the UK economy of the time, of the UK as the 'workshop of the world' and an economy dominated by primary and manufacturing industry (see also House 1978; Watson and Sissons 1964). This type of approach was continued by Hudson and Williams (1986) and Champion and Townsend (1990).

Since the election of the Labour administration in 1997, two other UK texts have been produced (Gardiner and Matthews 2000; Mohan 1999). These are illustrative of the way in which geography itself has changed radically with, for example, social polarisation and gender discussed. Many of the earlier texts, whether systematic or regional, would have embraced both the physical and human geography of the UK – about the relief and the resources of the landscape, as well as the socio-economic landscape.

In this book, rather like Mackinder's (1902) *Britain and the British Seas* and Demangeon's (1927) volume on *Les Isles Britanniques* in the series *Géographie Universelle*, we deliberately blur the regional and systematic approaches. But the structure and content of this book is very different to the earlier texts and reflects the changing nature of geography and the socioeconomic landscape of the UK. We are writing about the UK at the dawn of a new century, of a UK of economic and social divisions, of a postindustrial landscape and postcolonial cultural diversity. In Part 1, the chapters present a broad historical survey of geographical, socio-economic, cultural and political changes. As we aim to look holistically at the socioeconomic issues shaping the UK space and explore their wider impacts, we recognise the interconnectedness of spaces and places. This is reflected in the content of the substantive chapters of the book in Parts 2 and 3. These chapters each have a clear introduction and summary, with important theoretical principles explained in order to focus attention on the important issues in the subject area. Up-to-date further readings are given at the end of each chapter. We also include case studies, examples and revision questions.

Part 2 explores the theme of 'The UK in a period of change', through six chapters. Chapter 7 explores the complex issues of the demography of the UK. In Chapter 8 we look at the changing nature of employment and how this has impacted on the human geography of different parts of the UK. Patterns and trends in consumption and **leisure** make up the content of Chapter 9. Chapter 10 explores the related issues of health and well-being within the UK. The complex topics of culture and identity are discussed at some length in Chapter 11. Chapter 12 looks at the policy responses to a number of topical and often contentious issues. In Part 3, four chapters look at 'The UK: a society and state divided?' The linked themes of constitutional and political change are examined in Chapter 13, with particular attention paid to devolution and reform. In Chapter 14 we explore the geographical divisions in the UK in some detail by comparing the situation in the North East with that of the South East, treating London separately from the South East. Chapter 15 again explores contrasting circumstances by looking at the haves and have-nots in society. In the final chapter the geography of polarisation and division is discussed in terms of rural and urban areas.

THE UK IN A GLOBAL CONTEXT

2.1 INTRODUCTION

In this chapter we trace the changing and progressively circumscribed role of the UK in the world in the twentieth century. In particular we examine economic and political aspects, leaving a more detailed discussion of the internationalisation of culture to Chapter 5. The first section briefly traces the rise and fall of the UK's position in the world before turning, in the second and third sections, to its new relationships with Europe and a deepening and more intensively globalised world in the late twentieth century. The UK has, of course, been a major player in the internationalisation of economic and cultural processes as well as being subjected to these processes.

During earlier centuries, especially in the nineteenth century, the influence of the UK had stretched out to many parts of the world. Sir Halford Mackinder, appointed Reader in 1887 at what would be the first Department of Geography at the University of Oxford, and who later became a politician, analysed the implications of the closure of the world after four centuries of discoveries and territorial expansion by Western powers (Kearns 1993; Taylor and Flint 2000). This international vision, underpinned by the interests of the domestic power bloc consisting of an industrial bourgeoisie and a commercial aristocracy (Nairn 1977), has left a lasting influence to the present day. Thus the UK's contemporary understanding of its position as a player in the global system, and its vacillation over its relationship to the European Union, owes much to its earlier role as a hegemonic power and the global interests it developed.

2.2 THE RISE AND FALL OF THE UK

The UK reached the peak of its economic dominance in the mid-nineteenth century (Wallerstein 1983). It had emerged hegemonic from the competition with France and the Netherlands in the eighteenth century, leaving it free to pursue its colonial expansion. Its global penetration resulted from a complex pattern of settlement, direct rule and indirect administrative and political influence bolstered by economic power. Firstly, British emigrants settled lands where they displaced and almost eradicated indigenous populations. Its first colony was Ireland, followed by territories in North America and the Caribbean. The proceeds from the lucrative slave trade fed back into the British economy and enriched many port cities, such as Bristol and Liverpool, and country estates. The development of commodity markets underpinned the commercial success of London. Secondly, from the eighteenth century the British colonised the Indian subcontinent, which was its most populous and significant acquisition, and in the nineteenth century helped to carve up Africa. Thirdly, it exerted influence in the Middle East, often through indirect rule, and was a major force in the economic development of Latin America. Through these means it occupied territories, generated trade and spread its cultural influence, especially through the English language.

By the middle to late nineteenth century, the Empire was being brought into the midst of the metropolitan heartland. European ability to transport goods and peoples more rapidly and over longer distances led to a series of world exhibitions and fairs (Greenhalgh 1988). Rival powers used them to demonstrate their technological prowess and cultural superiority by collecting objects, customs and peoples from all over the world. The British were more concerned with showing off their technical skills – for example, the construction of the Crystal Palace in 1851. French exhibitions, on the other hand, were noted for their assemblage of cultural artefacts and peoples. Much booty, too, was plundered from weak empires or colonial possessions, which were thought not to be capable of appreciating what they owned and these supplied collections for newly created museums, such as the British Museum. This process of collecting has led to contemporary disputes over the ownership of artefacts. For example, the 'Elgin Marbles', which were bought and taken back to England from Greece by Lord Elgin in the first decade of the nineteenth century, are now claimed by Greece.

By the 1870s, the UK was being challenged by

Germany and the USA. Germany, now unified under Bismarck, was becoming a major industrial producer. The United States benefited from technological advances, such as the preservation of foodstuffs and transportation, thus presenting serious problems for European agricultural producers. The UK, unlike other continental states, pursued a policy of free trade. The conflagration of the First World War and the demands it put on Britain's economy finally revealed its weakening economy and the end of its hegemonic rule. It also demonstrated the upsurge of the USA, which had entered the war against Germany in 1917. The UK's heavy industrial base, such as the coal industry in South Wales, was badly hit in the 1920s and the Depression brought about the collapse of many economies in 1929. The 1930s heralded a period of protectionism, including in the UK. However, the UK, with its far-flung Empire, still retained the sense of being a world power.

The UK, though victorious, was effectively bankrupted at the end of the Second World War, which witnessed the emergence of the USA as the undisputed economic and military power, and it now began establishing an international regime based on the liberalisation of trade and commerce. At the Bretton Woods meeting in 1944, the USA laid out an institutional framework based on the International Monetary Fund and the World Bank. It also set about aiding Western European states through the Marshall Plan to build **Keynesian** welfare states (Overbeek 2000). The UK, especially under Ernest Bevin, the Labour Foreign Secretary and the architect of the Western post-war alliance, saw its future lying in a special Anglo-American relationship (Taylor 1990). Some have argued that this is one of the key reasons for the withdrawal from its colonies and military obligations.

The UK had ended the Second World War with its Empire intact, but one of the trade-offs for economic help from the USA and its support for a world role for Britain was easier commercial access to its colonial markets. The Empire, however, began to collapse after the war. The UK first granted independence to India, its most important colony, in 1947. African colonies too developed strong nationalist movements and demanded independence. In the 1950s, there began a process of evaluating the effectiveness of its military commitments that were increasingly relinquished to the USA. In 1956, **Suez** was only another, though significant, step in this military pulling back. In the late 1950s, it realised it could not sustain such an extensive military presence East of Suez and in the Middle East, as well as boosting domestic standards of living. The announcement of a full withdrawal from Africa came in 1960 with Prime Minister Harold Macmillan's speech on the 'winds of change blowing through Africa'. Soon afterwards, South Africa was expelled from the Commonwealth and this played a big part in the first application in 1961 to join the European Economic Community (EEC). The scaling down of its defence forces in 1968 further reflected its diminished capacity and appetite for a world reach.

In the early 1960s, after a period of post-war openness which corresponded to its continuing international vision of itself, the UK began to restrict the entry of its former colonial subjects. The vision of the colonies and white dominions after 1948, now called a Commonwealth, was that it was a group of multiracial societies with a historic link to Britain. The 1948 Nationality Act created the status of a citizen of the UK and Colonies, but with immigration to the UK it made it difficult to sustain the idea that a British identity was an exclusively white identity.

However, immigration legislation to control the number and source of incomers predates 1948. The 1905 Aliens Act constituted the first of a series of immigration acts in European states in the twentieth century. The Act was a response to racist sentiments and fear of economic competition against Russian and Eastern European Jews whose entry increased markedly from the 1880s. Until the passing of the 1962 Commonwealth Immigration Act, which put an end to mass immigration (from which those from the Republic of Ireland were exempt), students, political exiles and workers and their families entered relatively freely. Unlike many other European countries, permanent settlement and family immigration were allowed. The main form of contract labour was provided by the European Voluntary Scheme in the late 1940s to make good labour shortages in industrial sectors. The 1962 and subsequent Acts (1965, 1968, 1971) forced those seeking entry to obtain a limited number of employment vouchers. The 1971 Act gave patrials – those with a grandparent from the UK – special rights of settlement, and acknowledged the patently racist character of immigration legislation (see Chapter 7). Retrenchment in immigration flows partly stemmed from domestic racial tensions in a number of urban centres from the late 1950s. The politicisation of this issue was exemplified by the 1964 election in Smethwick, when the Labour candidate was defeated, and the vitriolic speeches of Enoch Powell warned against the dangers of the dilution of British identity.

The substantial scaling down of its defence forces in 1968, the closure against immigration well before other European countries, and its extremely sluggish economic performance were all contributory factors to creating a more insular UK whose role on the world stage was much diminished. This situation would be the prelude to a reorientation towards Europe.

2.3 EUROPEAN INTEGRATION AND UK RESPONSES

Of all the countries in the EEC, the UK had the most international perspective; it stood as offshore islands with special ties to the USA. This has undoubtedly shaped its attitude to joining the EC and the continuing divisions of how far it would be incorporated into pan-European institutions (Fay and Meehan 2000).

The first steps in the formation of the EEC came with the European Coal and Steel Community in 1951 (Box 2.1). The Community was launched by six countries in 1958 but without Britain, which had misjudged the will for cooperation. By 1961, its attitude to entry had changed as it rapidly decolonised. Its applications in 1962 and 1966 were vetoed by de Gaulle, France's President, who with some

justification saw Britain as largely supporting the USA. However, the departure of de Gaulle in 1969 and the reduction of the UK's overseas commitments helped ease entry. The UK's application for entry was eventually accepted in 1972 and it thus entered the EEC in 1973. The UK actually had very little trade with the Community, and by the early 1970s its companies had largely globalised. Its economic growth rates lagged far behind other states so that it entered as one of the poorest countries, yet, due to the workings of the Common Agricultural Policy, was one of the major contributors. It was thus not an auspicious time to be entering in tandem with a global downturn.

Budgetary issues were contentious right from the beginning. The Common Agricultural Policy, which had been designed for the interests of the original states some two decades before, swallowed up most of the budget. Instead of letting agricultural prices find their own levels, farmers' incomes were protected. Edward Heath, the Prime Minister at the time of entry, did bring about the implementation of European Regional Development Funds as partial compensation and later Harold Wilson attempted to renegotiate the unfavourable terms before holding a referendum. Other policies too made some impact. Equal opportunities legislation, enacted in 1975 and based on a fundamental article (119) in the Treaty, affirmed equality of pay and treatment for men and women. As we shall see in Chapter 8, though this helped within a few years to bridge some of the gap, women's earnings are still well below men's.

In the mid-1980s, in the context of deepening competition within the world political economy, discussions were held about directions the EEC should take. A social orientation was sidelined by a more neoliberal approach due to an increasing preoccupation with the need to be able to compete with the other two major trading blocs – the USA and Japan. This was accompanied by a measure of liberalisation in many European states, such as deregulation of working conditions and the encouragement of part-time and flexible contracts. The UK, of course, had been in advance of other European countries in pushing these kinds of working practices. It had also radically deregulated financial

Box 2.1 The UK and the European Community/ Union: key dates

1951	Creation of European Coal and Steel Community
1958	Launch of European Economic Community (France, Germany, Italy, Belgium, Luxembourg, Netherlands)
1961	First unsuccessful UK Application
1967	Second unsuccessful UK Application
1973	Entry of UK along with Ireland and Denmark
1974	Setting up of European Regional Development Fund
1980	First step towards European Monetary Union
1981	Entry of Greece
1986	Entry of Portugal and Spain
1987	Single European Act implemented
1985–90	Schengen Agreement and Convention
1990	UK enters Exchange Rate Mechanism from which it exits in 1992
1992	Maastricht Treaty of Europe, including rights of European citizens, signed
1996	Implementation of Schengen Accords with separate protocols for the UK, Ireland and Denmark
1997	Treaty of Amsterdam signed

markets. This ensured that the City of London maintained its world role and continued to attract large amounts of international capital. The Single Market Agreement, signed in 1986 and intended for completion by 1992, was supported by Prime Minister Thatcher in the belief that it was liberalising Europe. It was a framework driven by the largest industrialists who were global players.

Mrs Thatcher, however, opposed several key measures of further harmonisation, such as the European monetary union and the social chapter. The first was on grounds of loss of national sovereignty and the desire not to displace the US dollar as the single most important currency; the second that it would make labour less flexible and competitive with overseas suppliers. Her speech in Bruges in 1988 firmly rejected any federal union for Europe. The government also refused to participate in the Schengen Agreement concerning open movement across the borders of member countries that, it deemed, would breach British sovereignty. Schengen finally came into operation between nine countries of the European Union in 1997.

Until the end of the 1980s, debates on the orientation of the EEC revolved around the French view of a strategic regional organisation to regain economic autonomy and a British view that saw regional integration as no more than a contribution to the expansion of global capitalism (George 1996). From the 1990s and the reunification of Germany, it was Kohl, the German Chancellor, who moved to the fore of heightened integration. This was to fuel divisions within the Conservative Party, which was content with a common market but no more, and with many increasingly suspicious of a European Union (EU) dominated by German regulatory mechanisms and interests. The UK signed the Maastricht Treaty in 1992 but opted out of the social chapter. A significant aspect of the Treaty was that it codified the movement of EU citizens (defined as citizens of a member country) and their associated social and political rights. The new Labour government, elected in 1997, opted into the social dimension but has stayed out of the single currency which brought together eleven countries and came into effect in 1999. Other EU member states have subsequently joined 'Euroland'. Although less confrontational and desirous of showing itself to be a European player, the Blair government has not radically altered the nature of the UK's incorporation and hesitates over whether to join the single currency. The Conservative leadership has adopted a stridently oppositional stance to eventual acceptance of the euro and has even floated the idea of leaving the EU.

We can say that British governments, of whatever political hue, have adopted liberal positions in their approach to international political economy (Taylor 1990) and the worldwide reach of British influence; they have been wary of the regionalist and protectionist perspective of the EEC (George 1996). All British governments, with the exception of Edward Heath's (1970–4), have been Atlanticist to some degree, believing in the role of the USA as the rightful leader of the world system. With Thatcher, the rapport would become even stronger given the ideological similarity with President Reagan. The UK's vacillation over Europe has in effect reflected its indecision over its global position (Bonefeld 1999). It should be remembered that the UK has more multinational companies than any other country apart from the USA (see Chapter 3). This partly explains the paradox of why Britain, as the most neoliberal country in the EU, would seek to resist incorporation into the full European monetary union, which in itself is about tight monetary policy, deregulation and scaling down of the **welfare state** (see Chapter 6). Ironically, Britain would have easily qualified to enter the single currency in 1999 but chose not to do so for political reasons.

There is considerable debate about the nature of the state and its ability to maintain national sovereignty, especially in the face of the powerful forces of globalisation (see below). We would argue that although in flux, the state has not withered away and has in certain areas, such as immigration, reasserted its sovereignty. States have given up their power to the EU in some fields more than in others (Cerny 1990). The EU itself has adopted state-like powers and this has created a new geometry of state powers. The EU imposes guidelines and constraints

on what states can do, although the UK in particular has until recently refused to be a signatory to several conventions. Its role, however, is far more regulatory than directly interventionist or redistributional. Walby (1999) argues that its regulatory emphasis should not be equated as being without influence on policies and can be seen in areas such as equal opportunities and environmental policy. The EU has also passed several policies that have implications for UK social legislation, such as working hours, rights of mobility and attendant social rights for workers and their families enacted in legislation on European citizenship and human rights. In recent years an increasing number of cases have gone to the European Court of Human Rights in Strasbourg, which was created by the Council of Europe. The incorporation of the European Convention on Human Rights into British legislation as a Bill of Rights as of 2 October 2000 has already begun to have an impact on British legal rulings in areas of social legislation, such as gay rights. Bringing the Convention directly into the British constitutional system will make it much easier and less costly for cases to be heard.

2.4 GLOBALISATION AND THE UK

In the earlier period, when the UK commanded the world's stage, international connections, however extensive, tended to take place between nation-states. Hirst and Thompson (1996) have argued that the high era of commerce and trade between states occurred at the end of the nineteenth century. In the present period it has been argued that the scale of operation has shifted such that states are no longer the sole primary actors and that linkages between states have deepened and intensified in a number of fields (Dicken *et al.* 1997). The term 'globalisation' may thus be understood as referring to the processes, procedures and technologies – economic, cultural and political – underpinning the current 'time–space' compression or shrinking of the world.

Globalisation as a concept refers both to the compression of the world and the intensification of consciousness of the world as a whole. This development produces a sense of immediacy and simultaneity about the world. The processes and actions to which the concept of globalisation now refers have been proceeding, with some interruptions, for many centuries, but our consciousness of it is relatively recent. It has a material side in that the intensity, interaction and extension of many global processes are better organised and more extensive. Some of the major multinationals, especially those in resources and manufacturing, have a long history, for example BP Amoco and Shell. The latter's environmental damage in Nigeria has brought about considerable protest. They have often diversified and merged with other multinationals, for example in the case of Glaxo Wellcome, a leading multinational in the field of pharmaceuticals. Products are increasingly available from all over the world. Our computers are no longer assembled in the UK but in China or in Ireland; our clothes may be made in sweatshops in the UK or more likely in countries such as Morocco or China. However, the most spectacular development has been in the scale of globalisation in the financial services due to the increase in circulation of money after the oil crisis of 1973–4 and technological developments (Mohan 1999).

At the same time, globalisation can also be seen as a discourse and mythology which promotes a neoliberal vision of spatial, temporal and political relations in the present and the future which underpins an ineluctable global order against which there is no possible challenge. A world of flows and networks has supposedly displaced territoriality (Castells 1996). As a result of these changes, some writers have claimed that the world is now borderless, that places no longer matter and that the nation-state no longer plays a major role in economically, culturally and politically organising society (O'Brien 1992; Ohmae 1990). This view is not shared by all. Sassen (1991) argues that places still matter. She highlights the myriad of skilled and unskilled workers, many of whom are invisible, who ensure that services are delivered. Although cities may exhibit similar retailing national and international outlets, such as Boots, Prêt à Manger or Benetton,

they still retain some element of a distinctive feel. In particular metropolitan areas, especially the **inner cities**, are sites of cosmopolitan populations which often enable a variety of lifestyles and identities to flourish and the extension of new forms of citizenship.

As we saw earlier in this chapter, the UK, unlike many European states, had developed a strong sense of its position within a global reach. The decisive shift to neoliberal policies at the state level under the Conservative governments after 1979 also encompassed a positive attitude towards the emerging of global production and finance. In 1979, exchange controls were abolished and in 1986 financial trading in the City of London was deregulated – commonly known as the Big Bang. The UK has attracted a disproportionate amount of global finance to London since the opening up of money markets. The change to a Labour government in 1997 has not at all shifted support for an open economy with flexible **labour markets**. The UK must remain attractive to the inflow of capital, and Europe as a whole should model itself on the UK. This is the message Tony Blair, the Prime Minister, proclaims to more protectionist and wayward European states.

So far we have mainly cited writers and politicians who promote neoliberal versions of the impact of global processes, as well as globalists – those with an ideological slant on the positive advantages of globalisation. Thus globalists tend to draw attention to certain processes rather than others. The main domains they emphasise are:

- the economic (finance, manufacturing, transnational corporations);
- the cultural (food, music, film, youth culture, communications, including media and Internet);
- the political (international organisations, especially in relation to environmental issues and human rights, proliferation of sites of decision-making).

The picture, however, is much more complex and a number of writers have reached far more critical and nuanced understandings of the meaning and impact of globalisation in relation to different processes and kinds of states (Kofman and Youngs 1996). Globalisation is highly uneven, so that states exhibit considerable adaptability and variety in their responses to change and their capacity to mediate and manage international and domestic linkages. Others (Cerny 1990; Weiss 1997) have argued against the idea that advancing globalisation has diminished state capabilities, eroded cultural and institutional diversity, and uniformly homogenised everyday life. While money and finance have increasingly become 'global' in the sense that they circulate more freely, the same degree of movement does not pertain to production, trade or corporate practice. It certainly does not occur in relation to the circulation of people, where, on the contrary, states have tightened their controls, especially against those from the Third World and asylum seekers (Graham 2000). For example, in the past decade (1993, 1996, 1999) the UK has passed several draconian pieces of legislation against asylum seekers, making it more difficult for them to be granted refugee status (Schuster and Solomos 1999). The enactment of EU legislation in this field is not a matter of states giving up their sovereignty but rather collaborating to ensure that they do not have easier legislation than their European partners. On the other hand, states have made it easier for their own citizens to move within the EU and retain economic and social rights. They have made it easier for skilled labour to cross borders by recognising qualifications and degrees of other EU states, increasingly at the expense of those from the Commonwealth.

The relationship between the global and the local, or what has been termed 'glocalisation' (Robertson 1995), may differ radically in the degree to which a particular place is able to adapt to, convert, resist or submit to external influences. A firm may also feel the need to adapt to local customs. For example, McDonald's, which originally prided itself on providing the same product worldwide, has started to incorporate 'local' influences such as tikka burgers.

We should also bear in mind that attitudes towards an unfettered global economy may alter rapidly. There have been a number of protests against global capitalism since mid-1999. Many see the failure of the World Trade Organisation meeting in Seattle in November 1999 as a reaction towards the rule of the untrammelled economic forces against environmental and social considerations and regional and national cultures. One of the major celebrations held on May Day 2000 (1 May) represented a coalition of environmental and anti-capitalist groups (People's Global Action) in the symbolic and political heart of London and the UK (see Chapter 6). The protesters in Parliament Square, who daubed paint on the statues of a number of famous politicians, such as Sir Winston Churchill, were in part contesting the actions of organisations such as the World Trade Organisation, the International Monetary Fund and the World Bank. Interestingly they had selected a day which had different connotations, ranging from the internationalism of labour to an earthy popular tradition.

2.5 SUMMARY

We have shown in this chapter that the UK's historical tradition of an internationalist outlook that buttresses certain domestic interests partly explains the vacillation towards its full membership of the European Union. This has also made the option of the UK as offshore islands more attractive and one which allows it supposedly to maintain its own economic and political institutions. In effect both major political parties have embraced this attitude. At the same time, the UK is an active agent of global developments as well as feeling the effects of world-wide economic and cultural changes. The impact of global restructuring is locally uneven. As we shall see in Chapter 14 some localities and regions, such as the South East, have benefited from the concentration of capital, financial services, international movements of population and cultural dynamism; others, such as the North East, have lost economic activities that they have found it difficult to replace. Their fortunes are very much at the behest of multinationals, which may pull out on the basis of problems in the overall fortunes of the company.

 **PROCESSES OF SOCIOECONOMIC CHANGE**

3.1 INTRODUCTION

The British economy has been in 'crisis' since the late nineteenth century when Germany and the United States in particular made considerable efforts to industrialise. On the whole the state has not been used as an engine of growth for the UK economy as in other countries, for example with Germany. One notable exception occurred during the Second World War when the output of the entire British economy was geared to the war effort and government policy to increase domestic food supplies made agriculture the most efficient in Europe. It could be argued that global investment and the availability of imperial markets had enabled successive UK governments of all political persuasions to avoid any radical thinking about the continual decline of manufacturing during the first half of the twentieth century. The City of London, which was strongly attached to the Conservative Party from the 1880s onwards, had opposed the manufacturing lobby over free trade versus protection and financial measures such as exchange and interest rates. Imperialist ideology thus perpetuated industrial weakness in the UK. Even during the interwar years, despite the rapid growth of the consumer industries (such as motor vehicles and white goods) largely located in the South East, the UK's share of new consumer products in the world continued to fall.

At the end of the Second World War, although the UK had sustained some war damage, the level of destruction of infrastructure and production capacity was far less than on mainland Europe. It was therefore able to benefit for over a decade from reduced competition due to ravaged European economies, and the post-war labour shortage was met by migrants – first displaced European migrants and later from the former colonies and dominions, especially from the Caribbean and the Indian subcontinent. Moreover, after the war, the former priority given to overseas investments was re-established. The Labour Party, which had won a landslide victory at the 1945 election, set about winning its wartime commitments to full employment and social security and fair shares for

all, a package which all parties (including the Conservatives) adhered to in some form. The austerity programme was maintained in the post-war period in order to pay off wartime debts.

High rates of growth were recorded during the 1950s with rising wages and profits, but economic growth rates in other countries such as France, Germany and Japan were far higher. During the 1950s and early 1960s, the Conservatives, although maintaining a sociodemocratic consensus, did not devote much attention to a broad view of economic planning, the role of technology in an advanced capitalist society or pass legislation in keeping with social change. Economic policy was one of 'stop-go' and there was no educational reform (except for some expansion in the university sector), nor was a single hospital built during their thirteen years in office.

With the decolonisation of the 1950s (see Chapter 2) and reduced protection of British domestic and overseas markets, the full winds of international competition began to be felt. As a result of the above forces after 1961 British politics came to be dominated by economic issues. By 1966 the Labour government under Harold Wilson was forced to support an overvalued pound; its eventual devaluation in 1967 undermined Labour's economic strategy based on economic planning and a technological revolution and put severe limits on the financing of its other social goals, such as comprehensive education.

The decline in the UK share of world manufacturing continued (from 15.7 per cent in 1961 to 9.5 per cent in 1978) and this combined with a growing penetration of British markets by overseas producers. This is illustrated by the fact that 60 per cent of car sales were satisfied by imports by the end of the 1970s. The second Labour government of Harold Wilson pushed through a series of reforms in education, local government, and management of the economy in the 1970s, but neither Labour or Tory administrations in the 1960s or 1970s were able to reverse the economic decline of the UK. Thus the world accumulation crisis in the early to mid-1970s was confronted by a British economy still beset by poor levels of productivity and training.

The UK manufacturing sector had by now become far more centralised and internationalised (see Chapter 2). American multinationals owned about a fifth of the UK's visible exports while British multinationals made a third of their profits from abroad. The leading British multinational companies had largely written off the UK for their expansion strategies and were displacing their investment and jobs abroad. In some ways they were ahead of their time for European multinationals would do the same as part of the new international division of labour.

By the mid-1970s, the post-war political consensus had broken down, Prime Minister Edward Heath dismantled certain forms of state economic intervention and industrial bargaining but returned to the earlier corporatism in the face of union opposition, industrial disputes and strikes. The Labour government, first elected in 1974, found itself under pressure from a balance of payments crisis and was forced in 1976 to accept a loan from the International Monetary Fund with its inevitable restrictive conditions on public spending. This led to the 'winter of discontent' of strikes and industrial disputes in 1978/9, which led to public opinion turning against the government of Jim Callaghan. The Conservative Party, now under the leadership of Margaret Thatcher, was elected in 1979, offered a radical new approach to the economy, and sought to reduce the power of the trade unions. They adopted a monetarist policy and undertook a radical slimming down of a parasitic state as the road to economic recovery and the reversal of long-term national decline. Its populist social policies called for law and order, the return to family values, an attack on welfare scroungers and the reassertion of pride in national identity. The Conservative electoral victory, in June 1979, thus ushered in wide-ranging economic, social and political changes (see Chapter 6). In the remaining part of this chapter we examine the processes of economic change and dimensions of social change shaping the UK economy.

3.2 THE PROCESSES OF ECONOMIC CHANGE

The UK economy has been restructured involving the continued erosion of the manufacturing base (deindustrialisation) while the service sector has grown (tertiarisation) to dominate the economy (Bryson *et al.* 1999; see Chapters 8 and 9). The manufacturing base accounts for 20 per cent of gross domestic product (GDP), ranging from 29.3 per cent in the North East to 11.2 per cent in Greater London (Atkinson 2000). Economic restructuring has thus involved sectoral switches of capital (from manufacturing to **services**); geographical change in the location of investment (within the UK as reflected in the North–South divide, which is characterised by the concentration of investment in the South, as well as the movement of capital and production offshore); changes in the organisation of production (lean production, outsourcing, growth in the small firm sector, economies of scale) and the development of flexibility – especially on the part of workers – in the workplace. Restructuring has had a profound impact on the labour process (especially in the loss of male full-time jobs in manufacturing and primary industry) as well as on the division of labour (**feminisation of the labour market**) (Massey 1984).

During the momentous decade of the 1970s geographers turned to Marxian political economy as a source of theoretical inspiration in an attempt to explain the profound changes in the UK economy. The Marxian tradition does encompass a variety of approaches (Hudson 2000), but two geographers, David Harvey and Doreen Massey, have both played a pivotal role in developing the debate. Harvey brought Marxian approaches to the fore in geography (Harvey 1973), and Massey developed explanations of spatial uneven development in her restructuring thesis (Massey 1984). She argued that the geography of production and uneven development could only be conceptualised adequately through an analysis of the social relations of production, which are organised spatially, 'stretched out between areas', which she refers to as spatial divisions of labour (Massey 1984). She sought to identify the mechanisms in capitalist society which generated specific and unique outcomes linking theories of capitalist society to empirical outcomes (Mohan 1999).

One important aspect of economic restructuring has been deindustrialisation, the sustained decline in industrial activity (Hudson 2000; Martin and Rowthorn 1986). Employment in manufacturing in the post-war period peaked in 1966 at 8.5 million. Since 1966, there has been an almost continuous decline in the level of employment and output between 1973 to 1981. After 1981, output and investment recovered, whilst employment continued to decline and by 1999 manufacturing jobs numbered 4,289,000 (ONS 2000). The reasons for this erosion of the manufacturing sector are complex. Some argue that deindustrialisation (and reindustrialisation and tertiarisation) needs to be examined not in isolation but as part of a wider set of economic changes over time, and that episodes of economic development are linked to phases of major technological innovation associated with 'Kondratieff waves'.

Writing on deindustrialisation, Singh (1977) argued that the defining characteristic was a manufacturing sector that failed to sell enough of its products abroad to pay for its import requirements. Alan Townsend (1997: 80–1) adds several more features of deindustrialisation:

- manufacturing's declining share in all the big economies (including the UK) hastened by global competition and corporations investing abroad;
- redundancies in nearly all manufacturing industries in the UK recession of 1979 to 1982, which devastated many industrial towns through eliminating obsolete products and overmanning (Townsend 1983);
- the subsequent economic recovery, from 1982 to 1990, saw the number of manufacturing jobs increased only minutely;
- structural interdependence between national markets through internationalisation and multinational corporations has made the UK more vulnerable to changes in international demand.

A number of other explanations have been developed to account for deindustrialisation. These include external economic pressures such as the loss of competitiveness, the poor calibre of British management, as well as intractable labour relations (such as tightly demarcated jobs). The net result has been the sustained loss of millions of blue-collar jobs in both manufacturing and primary industry. State policies have also been blamed (for instance currency and interest rate policy), as well as issues of regulation and governance (Tickell and Peck 1999). But also wider political attitudes have played their part in the demise of manufacturing and primary industry, which are often termed 'sunset industries'. One particularly acrimonious 'battle' was the industrial dispute of the coal miners of 1984/5 (Hudson 2000). Today, the UK labour market is described as being one of the least regulated in the EU. UK employment law is said to offer UK workers less protection against redundancy (Harrison 2000).

Deindustrialisation thus has had a distinctive geography, with job losses concentrated in the industrial heartlands of the UK economy such as the coalfields and the West Midlands (see Chapters 14, 15 and 16). While some jobs have been lost due to their transfer to countries with lower labour costs, the majority have been lost due to technological change and automation and the drive for profits. Using Census of Population data for the period 1981 to 1991, the number of manufacturing jobs declined in Great Britain as a whole by 9.24 per cent (England recorded a decline of 9.57 per cent, Wales a 6.68 per cent decline and Scotland a 7.50 per cent decline). The restructuring of the UK economy (deindustrialisation, tertiarisation, the emergence of London as a key global node) has resulted in the growth in regional inequalities which have been articulated through a changing geography of economic prosperity and decline in terms of the 'North–South' divide (see Chapter 4). The economic restructuring has also had a political dimension expressed in the labour versus capital battle, and during the 1980s this battle with the unions was part of the Tory strategy for reshaping the UK economy.

Perhaps for the first time in more than a century, economic restructuring has resulted in young working people facing poorer economic prospects than their parents did at the same age. For poorly educated and low-skilled young men, in particular, the decline in urban manufacturing employment opportunities means that their employment opportunities are limited and the possibilities of gaining and retaining relatively well-paid work and therefore mainstream opportunities have declined in the final decades of the twentieth century (McDowell 2000). As the socially valued attributes of masculinity are so dependent upon economic participation in paid work, the loss of work for men is seen by many commentators as a crisis, and it has undoubtedly led to increases in poverty at the level of the household, as well as being associated with rises in suicide rates, health problems, domestic violence and divorce (Beatty and Fothergill 1996; Beynon 1999). The socioeconomic consequences of redundancy have been the focal point of two popular films of the late 1990s, *Brassed Off* and *The Full Monty* which were filmed in South Yorkshire. Today, all workers, blue and white collar, need to update skills in order to remain connected to the labour market and thereby sustain paid work until retirement.

Economic and social life in the UK is characterised by a number of dimensions of flexibility, including place within the labour market, the skills involved and the associated contractual arrangements (Pollert 1999). Over the last twenty years or so all employers (primary, manufacturing and services) have increased their flexibility by replacing permanent workers (blue and white collar) with those employed in non-standard employment (on a temporary or contract basis). The proliferation of these contracts for salaried workers, combined with downsizing and white-collar redundancy, has led Sennett (1998) to comment that a career (a succession of jobs of increasing levels of responsibility throughout a working life) can no longer be regarded as a 'well made road' (see also Beck 1992).

A second dimension of flexibility, 'place' flexibility, is an increasingly important feature of life for British workers (Henry and Massey 1995). Spatial mobility, especially for managers and professionals (scientists, engineers, and accountants), has typically involved inter-regional moves, but increasingly it involves moving internationally for relatively short periods as 'skilled transients'. 'Skilled transients' are now a major feature of global migration systems, and these flows are highly gendered (Hardill 1998). Housing is now a complex space; even though 'being together' remains a conjugal norm, housing as 'a unity' is disintegrating under the effects of increasing mobility demanded in the pursuit of paid work. There has been a growth in the number of 'commuter couples' who have two places of residence involving inter-regional and international weekly or less-frequent commutes (Green *et al.* 1999). The loss of manufacturing jobs in the industrial heartlands (such as the North East, the North West and the West Midlands) has led to the emergence of 'industrial gypsies', workers who accept fixed-term contract employment for manual and semi-manual work elsewhere in the UK and abroad (Hogarth 1986). One television programme that focused on this issue was the *Auf Wiedersehen Pet* series of the 1980s.

Women have increased their share of employment in virtually all industries and occupations, including the professions in the UK, leading Linda McDowell (1991: 417) to argue that, 'the feminisation of the labour market is amongst the most far-reaching of the changes in the last two decades'. This feminisation is largely the consequence of the growth of the service sector (tertiarisation) (see Chapter 8). About half of all women workers hold part-time jobs, many with non-standard employment contracts (see pp. 83–5). While the full-time pay gap has narrowed over the last years so that women earn an average 84 per cent of men's wages, female part-timers receive 58 per cent of the rate per hour of male full-timers (Ward 2000).

Some spheres of manufacturing industry have flourished since the 1980s, notably the new high technology industries (biotechnology, information technology, and materials) which tend to employ skilled workers often in small firms, especially in South East England along the 'Western Crescent', in East Anglia or in Silicon Glen. But some of these firms have experienced job-loss growth (an increase in the level of output but a decline in the number of jobs, due to productivity gains). The recession of the early 1990s not only adversely affected manufacturing employment (blue and white collar) but downsizing also became a feature of the service sector. In the last decade of the twentieth century skilled white-collar workers in producer services and financial services began facing an uncertain future, partly the result of the impact of new technology, flexibilisation and globalisation (Daniels 1999). This is still the case today.

3.3 DIMENSIONS OF SOCIAL CHANGE

During the post-war years, the social structure of the UK changed in a number of significant ways in terms of class composition, gender relations and ethnic and 'race' relations.

3.3.1 Class

Social class is a relational concept whereby the classes are seen as part of an overall system. Class can be defined in several ways. Though Marx recognised non-economic dimensions of class, he tended to focus on relations in the workplace and the market. The working classes, which had only their labour power to sell, are in opposition to those who owned the means of production (bourgeoisie). For

him society in the middle nineteenth century was becoming more polarised, with the middle strata being driven out with each economic recession. In contrast Weberian definitions focus more on status and see class as individuals who share the same life chances in the market. This approach yields a much less stark division into classes. It is not easy to measure class for it depends both on actual position within the labour process as well as on subjective consciousness. Occupational categories in the Census of Population are often used as surrogates for class. Class relations only entered into geographical analysis with the emergence of radical geography in the 1970s (Harvey 1973).

Class composition, as we have noted, has not remained static. The *Oxford Social Mobility Study* (Goldthorpe *et al.* 1980) showed that already by the late 1960s there was a great deal of fluidity in the intermediate zone and that the economic and social distinctions between manual and non-manual work were being eroded, due partly to new technology (Box 3.1). These classes are not the same as the more familiar ones used by the government (see Glossary). New forms of production requiring more administrators and managers, the expansion of the public sector, increasing levels of education and social mobility have all contributed to the development of a significant middle class (see Chapters 8 and 9). The middle class derives its basic

Box 3.1 Class divisions

Class 1: those exercising power on behalf of corporations and government and the traditional bourgeoisie
Class II: lower grade professions, administrators and high-grade technicians
These two classes constitute the service classes or the subaltern levels of the service class

Class III: routine non-manual
Class IV: workers on their own account
Class V: lower grade technicians and supervisors
Class VI: skilled manual workers
Class VII: semi-skilled and unskilled workers

Source: Goldthorpe *et al.* (1980)

income from salaries but has increasingly acquired wealth from housing investment and shares (see Chapters 4 and 15). The higher levels of the middle class, or what some have called the service class (Savage 1995), enjoy a high degree of autonomy at work though pressures of productivity since the 1980s have also led to considerable stress (see Chapter 8). This service class is unevenly distributed within the UK, and is more prominent in the South East, metropolitan centres and certain rural areas within commuting distance of metropolitan centres. It is the middle class whose interests are being increasingly catered for politically. 'New' Labour has tailored many of its policies to what it sees as the values of the middle classes, sometimes referred to as 'Middle England' (see Chapter 6).

The traditional working class has declined in number and has become fragmented as a result of deindustrialisation, new forms of production and the feminisation of the labour force (see pp. 80–5). The loss of jobs in the heavy industries and the shift into the routine service sectors since the 1970s have also weakened working-class communities especially. The thesis of the 'affluent' male worker in the 1960s argued that with the rise in wages many workers had acquired middle-class consumption habits and political affiliations. Though major distinctions still exist in styles of consumption, the working class partakes in the same world of consumption (see Chapters 8 and 9).

Class interacts with other dimensions of social stratification. A woman's class position was traditionally considered as emanating from her husband or father; thus women's relationship to the labour market and ownership of resources were obviously very different to those of men (Crompton and Mann 1986). Today, with women's greater independence, participation in the labour market, especially entry into service class occupations, and the increase in male unemployment, the picture of women's class position has become far more complex. It can no longer be treated as a derivative of male class positions. Their altered relationship to the labour market also raises the issue of whether class position stems from the individual situation or the household. So too may the experience of class differ for racialised minorities. Stuart (cited in Jackson 1994) has argued that black people's experience of class has been refracted through 'race' since racism has to a great extent limited their possibilities.

3.3.2 Gender and sexuality

Gender is part of everyone's identity, influencing how we think about ourselves, about other people, and about our relationships with other people (Blunt and Wills 2000). But just as identities extend far beyond the level of an individual, gender is also a social relation that positions men and women differently in different spheres of life. Both men and women have gender identities, which are often thought of in terms of masculinity and femininity. But 'gender' is usually distinguished from 'sex'. While sex – male or female – might be understood as a category based on biological or anatomical difference, gender is understood as a social construction organised around biological sex that varies over time and space (Blunt and Wills 2000). Thus, individuals are born male or female but, over time, they acquire a gender identity that is an understanding of what it is to be a man or woman. This gender identity is defined as masculinity or femininity. These gender identities are often naturalised, relying on the notion of biological difference, so that 'natural' femininity encompasses, for example, motherhood, nurturing, etc. However, the social relations between genders must be recognised as socially constructed and historically and geographically differentiated. There are therefore multiple masculinities and femininities (Laurie et al. 1999).

Following the period after the Second World War when women were exhorted to return to the home, to being 'homemakers', new possibilities opened up for women in the 1960s in education, the labour market and the ability to control their own reproductive body. The previously rigid boundary between the private and the public began to shift so that women became more visible in the public sphere (see Blunt and Wills 2000; also Chapter 8). By the beginning of the 1970s, a women's movement had formed in the UK and aimed to change gender relations radically (see Chapter 6). Feminist books, such as *The Female Eunuch* (Greer 1970), called for the end of women's subordination and for the personal to become political. This highlighted the need to politicise what happened within the home and domestic relationships. The first feminist articles were published in geography in the radical geography journal *Antipode* in 1973, the first textbook in 1984 (Women and Geography Study Group 1984); the journal *Gender, Place and Culture* has been published since 1994.

Gender identities intersect with other aspects of identity, such as social class, disability, sexuality, age and so on (see Laurie *et al.* 1999). To think of gender as a social construction has been politically enabling, allowing gender roles and relations to be destabilised and resisted by both men and women. So, for example, it is now unusual for a man to be the sole 'breadwinner' in a family, and more women are in paid employment (see Chapter 8). And yet, although some gender roles and relations may have changed others remain the same (see Chapters 8 and 9). While more women than ever undertake paid work outside the home, they remain disproportionately responsible for childcare and other domestic work and more likely to be employed in a part-time capacity and on lower wages than men (see Chapters 8, 15 and 16).

Since the late 1970s, and particularly over the course of the 1980s and 1990s, the work of feminist geographers has explored the connections between gender and geography and has challenged gender inequalities in both geographical discourses – knowledge of the world – and disciplinary geography (Blunt and Wills 2000). Feminist geographies address 'the various ways in which genders and geographies are mutually constituted' (Pratt, quoted in Blunt and Wills 2000: 91). Many areas of feminist theory and politics continue to identify gender in terms of difference from sex. But in recent years, the distinction between sex and gender has come to be questioned in important and challenging ways. Feminist writings have become more cultural and incorporated sexuality, exploring the materiality of sexed and gendered identities on a bodily scale and questioning the heterosexist matrix.

In the 1990s, feminist geographers have increasingly worked with poststructuralist ideas in their analyses of gendered spaces and the spatiality of gender. Poststructuralism represents a diverse field of work, which is concerned with the connections between power and the production of knowledge, the constitution and performativity of subjectivity, and the importance of difference (Blunt and Wills 2000).

Feminist and gay movements have both sought to change practices and attitudes in the home, the community and the workplace. Gay and lesbian movements have challenged the lifestyles imposed by patriarchal societies and have become highly politicised. Although gender relations have changed considerably in the past thirty years, there remains a great degree of sexism or discrimination and stereotyping against women (see Chapter 6).

3.3.3 'Race' and ethnicity

'Race' was first recorded in the English language in 1508 and denoted a class of things or persons. It was primarily in the nineteenth century, after two centuries of contact with other lands and diversity of peoples and with the rise of scientific thinking, that 'race' came to acquire the sense of a hierarchical division of the world's population on the basis of physical characteristics. Whilst colour has proved to be a key determinant in much Anglo-Saxon racial thinking, in Europe it looked more to myths of national origins, such as Aryan and Semitic. Racial categorisation was not only applied to exotic peoples but also to the Irish and to the working class well into the twentieth century. Disraeli's (1969) two nations were 'bred by different breeding, fed by a different food'. After the horrors of the Second World War when millions of Jews, gypsies and Slavs were killed on racial grounds, crude racial accounts became unacceptable but racist ideology certainly did not disappear. Racism can be defined as 'a system of beliefs which serves to identify a group on the basis of biological or some invariable categories and to then attribute other categories to it' (adapted from Miles 1989). Racist imagery is not monolithic or uniform; it varies according to class, region, ethnicity and gender (Husband 1982). Contemporary discussions of 'race' and racism in the UK often assume that racial thinking is narrowly linked with the immigration of colonial populations but, as we have noted in Chapter 2, it predates this post-war immigration.

It is common to speak of ethnic minorities when referring to migrants of colonial origins. Here again we need to be more careful in our usage of the term. 'Ethnicity' is a difficult term to define. The concept of ethnicity was developed in the plural structures of colonial societies where different tribal groups or migrants coexisted and often operated within separate economic and social worlds. In the British situation it refers to the strategic use of cultural distinctiveness in particular economic, social and political contexts. Although writers may be ambivalent in the use of 'ethnicity' as a euphemism or softer expression for 'race' (Jackson 1994), ethnic relations do in fact reflect prevailing power relations. The specific classification in the UK emerged from the dominant concern with the identification of groups of overseas origin seen to pose problems. The current classification tends to render invisible the presence of groups who do not fit into this situation but who display characteristics of ethnicity. This applies in particular to those groups who are classified under 'white' in the Census of Population, for example Eastern Europeans and Turks. The British categorisation of ethnicity (the ethnic question is not used in the Census of Northern Ireland) depends on colour and colonial origins. There is of course a subjective definition which emerges clearly in relation to Middle Eastern populations who may classify themselves either as 'white' or as 'other'. Other axes of ethnic identification may also be lost in the white category, such as national origin (for example the Irish) or religion (in the case of the Jews). There is also a problem in our conceptualisation of ethnic minority groups where there is a tendency to homogenise their class positions, national origins and religion, for example in relation to Asian groups (see Chapter 5).

The strategic deployment of cultural distinctiveness does not only pertain to groups of migrant origin whose territorial patterns can be highly dispersed. It also applies to religious groups such as the Catholics and Protestants in Northern Ireland who generally live in separate communities and often with their own institutions (see Chapter 16).

3.4 SUMMARY

In this chapter we have shown that the UK's economic decline continued in the post-war period, despite the growth in high tech industry, the service sector and London as a key node in the global economy. An earlier 'golden age' of high growth and near full employment and regional convergence has given way to slower growth, mass unemployment and regional divergence, as well as growing inequalities at intra-regional and intra-urban levels. We have also examined social change in the post-war period in three ways: in terms of class composition, gender relations and ethnic and 'race' relations. In Chapter 4 the social consequences of the UK's economic decline are explored.

AN ERA OF EXCLUSION

EXCLUSION

Polarisation and division

- **Introduction**
- **Recent debates on division**
- **Summary**

4.1 INTRODUCTION

At the end of the Second World War, there had emerged a level of consensus that social rights had to be extended so as to create a more inclusive society and bring the social exiles, or those too poor to participate in the community, into 'the fold'. The war itself – as indeed the First World War had – brought the different classes into closer contact than ever before, but this time massive change followed (Channel 4 1999). In 1945, the electorate, several million of whom (men and women) were involved in the war effort, voted massively for a Labour government and for radical change, with *all* Britons, regardless of status or class having 'a home fit for heroes'. The mood of the country in 1945 was thus for change, and a Labour administration set about creating a new order through such measures as the welfare state and education reforms. Sociologist T.H. Marshall (1950) saw the extension of social rights as a major means of class abatement and the culmination of over two centuries of struggles for different forms of rights.

It was these rights which characterised the notion of citizenship. Marshall (1950) defined this as full membership of a community where membership entails participation by individuals in the determination of the conditions of their own association in that community. Different rights developed historically with the extension of democracy in the nineteenth and twentieth centuries and the implementation of the welfare state. In the UK, the earliest type consisted of civil rights; that is, rights necessary for individual freedom (freedom of speech, thought and faith, the right to own property, the right to justice and ability to conclude contracts). Essentially these were the rights that signalled the development of capitalism. These rights were widely applicable by the nineteenth century as far as males were concerned. Political rights or the right to participate in the exercise of power as a member of a representative body or as an elector were, on the other hand, extended during the nineteenth century, again only to males. Social rights as a form of class abatement were the last to be gained.

By the end of the 1950s, the era of rationing and shortages was over and, with almost full employment, the UK seemed, to paraphrase Harold Macmillan, the Prime Minister of the time, never to have had it so good. Yet, by the 1960s, a number of social policy academics close to the Labour Party (such as Tawney and Townsend) raised the issue of the continuing existence of poverty in a period of greater prosperity. Townsend (1979) questioned absolute definitions of poverty (Rowntree 1937) which were outdated and failed to take account of the problems some people had in fully participating in society. He suggested one that was closer in tune to the concept of citizenship – poverty constituted a lack of resources that would enable a person to be able to participate in the normal expectations and customs of a society. This kind of definition also means that the indicators of poverty can change over time in order to embrace changes in society (Box 4.1). In the 1960s, Townsend used the example of not being able to afford a proper Sunday lunch as an indicator of poverty. The idea of a Sunday roast meal might not be so relevant today because of changes in family life and the way we gather together, and therefore is not so much an integral aspect of what we expect to be able to do normally. But Townsend's indicator of giving presents to near members of the family for birthdays or Christmas still holds, which would now also include the provision of some electronic material goods (televisions, telephones and microwaves are regarded by most people today as necessities not luxury goods), as well as perhaps an evening out (such as a meal/cinema).

By the late 1960s, policy measures were being developed to tackle poverty both nationally and locally. Part of the impetus was not just the material evidence of poverty but also the emergence of persistent racial tensions in some inner urban areas (for example, Labour's Urban Programme of the late 1960s). In the 1970s, a number of targeted policies were implemented in the field of education and housing and a more general area development focus to poverty policy as witnessed through the Community Development Programme (CDP Information and Intelligence Unit 1974). The financial problems of 1976 and the loan secured from the

**Box 4.1
Poverty:
key terms**

Deprivation. People are said to be deprived – materially and socially – if they lack the material standards (diet, housing and clothing) and the services and amenities (recreational, educational, environmental, social) which would allow them to participate in commonly accepted roles and relationships within society (Goodwin 1995). Deprivation is measured by the use of such indices as the Index of Local Deprivation (Lee 1999).

Poverty. This has been viewed in both *absolute* and *relative* terms. When talking about poverty researchers usually base their work on measures of deprivation rather than the identification of poverty by itself. The existence of deprivation is taken as a surrogate for the existence of poverty. The geography of poverty is complex, embracing the unemployed, those on low pay or in insecure work, the sick, the elderly, the unskilled, and some minority ethnic groups through the operation of labour and housing markets. Absolute poverty assumes that it is possible to define a minimum standard of living based on a person's biological needs for food, water, clothing and shelter. The emphasis is on basic physical needs and not on broader social and cultural needs. Seebohm Rowntree's studies of poverty in York in 1901, 1936 and 1951 used such an approach to poverty (Rowntree 1941; Rowntree and Lavers 1952). But another way of viewing poverty is of relative poverty, which goes beyond basic biological needs, and is not simply about a lack of money but also about exclusion from the customs of society. Relative poverty (Townsend 1957) is about social exclusion imposed by an inadequate income as was noted in *Faith in the City* (Report of the Archbishop of Canterbury's Commission on Urban Priority Areas 1985: 195): 'poverty is not only about shortage of money, it is about rights and relationships; about how people are treated and how they regard themselves; about powerlessness, exclusion and loss of dignity'.

Social exclusion. This is a 'broader concept than poverty, encompassing not only low material means but the inability to participate effectively in economic, social, political and cultural life, and, in some characterisations, alienation and distance from the mainstream society' (Duffy, quoted in Walker and Walker 1997: 8). The concept of social exclusion developed in France in the 1970s, and has been adopted by the EU and the Labour administration of Tony Blair.

Social polarisation. This refers to the widening gap between rich and poor in advanced capitalist economies and has been explored in a number of ways. One is based on the geography of social divisions (Champion *et al.* 1987), another is the global cities thesis (Castells 1989; Sassen 1991). A third approach emphasises social and spatial divisions in the UK highlighting the polarisation between households containing non-working ('work-poor') couples and those containing dual earner ('work-rich') couples (Woodward 1995), including the underclass (Robinson and Gregson 1992).

The underclass. This concept is now widely used in popular, political and academic discourse. At the core of debates is the notion that a new 'class' is emerging (or now exists) (Robinson and Gregson 1992). The underclass is characterised by a persistent, even permanent and intergenerational poverty, marginalised by spatial concentration and which is culturally distinct and separate.

International Monetary Fund (IMF), which imposed financial strictures on government spending, marked the beginning of the end of this post-war era. The new financial reality resulted in the tightening of welfare expenditure, which was reinforced by the 1979 electoral victory by the Conservatives.

In the 1980s, the discussion of poverty turned increasingly to the notion of polarisation (social and spatial) and to the shrinking portion of the UK cake held by the poorest. Income polarisation was also compounded by a number of policy measures introduced in the 1980s:

- a reduction in the level of income tax for high earners;
- changes to inheritance tax and capital gains tax;
- increasing use of indirect taxes;
- encouragement of increasing profit margins;
- an end to minimum wages legislation through the abolition of the Wages Councils;
- reduction in and removal of welfare payments such as to 16- and 17-year-olds;
- the abolition of earnings-related unemployment benefit;
- deregulation and the privatisation of a whole host of public sector activities, including the public utilities and so on.

At the same time those who were homeowners made substantial capital gains (Hamnett 1999).

A strong spatial dimension to polarisation was given through analyses on an increasing North–South divide, which received much publicity shortly before the general election of 1987 (Champion *et al.* 1987; Woodward 1995). In general, it could be said that academics were showing that polarisation and social disparities were growing between those who had benefited from the measures of the successive Thatcher administrations and those who had lost out, while the Thatcher government at the time tried to deny the excesses of **Thatcherism**. It also attempted to suppress evidence of the relationship between poverty and ill health (see Chapter 10), as for example with the Black Report, which had been commissioned by the previous Labour government.

In this chapter we examine the impact of economic and social changes since the 1970s that are creating a more socially polarised society. A number of axes of division are identified such as **'work-rich'** and **'work-poor' households**, homelessness and rootlessness, alienation and crime, rural/urban divisions, regional and ethnic divisions, environment and transport, and these are explored at greater length in subsequent chapters.

4.2 RECENT DEBATES ON DIVISION

Some recent research on the distribution of wealth and poverty in the UK has been produced under a Joseph Rowntree Foundation research initiative. This research highlighted that the number of people living in households with under half the national average income fell between the early 1960s and 1970s from five million to three million, but then rose to eleven million in 1991, to a point where one in five households now lives on under half the national average income (Woodward 1995).

There has also been some empirically based conceptual work on social and spatial divisions in the UK, and the idea of a polarising nation at the level of the household, which was first developed by Ray Pahl (1984). A process is identified based on distinctive work practices, creating a growing polarisation between households containing non-working ('work-poor') couples and those containing dual earner ('work-rich') couples (see Chapter 8; Pinch 1993). This thesis is more concerned with the amount of wealth in each category. Moreover, Hills (1995) also notes that the marked rise in the dispersion of male earnings is now wider than at any time for the period for which we have records.

The number of individuals under 60 living in households without paid work has more than doubled – from 4.1 million, or 8 per cent, in 1979, to 9.4 million, or 19 per cent, in 1993/4 (Pile and O'Donnell 1997: 32). This has been accompanied by a widening gap in the incomes of households in paid work and those out of paid work. The life chances of individuals are also influenced by the way that economic **restructuring** affects other members of the households to which individuals belong (Pinch 1993). If one partner becomes disconnected from the labour market and begins to receive benefits, the probability of his/her partner also becoming disconnected is high because of the working of the benefit system, in what has been described as the **unemployment trap**.

Research has also focused on the underclass, which is a rather chaotic conception but has been significant in the social polarisation debate in the UK. The term has developed as a distinctly Anglo-American concept (Robinson and Gregson 1992). There are two broad approaches to the underclass debate. One approach, which is usually associated with the political 'right', such as American political scientist Charles Murray, sees the underclass primarily as a 'cultural' phenomenon, characterised by high rates of illegitimacy, of crime and of dropping out of the labour market. In the USA, the underclass debate has a strong racial dimension. The other approach is associated with the 'left' such as the Labour politician Frank Field. Field sees the underclass as a 'structural' phenomenon, the result of economic and social changes, arguing that a new 'class' is emerging (or now exists) because of de-industrialisation combined with technological change. The underclass is said to mean something more than the poor; a structurally separate and culturally distinct group is being created, separated from the rest of society in terms of income, life chances and aspirations. The underclass is characterised by a persistent, even permanent and intergenerational, poverty; it is marginalised by spatial concentration and those in it are culturally distinct and separate.

During the 1990s, the concept of social exclusion gained increasing acceptance. Though its origins are European, we argue below that it is not that far removed from Townsend's (1979) definition of poverty. The difference is the vocabulary; the word 'poverty' still smacks of something starker than social exclusion. The traditional concept of poverty contrasts 'the poor' with 'the non-poor' as if the two groups comprised fixed classes in a hierarchically organised society. The poor do not succeed in the labour market, in a rather Victorian liberal view of society, and the poor become dependent on the state. Poverty, as a concept, has been used in ways which encourage the practice of 'blaming the poor', while social exclusion emphasises society's role in excluding certain people from full participation. Poverty may be both a precursor of social exclusion and a consequence of it.

Academic and policy debate has changed over the last decade largely due to the influence of the EU, and the term 'social exclusion', which, like the underclass, has been described as a chaotic conception (Samers 1998), is replacing 'poverty' in the academic and policy literature. Poverty and social exclusion, as Room (1995) has argued, differ in their intellectual and cultural heritages, with poverty being rooted in the liberal tradition of Anglo-Saxon societies while social exclusion reflects the conservative and social democratic legacies of continental European societies, especially France.

In France, social exclusion has a political history dating to the 1970s. Social debate at the time focused on those people and places that society had 'left behind' and 'cut off' from the mainstream of ordinary national life. Social exclusion described the condition of people living in the massive French suburban housing developments, *les banlieues*, who experienced lower than average incomes, higher than average rates of crime and poor quality housing. Social exclusion rapidly became popular political rhetoric in France and across Europe and was ultimately incorporated into EU policy for tackling issues of poverty and deprivation and therefore spatial exclusion (Hague *et al*. 1999).

Social exclusion as a concept–process, Samers (1998) argues, could be divided into processes associated with material exclusion and those associated with discursive exclusion (Box 4.2), and that there is not a social exclusion but many social exclusions. The EU assumes that every person within its territory has the right 'to participate in the major social and occupational institutions of the society' (Room 1995: 6). When such participatory rights fail, and people are disadvantaged in educational, occupational, income, medical and other standard opportunities, social exclusion has occurred.

**Box 4.2
Forms of
social
exclusion**

Material exclusions
- Exclusion from a job or 'primary sector' jobs (see Chapter 8)
- Exclusion from social services (including health and transport services and self-help organisations)
- Exclusion from adequate housing (both state and private sector)
- Exclusion from political (civic) participation (in local elections and/or regional and/or national elections, local cultural and social organisations)
- Exclusion from 'adequate' social contacts

Financial exclusions (banking; access to loans)
- Exclusion from (especially prestigious and/or well-funded secondary) education
- Exclusion from employment training
- Exclusion from 'adequate' consumption spaces
- Exclusion from recreation and leisure (activities and spaces)

Discursive (material) exclusions
- By academics (the question of representation; social 'invisibility' in reports and surveys)
- By government and policy-makers (social 'invisibility' in reports and surveys; inability of 'detection')
- By housing, social service and immigration authorities (racism, intimidation)
- Exclusion from schools, universities, and other educational authorities (racism, social categorising)
- Exclusion by the media (television, newspapers, etc.)

Source: After Samers (1998: 126)

In order to examine these problems and how member states were dealing with them, the EU established the European Observatory on Policies to Combat Social Exclusion in 1990. In 1997, Labour's Social Exclusion Unit (1998) adopted a similar remit for the UK. The SEU places emphasis on **'joined up government'** with **'joined up thinking'** to tackle social exclusion. Social exclusion may therefore be seen as a denial (or non-realisation) of the civil, political and social rights of citizenship. In its first two years the SEU has published reports into school exclusion and truancy, homelessness and sleeping rough and problem housing estates, and teenage parenthood and the problems facing 16–18-year-olds. As the Secretary of State for Education put it:

> There have been too many people consigned to a social scrapheap in the past. In addition to absolute poverty, we must eliminate the poverty of aspiration, which corrodes so many of our communities. So often it is compounded by restricted access to housing, lack of awareness of health issues or low educational expectations. We cannot tackle this systematic disadvantage by easing its symptoms. Instead we need to ensure that all policy is underpinned by the basic goal of guaranteeing equality of opportunity to all our people.
>
> (Blunkett 1999)

Social exclusion is thus seen as a dynamic process of being shut out, fully or partially, from any social, economic, political and cultural system which determines the social integration of a person in society. It is a problem that affects people rather than places *per se*, and there are different forms of exclusion: not just those on low income or in social housing, but it also embraces exclusion from full citizenship and participation in all aspects of society (Byrne 1999). Social exclusion is seen as a process with outcome measures such as poverty (Room 1995); for example, institutional measures that serve to exclude, such as the **poverty trap**, or linkages between housing and employment. It is a useful focus for the analysis of disadvantage, in that it allows emphasis to be given to the part place and space play in exclusion, in the dynamics of areas, through movements in and out of populations and the concentrations of poor people in certain neighbourhoods.

4.3 SUMMARY

During the post-war period debates and policies shifted in the 1980s from concerns about poverty to the notion of polarisation (social and spatial), and to the growing divide between the haves and the have-nots in the UK. In the UK today one in five households now lives on less than half the national average income. During the 1990s the concept of social exclusion gained increasing acceptance in academic and policy circles, the latter since the election of the Labour administration of Tony Blair.

CULTURAL CHANGE

- **Introduction**
- **Post-war cultural change**
- **Ethnic identities and cultures**
- **Summary**

5.1 INTRODUCTION

As we have already noted in Chapter 1, and discuss at greater length in Chapters 11 and 13, the United Kingdom emerged over several centuries out of strategic considerations and conquests. The multiplicity of cultures and identities inherent in a multinational state makes it difficult to analyse cultural change. Chapter 11 explores the very complex nature and scope of **culture** and identity in the UK and its component parts, but here we concentrate on cultural change in the post-war era from a UK perspective – an era when technological and socioeconomic change allowed a truly 'British' culture to emerge for the first time.

5.2 POST-WAR CULTURAL CHANGE

The UK has changed considerably since the end of the Second World War. Strong class cultures continued to divide society, although there had been a greater degree of mixing during the war (see Chapter 3). A high culture of literature, theatre and classical music was the preserve of the upper and solid middle classes. Although the early post-war years were typified by austerity, the newly elected Labour government strongly supported the dissemination of cultural amenities. John Maynard Keynes, the famous economist, set up the Arts Council in 1947. The newly acquired right of local authorities to use some of their rates on cultural activities led, for example, to the opening of a number of provincial theatres and the expansion of public libraries. By the middle of the 1950s, and the end of rationing, a number of artistic movements turned a critical eye to British society. John Braine, in the book *Room at the Top* (turned into a film), examined the loss of innocence in achieving social mobility. Alan Sillitoe, in *Saturday Night and Sunday Morning*, set in Nottingham, wrote about seduction, the mundaneness of everyday life, and of abortion and violence. While Hollywood films had for a long time been a major staple of British cinema-goers, the mid-1950s saw the transfer of highly successful rock

and roll films. From the late 1950s, a kind of cultural revolution leading to greater permissiveness and openness altered the lives of the general public and specific groups. As we also saw in Chapter 2, the UK population had by the 1960s become more diversified with the immigration of refugees in the 1930s, Eastern European workers after the war and colonial migrants from the late 1940s.

These developments raise issues about the nature of cultural change, amongst which are its class basis and geographical variation, the impact of Americanisation and closer contact with the European Union, the role of traditional institutions, such as religious bodies; changing nature of gender relations, sexuality and the new permissive society; the effect of new technologies, such as television and later the Internet in the 1990s; and the diversification of the UK population.

In this chapter we examine how the above processes influenced cultural change and the dynamic relationships between different cultures. First, however, it is necessary to clarify the meaning of culture. It is, of course, a term which has many meanings (Williams 1981), some of which are extremely broad and cover the whole way of life; others are more narrowly focused on the production of cultural artefacts and practices associated with them (Marwick 1991). The study of British cultures began in the 1950s with the work of Richard Hoggart (1957) and Raymond Williams (1958), Professor of Literature at Cambridge University and of Welsh origin. The interest in cultural change was continued in the 1970s by Stuart Hall and his colleagues at the Centre for Contemporary Cultural Studies (CCCS) at the University of Birmingham. Their study of the plurality of cultures was characterised by Marxist analyses using categories of social class and ideology (Clarke *et al.* 1979), gender and youth and subcultures (Hall and Jefferson 1976; Hebdige 1979). However, the CCCS did not include analyses of ethnic cultures or questions of **'race'** or geographical region. The new cultural geography, which developed in the UK in the 1980s (Cosgrove and Jackson 1987), located culture in the everyday and not just in distant lands or high culture. It initially focused on studies of symbolic landscapes (Cosgrove 1985), the spatial constitution of specific cultures, the politics of culture and the influence of gender, sexuality and ethnicity (Jackson 1994). In the 1990s, geographers have turned their attention to topics such as consumption (Bell and Valentine 1997), youth cultures (Skelton and Valentine 1998), place and individual identities (Crang 1998).

Our approach to the definition of culture is a relatively broad one in which culture is a system of shared meanings which people who belong to the same community group, or nation, use to help them interpret and make sense of the world (Hall 1995). The term includes the social practices that produce meaning as well as the practices which are regulated and organised by those shared meanings. Although cultures may develop in specific places, as with local and regional cultures, national cultures are largely produced and maintained through more national-level institutions. It is not necessary to have proximity though it is still common to think of cultures as if they depended on the stable interaction of the same people, doing the same sorts of things, in the same geographical locations. However, the significance of geographical proximity has lessened with the impact of communications and individual and group mobility.

A national culture does not necessarily express a unitary feeling of belongingness, but may represent what are in fact real differences as unity in order to produce, through an ongoing narrative of the nation (education, media, popular culture, heritage industry), identification across the divisions of class, region, gender, 'race' and the unevenness of economic development. This is especially the case in a multinational, **multiethnic** and **multicultural** state, like the UK, with its complex layers of multiple identities (see Chapter 11). Elites who play a major role in shaping national culture often tend to operate in extended geographical circuits, which in recent times have been increasingly internationalised. Many far-reaching changes have either started off in metropolitan centres or been adopted by higher social classes.

The cultural practices of the UK population began to shift quite markedly in the late 1950s. What happened between the late 1950s and the early 1970s

was not a revolution but a transformation in the opportunities and freedoms available to the majority as a whole and to distinctive groups and individuals within that majority (Marwick 1991, 1996). From the late 1950s economic recovery in Britain, and internationally, led to a substantial increase in the standard of living and the acquisition of consumer goods. For example, by 1961, 75 per cent of homes had a television and the average car cost a lot less than it had done previously. An autonomous youth culture, cutting across classes, began to develop. The country acquired an image of 'Swinging' Britain, which in reality was far more a London phenomenon. Places such as Carnaby Street, near Soho, and Kings Road, Chelsea, were associated with trendy fashion, lively nightlife and a permissive atmosphere.

In contrast, an indigenous rock and roll music emerged from the provinces. Liverpool, at one time in the early 1960s, hosted 300 bands, amongst which was the Beatles. Their popularity worldwide put the UK on the music map. Interestingly, it was Abbey Road in North London (the location of their recording studio) that took over as the landmark. So culturally, the UK both took products from abroad, especially the United States, and also became a major exporter of music and fashion. The 1960s was a period of increasing cultural interchange, including European films and design. In the past the UK had consumed Hollywood romances and musicals but now British producers stood out for their special effects films such as the James Bond or *Star Wars* films. Marwick (1991, 1996) has emphasised the weakening of authority and the increase in permissiveness during this period. Though class still dominated, in general, social mobility increased, partly through the expansion of higher education. Plummy accents did not dominate the public sphere to the same extent as they had done previously. Establishment institutions, such as the Church of England, had difficulty in maintaining the numbers attending their churches, along with other Christian denominations. On the other hand, attendance in non-Christian religious institutions, such as Muslims, Sikhs and Hindus, rose markedly from 1970 to 1990 (see Chapters 9 and 11). There are considerable geographical variations in Christian religious observance, with particularly high rates of church membership and attendance in Northern Ireland and parts of Wales.

A 'legislative moment' stretching from the late 1950s to the early 1970s altered the parameters of what one could do and how one could live one's life. This would include such liberal reforms as the Obscene Publications Act (1959), which revised the criteria for censorship, the Abortion Act (1967), permitting abortion in certain circumstances in Great Britain, and the Sexual Offences Act (1967), which allowed homosexual acts between consenting males over the age of 21 years in England and Wales. Prior to this change homosexual acts had to be conducted in secret. The availability of contraception in the mid-1960s and later abortion freed women from the fear of an unwanted pregnancy. The Equal Pay Act and the Matrimonial Property Act (establishing that a wife's work, whether as a housewife within the home or as a money-earner outside it, should be considered as an equal contribution towards creating the family home, if as a result of divorce, it had to be divided) were passed in 1970. Permissiveness also describes a sociological shift, a consequence of the post-war economic boom which encouraged a kind of controlled hedonism and a democratic liberalisation in the name of capitalism (Weeks 1985). One of the most significant shifts was the change in attitudes towards sexuality and sexual behaviour. Feminist movements reacted against the masculine preoccupations of the swinging sixties and made demands for new gender relations, rights and provision of services. Germaine Greer's *The Female Eunuch* became a bestseller in 1970, while a more cautious feminism was promoted through the pages of *Cosmopolitan* magazine, the first issue of which appeared in 1972 (Marwick 1991). However, the acceptance of more liberal attitudes and the continuation of a more authoritarian world-view varied considerably according to social, economic, political and geographical background (Ahrendt and Young 1994).

Can we speak of a single culture in the UK? As we saw in Chapter 1, and discuss in greater detail in Chapters 11 and 13, the United Kingdom is largely an artificial creation and comprises a number of distinctive cultures. Composed of different territorial units with their own languages, traditions, histories, identities and cultures it has long been in a state of cultural and ethnic flux. This has been boosted by substantial immigration in the nineteenth and twentieth centuries. As an imperial power, it has both influenced and been shaped by other societies and absorbed their cultural practices in distinctive ways. The early onset of industrialisation and international trading meant that many local customs were lost. Today, these external influences continue to transform national, local and sectoral cultures. These meanings and practices are reproduced, sustained and transformed through social and political institutions, such as schools, media, the heritage industry, as well as in everyday social relations. Of course, this will vary throughout the UK, especially in the minority nations – Scotland, Wales and Northern Ireland. Even within the **hegemon**, England, there are areas where regional cultural identity remains strong, most notably the North East and Cornwall (see Chapters 11 and 13).

Each of these institutions, and the values they disseminate, operates within a different space. For example, educational systems differ between Scotland and England and Wales, and now with the **devolution** of political power, Scotland is pursuing a distinctive policy of supporting students in higher education. Media, on the other hand, are international, national and local. Technological advances have meant that international media are easily available. We should remember that the UK itself is the source of much international media through the BBC and private companies, which sell their products globally. News bulletins highlight the mixture of international, national and local although international news is likely to be filtered through a domestic lens. At the same time local and community radio stations serving particular groups have multiplied. The Internet revolution has made available alternative sources of information and discussion, and has allowed minority cultures to flourish, especially among diasporic communities, although accessibility is still highly uneven (see

Chapter 9). Protest groups are able to organise more easily through the Internet (see Chapter 13).

The role of internationalisation and globalisation needs to be taken into account in the way it has replaced traditional ways and forms of consumption. In areas of everyday life, such as food, the nature of change has revealed different rhythms influenced by Britain's role in the world, drawing on foodstuffs from its overseas colonies. Food provides an excellent example of changing tastes. What had once been only available in the main metropolitan centres became widely disseminated in the 1980s. New fast food fitted well into British lifestyles, for example, short lunch breaks, street cultures, and changing family structures. The fish and chip shop diversified its range or was replaced by kebabs and Chinese takeaways as well as supermarkets offering a range of convenience and ethnic foods. New patterns of food consumption stemmed also from travel abroad and the presence of immigrant groups. Some migrant groups, such as Asians, Chinese, Greeks and Turks, have heavily invested in food outlets (see Chapters 9 and 11). There have been a number of geographical studies of the culture of food consumption (e.g. Bell and Valentine 1997).

5.3 ETHNIC IDENTITIES AND CULTURES

As we explain in Chapter 11, the UK has always been a multicultural state, with a variety of languages – two based on Germanic languages, the rest Celtic in origin – and cultural traditions. As we saw in Chapter 3, ethnic and racial identities became more salient in the post-war years due largely to immigration into the UK. Much of the immigration into the UK between 1950 and the mid-1970s was of non-white people from former colonies, especially the Caribbean and the Indian subcontinent (see Chapter 7). There are thus a wide variety of ethnic identities and cultures within the nations of the UK – some of these predate the predominant Anglo-Saxon cultures, others are much more recent.

Ethnic identity is not static; it is strongly influenced by the position of groups within a society and may be used strategically to obtain more material and cultural resources. It may also serve as the basis of political mobilisation to demand economic, social and cultural rights. There is no single definition of ethnic identities (Peach 1996). They are shaped as much by the way a group constitutes itself as by the perception of the dominant society. For example, at the outset of immigration the dominant society collapsed those originating from South Asia into a single Asian group, despite the fact that this population is very diverse in terms of religion, region and country of origin and caste (see Chapter 7). In recent years, with the rise of a more militant Islam, the creation of organisations and increased economic polarisation amongst Asians (Modood 1997), there is some recognition of its actual diversity. For many years, the Irish, although the largest immigrant group, were not recognised as an ethnic group. As a result of much lobbying and greater awareness of their relatively disadvantaged position in British society, they have now been included as an ethnic group in the 2001 Census of Population. Racism may be a strong force in encouraging ethnic identification – for example amongst African-Caribbeans (Jackson 1994). For others, ethnic identification involves a stronger espousal of Islam. The assertion of ethnic identity is not necessarily a response to the situation within the UK alone, but also to developments in the country of origin; for example, the assertion of Sikh consciousness in the UK, especially after the burning of the Amritsar Temple in India in 1984.

Many groups maintain links with their country of origin and other sites of migration; that is, a **diaspora**. Diasporas forge new relationships between culture, identity and place (Hall 1995). Gilroy (1993) has traced how Caribbean cultures have been shaped through the multiple migrations and links across the Atlantic since the time of slavery. Hybrid cultural forms are also produced through the mixing of ethnic and dominant cultures within the state as in music, fashion and design (Hall 1995; Chapter 11).

This does not necessarily mean that those with an ethnic identity uphold specific cultural practices such as dress, religion or language, although they may invoke customs at particular times. As Modood

(1997) noted, the most common expression of ethnicity is not what people do but what they say or believe about themselves. Some Asians are increasingly of the view that clothes are not a necessary function of their ethnic identity. On the other hand, their Caribbean counterparts may increasingly stress their identity through clothes and hairstyles (see Chapter 11). For many, religion is far more important in reinforcing identity and solidarity amongst ethnic minorities than the population at large. It is important amongst South Asians in how they describe themselves. Amongst Caribbean women, the percentage attending church is much higher than for the white population.

5.4 SUMMARY

There has been considerable cultural change in the UK during the post-war period. Liberalisation of attitudes and less rigidity between classes, as well as the emergence of distinctive youth cultures, have been some of the notable features. International influences, as well as the cultures of migrant groups, have also profoundly altered culture in the UK. Although the influences of the 'new multi-culturalism' pervade much of the UK, most evidence of this and the biggest impact is to be found in a few urban centres of England (see Chapter 7). At the same time, regional and local cultures have on the whole become versions of the national or inter-national, transatlantic culture. But cultural systems within the minority nations have retained a high degree of distinctiveness in the face of indifference from central government and the London-based media and cultural establishment. These are taking new forms as power is devolved to Scotland, Wales and Northern Ireland, while at the same time rekindling often long-forgotten cultures and identities in parts of England.

POLITICAL CHANGE

6.1 INTRODUCTION

For a state with no written constitution the UK has a relatively robust polity which, with the exception of a few crises, such as the Cromwellian Republic and Irish secession, has been remarkably resilient compared with many of its major European neighbours. For sure, there have been periods when elements of the constitution, such as the Monarchy or the House of Lords have come under threat, but, until recently, the political system and constitutional machinery have remained largely intact in a relatively successful, though extremely anachronistic mode of governance. The very conservative nature of the UK polity is reflected in the largely bipartisan nature of the political party system and the undemocratic and deferential nature of the Monarchy and upper chamber. That said, there have been a number of changes in the political and constitutional structures and systems that are worthy of brief consideration. Thus, this chapter looks at changes that have allowed the development of the democratic system through extension of the franchise, the increase of the power of the House of Commons over the House of Lords and the Monarchy, the increase in the power of the Prime Minister and Cabinet over the Commons and the developments that made the UK one of the most centralised states in the developed world. We also consider the development of the current party system, the geographical devolution of power and finally some of the wider political issues that have affected the UK in recent years.

6.2 CONSTITUTIONAL CHANGE

For much of the UK's history, the governance of the state has been largely in the hands of the privileged few. Money, title, blood ties, position and privilege dictated who should be in the legislature and who should put them there. Although the power of the Monarch declined, there was as much political muscle and clout in the House of Lords as in the Commons. Outwith Parliament, the Establishment was very much a mirror image of both Houses. With industrialisation came representation of the 'newly monied', and the gradual enfranchisement of certain sectors of the community. A series of reforms from 1832, 1867 and 1884 gradually extended the franchise, but there was still no place for the working-class man until the widening of the franchise and the development of the nascent Labour Party in the latter years of the nineteenth century. It was only in 1918 that all men over age 21 got the vote.

One serious omission in terms of the franchise was the complete rejection, since 1867, of any notion that women might wish or have the right to vote. By 1903, the Women's Social and Political Union embarked on a campaign of civil disobedience, and later violent means, to win the right for women to vote. The work of the militant suffragettes largely failed to win general support among the population – women and men. It was only after the First World War and the Reform Act of 1918 that some women – those over age 30 and who were ratepayers or the wives of ratepayers – became entitled to register to vote. Women had to wait until 1928 for parity with men.

The power of the unelected House of Lords was first seriously challenged by Prime Minister David Lloyd George in 1911. A major constitutional crisis occurred as a result of the Lords rejecting a mildly redistributive Budget. Through guile, intrigue and threat, Lloyd George managed to curb the powers of the second chamber. This meant that while sovereignty remained in the three branches of government – Commons, Lords and Crown – the real power became concentrated in the only democratic element, the House of Commons. That said, from the middle of the nineteenth century, an increasing amount of power was being exercised by the office of Prime Minister and his Cabinet. This continued with the election of such powerful characters as Lloyd George.

Constitutional changes since the Second World War have been no less dramatic. The franchise was extended to those aged between 18 and 21 in 1968. The powers of the House of Lords were further curbed in 1949. The House of Lords Act of 1999 removed the right of most hereditary peers to sit and vote in the House, but an amendment allowed ninety-two hereditary peers to remain until the House is fully reformed. In January 2000, a Royal Commission report was published and offered three options for composition involving varying numbers of appointed and elected members. Furthermore, the present government has set in train a revolutionary programme of devolution (see pp. 43–4). The Monarchy, however, has gone relatively unchallenged by the main political parties, though support for a republican stance is more openly admitted than previously. The UK joined the European Economic Community (EEC) in 1973, and this was confirmed in a referendum in 1975, when the electorate, in its first ever referendum, voted by 67.2 per cent to remain in the EEC. The UK's membership of the European Union (EU), as the EEC became, remains a major contentious issue in UK politics (see Chapters 2 and 13).

6.3   POLITICAL CHANGE

Throughout much of its life the UK has had a political system that has mainly consisted of two large parties dominating policy and political debate, with several smaller parties struggling for a place in the **first-past-the-post** electoral system. During much of the nineteenth century, the administrations varied between different forms of Tory or Liberal governments. One of the most far-reaching changes in terms of the relative position of political parties came in the early part of the twentieth century when the working-class Labour Party displaced the Liberals as the main opponent of the Conservative Party. The Labour Party started as the Labour Representation Committee in 1900 – an amalgamation of working-class trade unionists and middle-class intellectuals. By 1906, twenty-nine of its candidates were elected to Parliament. By 1924, with Liberal support, the Labour Party was able to form its first, albeit short-lived, government. The second Labour administration (1929–31) was again reliant on Liberal support. The Liberal Party began its decline as the main opposition to the Conservatives in the interwar years.

Although the UK has generally been dominated by the two-party system, there have been situations when parties are in a state of flux or the electorate has not been able to make a decisive choice at the ballot box. Thus, no single party can form a government. Also, at times of crisis, such as around the world wars, coalition governments, 1915–22 and 1940–5, have been deemed expedient. Similarly, during the interwar global economic depression, three 'National governments' survived from 1931 to 1940.

In the post-war decades, party politics has been dominated by the Conservative and Unionist Party and the Labour Party. These have generally been able to form viable governments; occasionally they have had to rely on other parties for support. The Conservatives, for example, have often relied on the various unionist MPs from Northern Ireland. The Labour Party, as we have seen, has sometimes had to rely on the Liberals for support, either informally or in formal pacts such as the Lib–Lab pact of 1977–8 (Sked and Cook 1990). Also, during the 1950s and 1960s, British politics operated under a 'two-and-a-half party system' (Miller 1998: 175); that is, the Conservatives and Labour, plus the Liberals. While the UK Parliamentary system still tends to run like this, with the Conservatives currently in opposition and Labour in power, this has broken down in Scotland and Wales. In Scotland, the Scottish National Party is the second party, with the Conservatives third and the Liberal Democrats fourth. In Wales, there is a similar situation with Plaid Cymru as the second party (*The Economist* 1999a).

In the 1980s, a split developed in the Labour Party over British membership of the EEC and other issues. This led to the creation of a new party in British national politics, the Social Democrat Party, which subsequently merged with the Liberals to form the Liberal Democrats. The Liberal Democrat Party, which fought the 1992 election, was formed in 1988 as a merger of the Liberal and Social Democrat parties, which fought the 1983 and 1987 general elections as the Alliance (Pearce 1998). The fortunes of the Liberal Democrats and other smaller parties are not helped by the UK's electoral system.

The first-past-the-post electoral system, based as it is on simple majorities, has always tended to favour the bigger parties. However, in recent years, elections to the European Parliament have employed alternative voting systems. The Scottish Parliament and the Welsh and Northern Ireland Assemblies also used, for the UK, innovative systems. The London mayor and Assembly were elected using new procedures. There are also plans to use an alternative system for Westminster elections, after first being suggested by the Electoral Reform Society as early as 1884 (Davies 1999). Whatever system is adopted, it is hoped that more of the electorate can be encouraged to become more active. There has been a continual decline in turnout at recent elections at all levels. Membership of political parties is also in decline. Even the 1997 election, which generated a lot of media and popular excitement and saw a fundamental shift in the way voters behaved (Norris and Evans 1999a), had a relatively low turnout.

Mohan (1999) has identified three major post-war phases in the political economy of the UK: one nation politics, from 1945 to 1975; two nation politics, from 1979 to 1997; and the 'third way' of 'new' Labour. The first phase was one of consensus government during which Keynesian approaches to management of the economy were applied. These sought to encourage steady economic growth through strong government control of the economy and vigorous public spending programmes. This era was also characterised by a commitment to a mixed economy with a significant amount of public ownership of key industries, as well as the promotion of the welfare state. Another feature of this period, noted by Mohan (1999: 36), 'was the attempt systematically to reduce spatial inequalities'. Although this is viewed in retrospect as an era of consensus there were nevertheless considerable differences between Labour and Conservative in terms of ideologies and policies – indeed much more than there is now. Some would go as far as to describe the notion of consensus as a myth (Jones and Kandiah 1996). However, regardless of the party in power, government was not characterised by the confrontational, dictatorial, exclusionist and centrist policies that were so typical of the two-nation style of the Thatcher–Major era.

The period between 1976 and 1979 was a short interval during which management of the economy was largely beyond the control of the government. Inflation was growing at alarming rates, as was unemployment, yet productivity was decreasing. The government of Jim Callaghan was forced to seek a US$3,900 million loan from the International Monetary Fund to protect the pound. To secure such a loan, further cuts in public expenditure were required and taxes were further increased (Sked and Cook 1990). Thus, although there was an element of consensus during this brief period in political terms, especially since Labour had to align with the Liberals in a formal pact, the economic crisis marked the end of the post-war consensus and abandonment of the Keynesian economic policies of the earlier period.

The period of two-nation politics was characterised by a neoliberal approach (see Chapter 2) to economic management. Mrs Thatcher was a keen disciple of the American economist Milton Friedman, whose monetarist theories suited her policies. The aim was to roll back the state, cut public spending and let markets regulate the economy. The government would adopt strict fiscal measures to keep inflation low, but at the same time reduce state 'interference' in the economy. This led to deregulation, privatisation, control of trade unions and reduction in direct taxation. The results were a rise in unemployment and increasing job insecurity (see Chapter 8), a widening of the divisions in society (see Chapter 15) and a serious undermining of the integrity of the Union (see Chapter 13). Further, the Thatcher era was seen as one in which spatial inequality in the UK increased (Mohan 1999: 36) during which 'it seemed that spatial inequalities were deliberately used as an element in a political strategy'. Increasingly the Tories were seen as a party of the South and as such voters became more polarised by region. During the Thatcher era, the UK became one of the most centralised states in Europe. Tiers of local government – the Greater London Council (GLC) and the metropolitan counties – were abolished and the powers of the remaining elements were curtailed. Many of these powers were given to unelected **quangos**. This 'quangocracy' depended on political patronage and as such helped to cultivate the image of sleaze, corruption and disdain that was the final undoing of the Thatcher–Major project.

Mrs Thatcher personally, as well as her governments, was particularly disliked in Wales, Scotland and the Nationalist community in Northern Ireland (King *et al.* 1993). Nothing upset the Scots more than the Thatcher government's insistence on introducing the community charge (poll tax) – a new way of funding local government – in Scotland in 1988, a year before the rest of the state. This led to mass civil disobedience, largely unreported in the 'national' media. In a short space of time a remarkable 700,000 summary warrants were issued for non-payment of the tax (Devine 1999a). Once introduced into England, the community charge was soon scrapped after riots in London.

Thus the UK entered the third post-war political phase in which Tony Blair and 'new' Labour have trodden a 'third way'. Mr Blair's much publicised 'third way' is heavily influenced by the thinking of Anthony Giddens (1998) and aims to steer a path between the policies of the consensus era and the extremes of Thatcherism. Yet, in economic terms, the 'third way' seems to have adopted more of the policies of the latter than the former. Indirect taxes have increased while direct taxes have been kept low. This tends to be more burdensome on those with low incomes, thereby perpetuating social division, but is reckoned to please the middle classes upon whom 'new' Labour relied so heavily for its landslide victory. The Blair project has also continued the Thatcherite assault on the welfare state, particularly the benefits system. Further, 'new' Labour, so critical of the private finance initiatives (PFIs), and the privatisation programme generally, when in opposition, has embraced this method of funding capital projects with apparent gusto. The use of this funding method has spatial implications since private investors are unlikely to support projects in places where there will be little return on investment (Mohan 1999). PFIs are increasingly being used in more controversial areas, especially the NHS. The government has committed to twenty-

five major hospital schemes in England worth £2,200 million, and eleven in Scotland worth £408 million have been approved. These are designed, built, maintained and owned by the private sector and leased to the NHS (Pearce 1998). The private sector involvement in the building and running of prisons and other 'social' institutions has also been heavily criticised. That said, 'new' Labour has instituted many reforms that would have been anathema to Mrs Thatcher, such as devolution and the minimum wage.

Although the recent reforms of the Labour government have gone some way to reverse the centralising forces of the Thatcher–Major project, there has been increasing criticism of the continued growth of the Prime Minister's office. Mrs Thatcher was notorious for her dictatorial style within the Cabinet and willingness to run roughshod over the House of Commons. However, Prime Ministerial power has increased and the Cabinet Office, answerable only to the Prime Minister, has been the fastest growing government body (Davies 1999). Furthermore, the Prime Minister has become increasingly reliant on unelected advisers and press officers, or spin doctors, whose job revolves around manipulating policy and the electorate via the mass media. In policy terms the Labour administration has been dubbed 'Blaijorism' in that critics see little difference between the ideologies of the Major governments and that of Tony Blair. While Blair has delivered on devolution and some other manifesto pledges, the overall economic and social policies are not greatly dissimilar to those of previous Conservative administrations.

6.4 DEVOLUTION: THE END OF THE UK?

As Chapter 13 shows, the 'unitedness' of the UK has always been in doubt. From its earliest beginnings there have been those, mostly, but not exclusively, in the minority countries, who did not want to join the British Union – just as there are those now, mostly, but not exclusively, in the hegemon, who do not want further integration with the much newer

European Union. Those with the most vociferous calls for leaving the Union were those who got least out of it – the Irish. Since the Act of Union of 1800, which saw the UK expand, in 1801, to encompass four countries, Irish dissatisfaction has manifested itself in violent revolts, uprisings and political agitation leading to sometimes brutal reprisals by the UK state, and the subsequent development of some very sophisticated terrorist techniques by the various Irish secessionist groups. Irish terrorist groups have been operating in Great Britain at least since the **Fenian** outrages of the mid-nineteenth century. Indeed, despite 'Home Rule' for Ireland dominating much of the political debate in London during the later nineteenth century and the early twentieth, it was only after a bloody rebellion that Ireland (or at least just over three-quarters of it) finally withdrew from the British Union. The British–Irish War of 1919–21 is known as the War of Independence in Ireland (Davies 1999). The remnant, which became Northern Ireland, started life with its own devolved parliament at Stormont. But this was to prove as undemocratic, unrepresentative and as despised as the London government ever was, at least among the minority Roman Catholic population who aligned with the Republic and the Dublin government. The blatant discrimination in the areas of housing and employment, especially, did much to fuel the modern 'troubles' dating from the late 1960s, and which have cost the citizens of the UK so much in terms of life and limb, to say nothing of the vast sums of money that could have been better spent. The efforts of politicians in both parts of Ireland, Great Britain, the USA and elsewhere finally led to a peace process – albeit fragile – and the return of devolved government in Northern Ireland, but this time on a much more democratic basis (see Chapter 13).

In Scotland and Wales, secessionist movements have not been as active or as violent as those in Ireland. Nevertheless, some violence, mostly anti-English, has occurred. However, most secessionist energies have gone into democratic endeavours through the vehicles of the Scottish National Party and Plaid Cymru. These have had varying degrees of success at the ballot box, but both helped galvanise nationalist sentiment during the long period of Thatcher–Major governments (1979–97), when both of these minority countries felt aggrieved and neglected by these 'English' governments, given the low representation of the Conservatives in Scotland and Wales. The Blair government, however, embarked on a promised programme of devolution. This has happened, and the Scottish Parliament and Welsh Assembly are now operating in Edinburgh and Cardiff, respectively. Part of the devolutionary programme included the setting up of a Greater London Authority, composed of an elected Assembly and an elected mayor. This has also occurred. With devolved governments in Wales, Scotland and, once more, in Northern Ireland, there have been more calls for an English Parliament. Thus far, however, the government has no plans for such a body, seeing the Westminster apparatus as sufficient for the legislative needs of England as well as the UK as a whole. However, with an assembly in London, as well as the minority countries, a number of English regions, especially those peripheral to the South East, are gearing up for directly elected regional assemblies, which central government is keen to see created where there is a demonstrable local desire for them (see Chapter 13).

Given the political structure of the UK, a supposedly unitary state composed of constituent countries, what is surprising is not that secession and devolution have taken place but that the state has survived as long as it has. As Davies (1999: 727) points out, 'British democracy operated from the start on the understanding that a handful of southern English counties decided everything.' That could not last indefinitely, and the break up of Britain (Nairn 1977), that started so violently in 1919, might soon be completed by peaceful means.

6.5 POLITICAL IDEOLOGIES, PRESSURE GROUPS AND IDENTITY POLITICS

The discussion thus far has been largely concerned with the more formal aspects of government and governance in the UK. The whole notion of party politics and the nature of the constitution are centred on different political ideologies – differing opinions and viewpoints about how the state should be governed and organised. All of the debates surrounding such governance have been informed by pressure groups or special interest groups acting outside the formal structure. Some of these groups have formal links with political parties or governmental organisations. Large companies, professional organisations and labour unions donate money to political parties and expect policy measures in return. Historically, private companies have tended to support the Conservative Party – the traditional defender and supporter of capital – while trade unions supported the Labour Party – the traditional defender and supporter of labour. However, since the Labour Party's shift to the right, there has been something of a breakdown of this pattern, especially with business and some wealthy individuals supporting Labour. Other pressure groups have also tended to switch allegiance according to political expedience. Box 6.1 shows just a few of the bewildering array of pressure groups. These pressure groups can have a significant impact on government policy and affect the prevailing political ideology. As we have already seen, interest in formal politics has been declining so that these pressure and issue groups are becoming a more important part in the overall political process of the UK. More citizens are involved in pressure and protest group activity than in political parties. For example, more people are members of the Royal Society for the Protection of Birds than of the Labour Party (Grant 1989).

Many pressure or issue groups, however, have no party political allegiance or are apolitical in a formal sense, while others are deeply opposed to the whole formal political process. There are thus much wider political issues than the constitution and party politics to consider when we look at political change in the UK. One of the most successful of campaigns has been that of the 'women's liberation movement', which was seen as second wave feminism following the first wave of the suffragettes. The first British women's rights group was formed in Hull in 1968 (Black 2000). In fact, there were several women's movements. This was a disparate and eclectic range

of long- and short-term groups with a variety of agendas and methods. They encompassed feminists with far left-wing and anti-male leanings seeking to remodel society to less politically inclined single issue groups. The 1970 national conference of the Women's Liberation Movement agreed four main demands: equal pay; equal education and opportunity; free and automatically available contraception and abortion; extension of nursery provision (Black 2000).

The liberation and licence expounded by some women's groups and explored in Chapter 2 was seen as a threat by the so-called moral majority. The 'Clean-up TV' campaign, started by Mary Whitehouse in 1964, was but one manifestation of this liberalisation. The women's movement has had some successes, such as access to family planning, easier divorce, equal opportunities legislation, changes in inheritance legislation, and so on. But as we show in Chapter 8, women have still not reached parity with men in terms of pay and career progression. Nor has there been a great deal of change in the arena of government. Despite the relatively large number of female MPs returned in 1997 – the much-hyped 'Blair's Babes' – party politics and the Establishment in the UK are still dominated by men.

Associated with the demands of women's movements and the increasing liberalisation of society was the increase in legislation towards equal opportunity. Women and non-white groups were constantly discriminated against in the workplace and elsewhere. It was quite common for employers to advertise jobs for men only or for whites only. Women and non-whites would routinely receive lower rates of pay and poorer conditions than white men. Pubs would admit men only to public bars, while reserving lounge bars for 'ladies and couples only'. Signs in guesthouse windows advising 'No dogs, no wogs, no Irish' (or similar sentiments) were not an uncommon sight. Growing pressure to prevent such blatant discrimination led to the introduction of Race Relations Acts in 1965, 1968 and 1976, the latter being the same year the Sex Discrimination Act was passed. Formal organisations to monitor discrimination were set up in the form

Box 6.1 Some pressure and protest groups in the UK

Group	Type	Issue
AA (Automobile Association)	Pressure	Roads lobby
Alarm UK	Protest	Environment/anti-roads
Amnesty International	Pressure/protest	Prisoners/torture
BMA (British Medical Association)	Pressure	Doctors' interests
British Hauliers Unite	Protest	Roads lobby
CAMRA (Campaign for Real Ale)	Pressure	Heritage
CBI (Confederation of British Industry)	Pressure	Industrialists' interests
CND (Campaign for Nuclear Disarmament)	Pressure/protest	Nuclear weapons
CPAG (Child Poverty Action Group)	Pressure	Children's interests
CPRE (Council for the Protection of Rural England)	Pressure	Environment
Farmers for Action	Protest	Farmers' interests
Friends of the Earth	Pressure/protest	Environment
Greenpeace	Pressure/protest	Environment
NUT (National Union of Teachers)	Pressure	Teachers' interests
National Farmers' Union	Pressure	Farmers' interests
National Trust	Pressure	Environment
RAC (Royal Automobile Club)	Pressure	Roads lobby
Road Haulage Association	Pressure	Roads lobby
RSPB (Royal Society for the Protection of Birds)	Pressure	Environment
Shelter	Pressure	Homelessness

of the Equal Opportunities Commission and the Commission for Racial Equality in 1975 and 1976, respectively. Although these bodies have been successful in abolishing some of the more obvious aspects of discrimination, there continues to be a great deal of racism and sexism in British society.

Equality of opportunity and legislation to tackle discrimination against another large group in British society have been slower in coming. Homosexual

acts between men have long been outlawed in the UK. Not till the permissive period of the 1960s did legislation permit homosexual acts between consenting males over the age of 21 years in England and Wales – though 1979 in Northern Ireland and 1980 in Scotland (see Chapter 5). Although lesbianism has never been illegal and never attracted the same opprobrium as male homosexuality, it was nevertheless behaviour that was not openly tolerated. From small beginnings, and often at considerable cost to individuals in terms of physical assault, job loss, intimidation and so on, a 'Gay Rights' movement grew up. Increasingly, men and women declared their sexuality openly and 'came out'. Because of the activities of this movement, made up of a number of groups and individuals, societal attitudes to homosexuality and other forms of sexual expression began to change. However, the homosexual cause suffered something of a setback with the onset of the HIV/AIDS epidemic, when the tabloid newspapers and other right-wing commentators dubbed it the 'Gay Plague'. This tragedy only served to increase solidarity and resolve and led to the setting up of the HIV/AIDS charity the Terrence Higgins Trust, as well as widespread campaigning for increased research into the disease, most notably from people in the entertainment industry. The pressure group Stonewall is one of the best known of the groups campaigning for homosexual and bisexual rights and issues.

The 'Gay Rights' movement has had a number of successes and it is increasingly common to see 'out' homosexuals in a wide range of prominent positions, including the government. 'Gay Pride' parades are now a common feature of the cultural calendar of many British cities. Television programmes, such as *Rhona* and *Queer as Folk*, show lesbians and gays in mainstream situations and not the caricatures prevalent in the *Carry On* films or *Are You Being Served* era. Nevertheless, tolerance of homosexuality is not universal, and young homosexuals still feel forced to leave rural areas and small towns for the anonymity and gay scene in metropolitan areas. This concentration in certain areas is not lost on politicians, some of whom actively court the 'gay vote', or the business community who value the 'pink pound'. The recent controversy over Section 28 has demonstrated how widespread anti-homosexual feeling is in the UK, particularly in Scotland. There are still a number of unresolved issues, including the harmonisation of homosexual and heterosexual consent – from 21 to 16. And it was only the threat action from the European Court of Human Rights that forced the British military to address the issue of homo-

sexuality in the armed forces. The question of equality is closely linked with the rise of identity politics most notably around gay rights, disability action groups and so on, rather than the more traditional class-based party politics (Knox and Pinch 2000).

Perhaps the area in which pressure or protest groups have been most visual and vociferous in the UK in recent decades has been that of the environment (see Chapter 12). The environmental movement is now a major social and political phenomenon. Even in the early 1980s around one in ten adults belonged to one of the myriad environmental groups (Lowe and Goyder 1982). Some of these groups are professionally organised and global in scale, others are more *ad hoc*. Norris (1997) recognises three types of environmentalism. First, there are many groups concerned with traditional issues like the National Trust or the Council for the Protection of Rural England (CPRE). These tend to be conventional in terms of lobbying politicians and policy-makers. Second are the more radical green organisations such as Friends of the Earth or Greenpeace. Although these are also professionally organised, and pursue their ideals through conventional channels, they have embarked on or supported high-profile campaigns on single issues. Newer and more radical groups, such as Earth First! or Alarm UK, are even more dependent on direct action to address single issues (Anderson 2000). Third, there are the anti-nuclear groups which are largely single issue but have a wider environmental ethos. The best known of these is CND (Campaign for Nuclear Disarmament), which was founded in 1958 but is as much a moral crusade as a defence of the environment. CND is a nationwide organisation but has been particularly active in specific areas such as the west of Scotland, where the main nuclear deterrent, the Polaris submarine, was based at the Holy Loch, and at Aldermaston in England. Another well-known anti-nuclear protest movement was that at Greenham Common, where mostly female protesters demanded the removal of American Cruise missiles. These protests tended to be very specific spatially but were based on global issues. The Greenham Common women were more like the newer and often anarchic movements, best charac-terised by the **ecowarriors**, such as Swampy (see Chapter 16), whose frustration with conventional political machinery has led to what has been termed DIY politics (McKay 1998). Much of the radical protests related to wider environmental issues are a response to the 1995 Criminal Justice Act, which gave police new powers to prevent peaceful protest and trespass on public land. It is not only the more radical elements of society that have resorted to DIY politics, as the blockades of oil refineries in September 2000 demonstrated. An alliance of farmers, in the form of Farmers for Action, and road hauliers – two of the most heavily subsidised groups in the UK – created a situation whereby the social and economic well-being of the community was put under threat, and which resulted in shocking and pathetic instances of widespread panic, greed, theft and violence as motorists sought petrol.

As with the older movements, such as CND, the newer groups' attentions often have a very local geographical focus yet are part of much wider national and global issues. This local–global dimen-sion has been seen in the protests against live animal exports in 1995; protests in support of striking trade unionists in which the radical group, Reclaim the Streets, participated with Liverpool dockers; and the demonstrations over the expansion of runways at Manchester airport (McKay 1998). The more recent protests and direct action against genetically modified foods has also highlighted the local–global nexus in environmental issues. Both the more established, respectable, traditional lobbying organ-isations and the newer, radical, grassroots groups have made use of that most modern of technologies, the Internet, to improve their performance. The Council for the Protection of Rural England, Friends of the Earth and Greenpeace, for example, have well-organised web pages to communicate with the public. The newer groups have made more use of the Internet to communicate with each other and with their members to organise protests and the like. This was best exemplified in the recent anti-capitalist protests in London and elsewhere. The farmers and hauliers, referred to above, also used the Internet and mobile phones effectively to co-ordinate their efforts. Although the main reason for the success of the latter campaign was due to inaction by the police, in stark contrast to the apparent overreaction by police in other protests such as Reclaim the Street marches, anti-capitalist demonstrations or miners' strikes.

A more mainstream and national organisation that has made a big impact recently on wider political issues is the Countryside Alliance (see Chapter 12). Although this purports to be a body to defend the countryside it does not by conventional standards have environmental issues at its core, with its pro-hunting, anti-raptor and anti-rambler stance (*The Guardian* 2000a). This is largely a pro-hunting body sponsored by landowners and other vested interests, but it does have significant funding and political clout.

Most of these movements are characterised by the heterogeneity of those involved. Single issue politics tends to unite people from a wide range of backgrounds and from all parts of the UK, though education and age appear to be better predictors of support for environmentalism than gender, class or income (Norris 1997). The environmental issues and the green movement have galvanised some supporters into pursuing their goals through formal mechanisms such as the Green Party. The Greens have had little success in national elections but fare rather better in local elections. However, in the 1989 European elections almost 15 per cent of the vote went to the Greens and since then they have taken over from the Liberals as the recipients of protest votes. Support for the party comes mainly from the South of England, where environmental pressure is most keenly felt, and is weakest in the North, Scotland and Wales (Pattie *et al.* 1991).

6.6 SUMMARY

There have been a number of important political changes in the UK this century and especially since the Second World War. Universal suffrage is now considered as a given and so taken for granted that turnout at all elections, as well as interest in party politics, has declined, at least among some groups, to alarmingly low levels. Attempts to reform the voting system might go some way to reverse this trend. It might also, as has happened in the Scottish Parliament, lead to a less confrontational and more cooperative form of government. The House of Lords, first seriously challenged by the Welshman, Lloyd George, has been all but fully emasculated by the Scotsman, Tony Blair. A truly effective second chamber, however, is still a work in progress. Reform of the Monarchy is a work yet to be started. Devolved governments in Scotland, Wales, Northern Ireland and London have been put in place. As yet, England is not to be afforded the same consideration, although regional devolution is likely to spread beyond London to other English regions. The impact of the EU on the polity of the UK is of perennial interest and is likely to remain one of the major political issues of the coming decades.

The decline of interest in party politics is not merely a reflection of the switch of the Labour Party to the right, leading to little in the way of ideological difference between the two main UK parties, it is a reflection of the growth of single issue politics and frustration with the *status quo*. This has led to the development of some large-scale, populist movements, which do not seek to influence the governance of the UK by altering the constitution or influencing party policy the way the suffragette or trade union movements did, but hope to challenge much deeper notions of the way the political economy is organised.

PART ②

THE UK IN A PERIOD OF CHANGE

THE UK POPULATION

7.1 INTRODUCTION

A fundamental aspect of the human geography of any area is the human population. Like all animal populations, human populations have their own distributions, dynamics, structures and problems. The population of the UK is just over 59 million, a considerable increase from the 38.2 million of 1901. This makes the UK the world's nineteenth most populous country – exceeded in Europe only by Germany and Russia.

Even in an economically advanced country like the UK, where population dynamics are relatively stable, aspects of the human population are continually changing. It is important to look at patterns and trends in population dynamics in some detail, since all of the other issues dealt with in this book relate to population, and are in turn affected by population. Aspects of work are age and gender specific, as are patterns of consumption and leisure. Local age and sex structures influence patterns of health and well-being, as well as education, welfare and housing provision. Past migration trends can influence patterns of culture and identity. Migration also interacts closely with the labour and housing markets and can influence political decisions.

In this chapter we look at the demographic elements of the population – fertility, mortality and migration – which result in distribution and change in terms of growth and decline, as well as change in the composition of the population. We also look at some of the sociodemographic characteristics of population. These are issues which, as well as being related to purely demographic aspects of the population, are also related to some legal matter (for example, international migration and marital status), or to public policy (for example, ageing, extra-marital fertility, household composition and ethnicity).

7.2 POPULATION DISTRIBUTION

The UK is a very densely populated country at 243 persons per square kilometre (ppsk), which is well above the European Union average of 117 ppsk. Leaving aside the smaller states of the global system, such as Monaco or Singapore, the UK is the seventeenth most densely populated country in the world. However, this population is not evenly distributed within the UK. England has the bulk of the population – over 83 per cent in just over half of the land area. This makes England, at 378 ppsk, the seventh most densely populated country in the world, ranked between the Netherlands and the Lebanon. At only 66 ppsk, Scotland has relatively low density, much like that of Egypt or Burma (Table 7.1).

One characteristic of the UK population is how urbanised it is. As the oldest industrial country, the UK has long been urbanised. England's population, for example, became more urban than rural at some

Table 7.1 Population, density, distribution and change, UK and component countries, 1997

Area	Population thousands	Persons per square kilometre		Percentage of UK population	Percentage change 1981–97
UK	59,009	243	(17)	100.00	4.7
England	49,284	378	(7)	83.51	5.3
Scotland	5,123	66	(109)	8.68	−1.1
Wales	2,927	141	(29)	4.96	4.0
Northern Ireland	1,680	124	(33)	2.85	8.9

Source: ONS 1999a, *Regional Trends 34*, Sub-regional tables; CIA 1999, *World Factbook*
Note: Population estimates, based on 1991 Census of Population. Figures in parentheses show global density rankings.

point between 1841 and 1851, and even in 1801, 35 per cent of the population was classed as urban (Coleman and Salt 1992). Scotland is the most urbanised of the component countries (Compton 1991), and has been so since the seventeenth century (Wrightson 1989). Scotland also has the greatest disparity of population distribution. Glasgow, for example, has a density of 3,494 ppsk, while Highland District has only 8 ppsk. Northern Ireland remains the least urbanised part of the UK, but the main foci are the cities of Belfast and Derry.

There are many lenses through which we can look at population distribution, but two main phenomena are of particular relevance – physical geography and economic history. By comparing a map of the physical geography of the UK with a map of population distribution, it is clear that some areas are more favoured than others. Thus, the Highlands of Scotland, the Scottish Borders, the mountains of Wales and the Lake District are sparsely populated. Not only do these areas present a harsh environment, but travel has also been difficult in these areas. Some areas are more difficult to access from population centres than others. For example, the Isle of Wight, situated in the densely populated south of England, is more densely populated than any of the Western or Northern Isles, which are very peripheral to the rest of the UK. Thus the population of the UK is not only concentrated in lowland and coastal areas but is clearly over-represented in the area closest to continental Europe. In fact, the five regions that make up the 'South' (South East, East, London, South West, East Midlands) house 51 per cent of the UK population on 33 per cent of the land area.

The concentration in lowland and coastal areas is linked to the history of agriculture and trade. The industrialisation of the eighteenth and nineteenth centuries saw a shift to what had been relatively sparsely populated areas where raw materials could be found. This led to high concentrations of people in specific areas – Merseyside, Greater Manchester, South Yorkshire, Greater Birmingham, South Wales, Tyneside, Teesside, Clydeside. The uneven distribution of the UK population 'is a legacy of the UK's pioneering role in the industrial era and the associated rapid growth of factory-based settlements at a time when personal mobility was very restricted' (Champion 2000: 176). Although personal mobility has now increased and the *raison d'être* of the older industrial areas has long gone (see Chapters 8 and 14), one of the key characteristics of the UK population is how reluctant many people are to move, compared to more spatially mobile societies such as the United States or Australia. Inexorably linked to population distribution is population change, which is now explored.

7.3 POPULATION CHANGE

Between 1981 and 1997, the UK population increased in overall terms. That is to say, more people were born than died, plus more people entered the country than left. Within the UK, all countries, with the exception of Scotland, have gained population over the last three decades (Table 7.1). Scotland, like Ireland, has had a long history as a net exporter of people, both to other parts of the UK and overseas, though there appears to have been a slight reversal in that trend recently. That said, the projection is for a fall in that country's population from 5.128 million in 1996 to 5.048 million in 2013 (*Population Trends* 1998). The North East and the North West have also been regions of traditional population loss. However, while the exodus of people is predicted to slow down in the North West, it seems that the North East will continue to lose people. All the other regions of England have a bright future in terms of population gain, especially the East, South East and South West (Table 7.2). This growth is not without its problems, of course, as homes will have to be found for these people and such growth is proving to be a nightmare for planners and policy-makers (see Chapter 12). This is tangible evidence of the North–South divide discussed elsewhere (see Chapters 14 and 16).

One of the main components of population change is natural change – the interplay of fertility and mortality, where natural decrease is experienced when deaths are exceeded by births, or natural increase is experienced when births exceed

Table 7.2 Percentage population change, regions, 1971–81, 1981–91, 1991–7, 1996–2021

Area	1971–81	1981–91	1991–7	1996–2021
UK	0.8	2.6	2.1	—
North East	−1.6	−1.3	−0.3	−3.5
North West	−2.4	−0.8	0.1	1.1
Yorkshire and the Humber	0.3	1.3	1.1	3.3
East Midlands	5.5	4.7	3.0	9.2
West Midlands	0.8	1.5	1.1	1.8
East	9.0	6.1	3.6	12.2
London	−9.6	1.2	3.4	9.4
South East	6.1	6.0	3.6	12.8
South West	6.6	7.7	3.4	12.6
England	0.9	3.0	2.2	6.9
Wales	2.7	2.8	1.2	—
Scotland	−1.1	−1.4	0.3	—
Northern Ireland	−0.2	4.5	4.5	—

Source: ONS 1999a, *Regional Trends 34*, Table 3.1; *Population Trends* 1999, Table 1, p. 2

deaths. Because of very low current fertility levels, natural change does not have the major impact on local population dynamics it once did. That said, there are still variations in natural change from place to place. Some regions have experienced natural decrease while others have experienced natural increase. The large elderly population of the South West, for example, means that mortality will be high and fertility low, in relative terms, and thus that region experienced a natural loss of 0.4 per thousand. On the other hand, London experienced natural increase of 5.8 per thousand because of its relatively young age structure and large ethnic minority groups (see Chapter 14). In a totally closed system, such as the Earth, this is the only element of change. But at spatial scales below this, migration will influence population change. The geography of population change is therefore complex, since multidimensional behaviours have to be accounted for, such as fertility decisions, the behavioural factors that influence mortality and the complex decisions that determine migration. Nevertheless, from Figure 7.1 it is clear that the areas suffering most population loss between 1981 and 1997 are the older industrial areas of the Welsh valleys, the West Midlands and northern conurbations, Clydeside and some of the peripheral islands. Many rural areas in the North have gained due to **net migration**, much of it from the losing areas. Apart from Bristol, Medway and Portsmouth, the South has shown a total gain in population.

These three components of change – fertility, mortality and migration – are now discussed in more detail.

7.3.1 Fertility

This is the 'in' or entry to the system. In crude demographic terms, people are born and are thus added to the stock of population. Human fertility is a very complex phenomenon and variations in fertility are difficult to explain (Graham 1998a). Conventional explanations include variation in structural features and demographic factors such as mortality (Andorka 1982), age structure (Coward 1986) or sex ratio (Wilson 1978). The structural impacts of migration on sex ratios and age structures can also have a direct effect on fertility (Creton 1991).

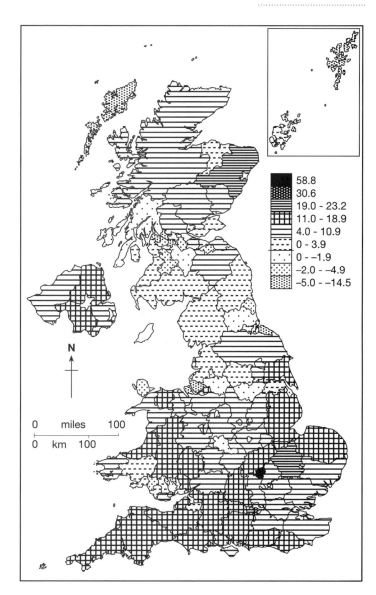

Figure 7.1 Percentage total population change, counties and Unitary Authorities, England and Wales; New Councils, Scotland; Health and Social Services Boards, Northern Ireland; 1981 to 1997

Source: Data derived from ONS 1999a, *Regional Trends 34*, Tables 14.1, 15.1, 16.1, 17.1.

As well as variations in demographic phenomena accounting for differentials in human fertility, socioeconomic variables have also been hypothesised to be crucial in accounting for spatial differentials. Thus income, housing tenure, education, religion, ethnicity, occupation and social class have been shown to affect fertility (Andorka 1978; Cooper and Botting 1992; Graham 1988, 1994a). Because these causal variables vary spatially, it should follow that there will be geographical variation in fertility.

Figure 7.2 does show a varied pattern of fertility, though the overall pattern is somewhat difficult to interpret. Here, the **total period fertility rate** (TPFR) is used. This takes account of the age structure of a population and is thus a more accurate measure than the more commonly used **crude birth rate**. At this level of analysis the high fertility of Northern Ireland is evident. Parts of industrial Wales also manifest high levels, as do some of the areas with significant non-white ethnic minorities. Closer

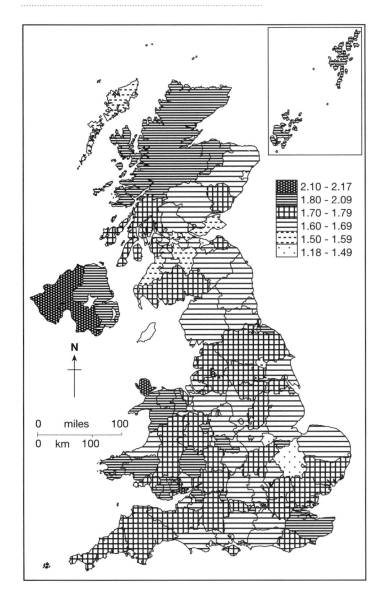

2.10 - 2.17
1.80 - 2.09
1.70 - 1.79
1.60 - 1.69
1.50 - 1.59
1.18 - 1.49

N

0   miles   100
0   km   100

Figure 7.2 Total period fertility rate, counties and Unitary Authorities, England and Wales; New Councils, Scotland; Health and Social Services Boards, Northern Ireland; 1997

Source: Data derived from ONS 1999a, *Regional Trends 34*, Tables 14.1, 15.1, 16.1, 17.1.

inspection of London boroughs, for example, shows high TPFR in Newham (2.68), Tower Hamlets (2.40), and Hackney (2.35), but in other Inner London boroughs fertility is very low, for example, Westminster (1.17), Kensington and Chelsea (1.25), Hammersmith and Fulham (1.37). As with high fertility, the areas with the lowest levels are widely dispersed, although the generally low fertility in Scotland can been seen in the map. Within England

and Wales, Armitage (1997) has noted the above average fertility in the 'manufacturing' group of local authorities, with correspondingly low fertility in the 'services and education', 'resorts and retirement' and 'most prosperous' groups. Not only is fertility higher in the 'manufacturing' areas, but most of this is concentrated among young women. These are the areas which often suffered economic and social deprivation as the industrial base declined, and which have also been shown to be the least healthy.

At a more refined level of analysis the lowest fertility is in Cambridge District (1.11) and the highest is in the Isles of Scilly (5.14). This latter figure is exceptionally high and is more akin to the TPFR found in many African countries. It is accurate but is based on small numbers of births and is thus subject to statistical artefact. What is striking is the number of areas (as used in Figure 7.2) that have TPFR above replacement level. Only Blackburn and Darwen in the North West, Torfaen in Wales, and Western and Southern in Northern Ireland, have a TPFR above 2.1 children per woman. Below this level a population is in decline (leaving aside the effects of migration). With a TPFR of 1.81 the UK fertility is somewhat below that. Of course, this is an average, the most fecund woman in the UK is a 40-year-old who has given birth to twenty children (Boseley 1999a). Of the four countries, only Northern Ireland, with a rate of 2.59, has an above replacement level (ONS 1999a). Fertility has declined since the baby boom peak of 2.93 children per woman in 1964 (Armitage 1997). This has clear implications for policy-makers in terms of provision of schools, health care and the like (Jackson 1998). The low birth rate, combined with improved **life expectancy at birth**, has much wider implications in terms of the ageing of the population (see pp. 65–6).

One of the most widely commented on aspects of fertility is that of teenage pregnancy. The UK has the highest teenage pregnancy rate in Europe, and, according to Tony Blair, this is 'appalling and it should be a matter of anxiety and concern to anyone who believes in the future of the country' (quoted in Rawnsley 1999: 8). In 1996, fertility rates among the 15–19 age group were lowest in the Netherlands at 4 per thousand and highest in England and Wales at 30 per thousand (Botting 1998). The North East is the region with the highest incidence of teenage births at 41 per thousand. The North West, York-shire and the Humber, Wales, Scotland and the West and East Midlands all have higher rates than the UK average. Here again is evidence of the North–South divide. As Botting (1998) has noted, most teenage births occur in local authorities

characterised as 'coalfields', 'manufacturing' and 'ports and industry'. For example, 'Dundee has an unfortunate reputation as the country's teenage pregnancy capital', at 69 per thousand (Wazir 1999: 18). Nottingham is another city with an unenviable rate of teenage pregnancies, with 100 in every 1,000 girls aged 13–15 getting pregnant (Scott 1999). There have been a number of well-publicised examples recently of 12- and 13-year-olds giving birth (e.g. Barnett 1999; Reeves 1999a).

Not all teenage pregnancies are unplanned, and some of the unplanned result in wanted babies. But it is among this age group that most unplanned and unwanted pregnancies occur. 'These pregnancies can have long term implications on the health and socio-economic future of both the mother and child' (Botting 1998: 19). The health risk associated with high levels of unprotected sex among these young women and girls is also a major cause of concern (Anning 1999; Arlidge and McVeigh 1999; see Chapter 10). Since 90 per cent of teenage mothers receive income support (*The Economist* 2000a) the government is keen to tackle this issue. There are proposals to use the benefits system to try to reduce the rate of teenage pregnancy, in line with methods which have proved so successful in reducing teen-age pregnancy in the USA (Summerskill 2000a). Other methods include free contraception provided at school, hostels for teenage mothers and reminding teenage boys that they will be pursued by the Child Support Agency to pay for the upkeep of children they father (Wintour and Bright 1999). However, the government is reluctant to follow the very successful Dutch model of full and frank sex education from a young age, for fear of upsetting Middle England (Bright 1999; Laurance 1999).

This is something of a paradox given that the average age of first birth has risen quite dramatic-ally in recent years. The mean age of mother at birth increased from 26.2 years in 1971 to 28.8 in 1997 (*Population Trends* 1999). Women are not only having fewer children but are starting families later. Although most births are to women aged 25–29, an increasing number are to women in the older groups. Birth among the forty and over age group is becoming more common, as the well publicised pregnancy of the Prime Minister's 45-year-old wife, Cherie Blair, demonstrated. However, although birth postponement is usually to do with career choices and financial security, the health of the mother and child can be compromised during pregnancy and birth among older women (Reeves 1999a). Advances in fertility treatments and obstetric care mean that women in their fifties and sixties can now have children. Again, this has caused comment among the media and policy-makers. It seems that when it comes to fertility women cannot win and certainly are not expected to have the same range of fertility as men (Walter 1999). There is a clear divide in attitude to children according to social background, and peer pressure has an important bearing on fertility behaviour (Adams 1999; Wazir 1999).

Despite the predictions of some population analysts (e.g. Andorka 1982; Hawthorn 1982) that fertility differentials would diminish or disappear in economically advanced societies, it is clear that, in the context of the UK, there is considerable variation in fertility at regional and district level. Indeed, these changes in fertility fit well with **demographic transition theory**, which postulates low but fluctuating fertility rates in advanced societies.

The geography of fertility is the result of complex multidimensional processes and it is difficult to argue with Jones (1990: 133) that fertility variation is much less dependent on place than mortality variation. Nevertheless, fertility behaviour can be influenced by the social and cultural environment, and fertility patterns themselves impact on the social, cultural and even physical geography of an area in terms of age structure and education, leisure, retail, housing and other provision.

7.3.2 Mortality

This is the 'out' or exit from the system. Mortality rates in economically advanced countries are relatively low. This, coupled with low fertility, has meant that the population is ageing and this is dealt with below. The decrease in mortality at all ages has led to an increase in life expectancy, which for males is 74.4 and females 79.6. Life expectancy in the UK

has increased by more than 30 years this century and is increasing by around two years every decade (Browne 1999a). The gender differential is accounted for by physiological factors – mostly hormonal – and by differences in behaviour and lifestyle. However, many of the dangerous occupations that once killed men either no longer exist (see Chapter 8), or else are much more strictly regulated. Further, more women are engaging in behaviours that puts their health at risk such as smoking, drinking, sexual promiscuity, dangerous driving and the like. There is thus expected to be a convergence in the sex differential in the future. Indeed, the Institute of Actuaries has noted that life expectancy is increasing faster for men and that the gender gap has narrowed by one year since 1980.

The most important cause of death is circulatory diseases (mainly heart attacks and strokes), which account for over two-fifths of all deaths. Cancer causes around 25 per cent of deaths and respiratory disease 16 per cent (Pearce 1998). Aside from road traffic accidents (especially among young men) most mortality is caused by chronic disease rather than infectious disease. 'Social environment and lifestyle have been the main driving forces behind ischaemic heart disease' (*Population Trends* 1997: 44), the most likely causal agents being diet (especially saturated fats) and smoking. Behavioural change, it would seem, could bring about a reduction in coronary heart disease. There is a very marked geography of mortality from various causes due to the complex and multidimensional nature of health and death, as the maps provided by Coleman and Salt (1992) and Dorling (1995) show. This is borne out by Figure 7.3. Here, we use the **standardised mortality ratio** to take account of the different age and sex structures in different areas. Otherwise, if the commonly used **crude death rate** were used, some areas with substantial elderly populations, such as the south coast of England or Cornwall and Devon, would show very high mortality rates. The areas with the highest mortality include the older industrial districts that suffer from multiple deprivation and unhealthy lifestyles (Coleman and Salt 1992). Apart from some urban areas, such as Luton, Medway, Slough and Portsmouth, the North–South divide in

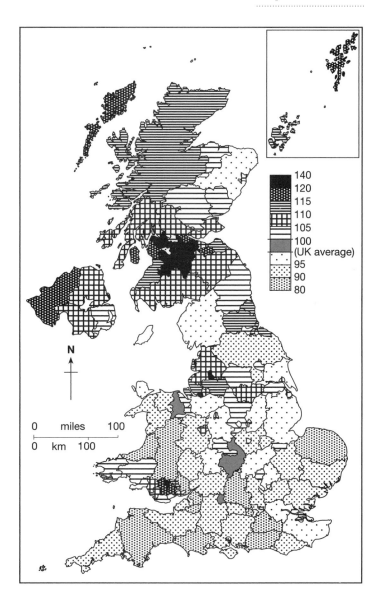

Figure 7.3 Standardised mortality ratios, counties and Unitary Authorities, England and Wales; New Councils, Scotland; Health and Social Services Boards, Northern Ireland; 1997

Source: Data derived from ONS 1999a, *Regional Trends 34*, Tables 14.1, 15.1, 16.1, 17.1.

mortality distribution in the UK is evident (see Chapters 10 and 14).

One aspect of mortality that is of particular interest is infant mortality. **Infant mortality rates** are particularly important when considering life expectancy and are directly related to fertility (Graham 1994b). The mortality of infants in the UK is now about as low as it is possible to get. The rate dropped from 17.9 per thousand in 1971 to 5.7 per thousand in 1998 (*Population Trends* 1999). Improvements in diet, housing, health of the mother, obstetrics, coupled with universal vaccination programmes, mean that infants no longer die in the large numbers that were once common. The virtual eradication of infectious diseases, such as measles, rubella, diphtheria, whooping cough, polio, chickenpox, and the like, means that accidents or congenital problems are the biggest causes of infant death.

Table 7.3 Percentage of births outside marriage, 1997; infant mortality rate, 1996–8; increase in number of households, 1981–2006; percentage of population aged 65 and over, 1997; percentage of households lone parent with dependent children, 1998; UK and statistical regions

Area	Births outside marriage	Infant mortality rate	Increase in household numbers (millions)[a]	Population aged 65 and over	Lone parent with dependent children
UK	36.7	5.9	5.03	15.7	6.6
North East	46.2	5.8	0.14	16.0	7.3
North West	41.1	6.4	0.38	15.7	8.2
Yorkshire and the Humber	39.8	6.6	0.37	15.8	6.5
East Midlands	37.8	5.8	0.42	15.9	5.7
West Midlands	37.5	9.7	0.38	15.6	6.5
East	32.2	5.1	0.63	16.0	5.1
London	35.0	9.2	0.62	13.1	8.4
South East	31.5	7.9	0.93	16.1	5.1
South West	34.7	5.4	0.57	18.5	5.4
Wales	42.8	5.7	0.21	17.3	7.4
Scotland	37.7	5.7	0.38	15.2	6.7
Northern Ireland	26.6	5.6	—	13.0	8.0

Source: ONS 1999a, *Regional Trends 34*, Tables 14.2–17.2, 3.19, 3.4, 3.20

Note: [a] Great Britain only.

Most infant mortality is now confined to the perinatal – stillbirths and deaths within the first week of life. The **perinatal mortality rate** is a sensitive indicator of the quality of a health service within an area (Farmer and Miller 1983). Despite the very low levels, there are still geographical variations in infant mortality (see Table 7.3). However, at this crude spatial scale it is difficult to determine a pattern. Although infant mortality is still related to conditions of the social and economic environments of mother and child, it seems that access to health services is also important (Graham 1994b).

7.3.3 Migration

This is the 'shake it all about' of the system where population is redistributed. Migration is normally studied from two aspects – internal and international. In advanced capitalist countries there are normally few formal limits on mobility. Now that both fertility and mortality rates are relatively stable, internal migration is the main component affecting population distribution and structure. The UK has had a long tradition of being involved in modern international migration streams. Parts of the UK have long been net exporters of people, especially Ireland and Scotland, and absorbed successive waves of immigrants. However, in recent years, along with other advanced capitalist countries, the UK has introduced ever-stricter controls on many types of international migration, especially from developing countries (Graham 2000). Because this is as much a policy issue as a demographic one it will be dealt with in more detail below.

That said, it is still a relatively important part of the demographic equation in the UK. In 1997, an estimated 92,000 more people migrated *to* the UK than *from* the UK, and there has been a net gain due

Table 7.4 Net international and inter-regional migration, 1981 and 1997

Area	International migration (thousands) 1981	International migration (thousands) 1997	Inter-regional migration (thousands) 1981	Inter-regional migration (thousands) 1997
UK	−80	30	—	—
North East	−10	1	−8	−6
North West	−9	12	−5	−10
Yorkshire and the Humber	−5	2	−5	−7
East Midlands	−5	3	5	11
West Midlands	−2	6	−12	−11
East	−9	1	17	20
London	−6	33	−32	−55
South East	−9	2	36	24
South West	−3	5	20	32
England	−58	65	1	−4
Wales	−8	7	3	5
Scotland	−11	−10	−1	2
Northern Ireland	−2	−1	−3	−3

Source: ONS 1999a, *Regional Trends 34*, Table 3.11

to international migration for the ten years prior to this (ONS 1999b), though this was not the case in the early 1980s (Table 7.4). Of the constituent countries, England receives the vast majority of international immigrants, and most of the gain is in non-British citizens while the loss is of British citizens. Both Scotland and Northern Ireland are net losers in the international migration equation. Most immigrants come to the UK for formal education and most emigrants leave for work-related reasons (Vickers 1998). Because of changes in immigration law, the UK now receives more immigrants from the EU and the USA than the New Commonwealth countries, though the Old Commonwealth remains an important source. For example, around 200,000 French live in the UK, mostly in London, ten times the number of ten years ago. They are the UK's seventh biggest immigrant group, and the sixth biggest source of immigrant labour (the largest is still Ireland) (*The Economist* 1999b). There are around 200,000 US citizens living in the UK, twice the number of ten years ago, despite the closure of many

US military bases (Reeves 2000a). This is seen as part of the Americanisation of British culture (see Chapter 11). Nevertheless, most of the UK's migrants accepted for settlement still come from Asia and Africa.

Though governments have attempted to channel population movements within the country through regional policy (see Chapter 14), these efforts have largely failed. Because fertility and mortality are so stable in economically advanced countries, most local population change occurs via patterns and trends in internal migration. Internal migration within the UK has increased, presumably as more and more people are 'getting on their bikes' and looking for work or better living conditions. The country can be considered 'a nation on the move' (Champion *et al.* 1996: 8). The broad regional differentials, in terms of net change between 1981 and 1997, can be seen in Table 7.4. In 1981, apart from London, the North–South divide is clear in that the northern regions lost more people than they gained. This pattern was repeated in 1997, though it is noticeable that Scotland had more people moving in than out, while the reverse was the case in England. As Champion (2000: 176) has noted, 'The principal feature of regional population change throughout the twentieth century has been the drift of population from North to South.' The North loses about 23,500 people every year to the South (Reeves 1999b). Such is the loss of population from peripheral areas, such as the Western Isles, that communities can easily become unviable.

Although the UK is a very urbanised country, since the late nineteenth century it has mostly been in the form of **suburbanisation**. The development of electric tramways allowed people, mostly the wealthy, to move out of the city to newly developed suburbs. This migratory trend increased with the advent of affordable motor cars. Suburbanisation is particularly problematic in the South East where efforts to reconcile the demand for new development land with demand for housing has become a major political issue (see Chapter 16). A more recent trend has been a tendency towards **counter-urbanisation**. Dispersal of people away from the large urban areas has occurred throughout this century but accelerated noticeably from about 1960 (Compton 1991; see Chapter 16). While people are moving out of cities there has been an increase in a return to city living by some people, notably the young better-off professional people who are attracted to city life and proximity to work and leisure (see Chapters 9 and 16). However, as the assistant director of the Council for the Protection of Rural England has noted, 'At the moment, city living is still seen as a pioneering thing, not the norm' (quoted in McGhie 1999: 18). The government is very keen to see more people remaining or returning to the cities but the current tax situation penalises developers who want to renovate old buildings as dwellings. It is much cheaper and easier to develop greenfield sites. The situation is somewhat different in Scotland where a large stock of high density, mixed quality tenement flats (along with more liberal liquor licensing laws) has ensured that the inner cities have remained viable and vibrant places, much like many Continental cities (see Chapter 9). Despite a trend towards counterurbanisation, the

UK remains the most urbanised country in Europe. 'The continuity of urban dwelling is arguably the most salient stable element' of the UK population (Compton 1991: 39).

7.4 POPULATION ISSUES

There are a number of issues directly or indirectly related to the main demographic phenomena discussed above. These are either a result of the main components of the demographic equation or affect the nature of the population in some way. They have all attracted the attention of the media and the policy-makers and are thus worthy of some further consideration here. They are also linked with the other issues discussed in this book.

7.4.1 Extra-marital fertility

One of the most newsworthy aspects of demography in recent decades is that of extra-marital fertility. Indeed, extra-marital fertility was relatively rare until the 1960s (Coleman and Salt 1992), subject, as it was, to social stigma and the pressure on couples to engage in a 'shotgun marriage' to legitimise the yet-to-be-born child. Since 1987, the Family Law Reform Act ended the legal discrimination against children born outside marriage and the terms 'legitimate' and 'illegitimate' no longer have a legal (as opposed to a social) meaning (Cooper 1991). However, while the 1987 Act removes the word 'illegitimate' from much of the legal framework it does not remove *all* of the legal differences following from the marital status of the child's parents (Stevens and Legge 1987). Within the UK different legal systems mean that definitional differences exist between Scotland and England and Wales (Illsley and Gill 1968).

Aside from the 'moral outrage' and welfare arguments that surround extra-marital fertility it is still worthy of consideration because of the associated high infant mortality, stillbirths and low birth weight (Botting 1998; Gill *et al.* 1970). Much of this is due to the fact that many extra-marital births are first births, that mothers seek antenatal care late into pregnancy and unmarried mothers tend to be drawn from the lower social classes (Benjamin 1965; Haskey 1995). Unlike teenage pregnancies, the UK's extra-marital fertility rate is not the highest in Europe. However, with 36.7 per cent of live births, it is still relatively high (Table 7.3). Extra-marital fertility in Northern Ireland is noticeably low, reflecting both the strength of organised religion and the traditional family. Otherwise the North–South divide can be seen again in terms of this aspect of human behaviour. In 1997, over half of all births in Nottingham, Hull, Liverpool, Blackpool, Redcar, Middlesbrough, Hartlepool, Merthyr Tydfil, Dundee and Glasgow occurred outside marriage (ONS 1999a).

It would seem that increasing marital breakdown, acceptance of lone mothers, postponement of marriage, increase in consensual unions and increasing social and economic independence of women will ensure that the trend in extra-marital fertility will increase and probably reach the proportions currently found in Scandinavian countries. Nevertheless, because extra-marital fertility is influenced by so many variable factors, which themselves vary over space, there will continue to be a geography of this aspect of human behaviour (Graham 1998b).

7.4.2 Ageing population

Like all countries, the population of the UK is ageing. Simply put, the average age is increasing and the percentage of the population in older age groups is expanding at the expense of younger age groups. In 1851, only 4.6 per cent of the population was aged over 65. By 1997 the proportion was 15.7 per cent, and by 2020 the proportion is predicted to be 19.3 per cent (ONS 1999c). Not only has there been an expansion of the elderly, those from retirement age to age 75, but the so-called old elderly, those over 75, have also increased as a percentage of the population (Thane 1989). More remarkable has been the increase in the number of centenarians. Before the Second World War, they 'were very rare and were more the objects of curiosity than anything else'. But since the 1950s,

numbers started to rise and by 1996 there were 6,000. Although the number is small, the rate of increase has been described as an 'explosion' (Thatcher 1999: 6). The number of centenarians is likely to increase tenfold by 2031 (*Population Trends* 1997). By the 2080s, at least one person a year will reach 116 in the UK. The policy implications of this are clear. As *The Economist* (2000b: 37) points out, 'Someone who retires at 60 and lives to 110 is an actuaries' nightmare.' Dependency ratios are increasing as more and more elderly must be supported by fewer people of working age. Although the elderly are spatially ubiquitous some areas have high concentrations (Table 7.3) and this puts considerable strain on local services. At a finer scale, the percentage of population of pensionable age ranges from 11.5 in Milton Keynes to 33.0 in Christchurch, in Dorset (ONS 1999a). This is largely the result of retirement migration, with increasingly large numbers of people retiring to seaside and rural areas such as the Scottish Borders, North Wales, the South West and the south coast of England.

One of the main issues surrounding ageing is the dependency burden – the non-active population as a proportion of the active. The UK's dependency rate is lower than some other EU countries – Italy, Spain and Luxembourg. However, the UK currently has one dependent person per non-dependent. Northern Ireland has the highest level of dependency, due to higher numbers of children rather than elderly. Scotland has much the same proportion as England. Wales and the northern regions of England also have high dependency rates, while the South East has the lowest (ONS 1999a). This increasing dependency burden is of particular concern from an economic perspective. The state pension is no longer seen as a viable way of sustaining ever increasing cohorts of pensioners. Although occupational and private pension schemes have increased in popularity, many young people are making no provision for retirement. 'Pensions are not a sexy subject when you're in your twenties', and 'those who fail to make sacrifices now risk joining an impoverished underclass in 20 or 30 years' time, consigned to living on incomes a fraction of those of an increasingly affluent workforce' (Brodie 1999: 2). Not all observers, however, view ageing in a negative way (e.g. Day 1995; Thane 1989) because of the economic and social contributions from this section of the community.

Even though the average age of the population is increasing and the dependency burden with it, the UK is still a very ageist society, especially when it comes to employment (see Chapter 8). For example, the proportion of men aged 50 to 64 neither in work nor looking for it has increased from 11 per cent in 1976 to 27 per cent in 1999 (Rennell 1999). However, this should change as the proportion of the work-force aged 25 to 34 falls by 17 per cent till 2006 and employers become increasingly dependent on workers 50 and over (MacErlean 1999a). The age-ing of the population is not lost on the leisure and consumption industries (see Chapter 9). The fact that the over 50s have 80 per cent of the UK's private wealth gives them considerable economic clout. The fact that there are increasing numbers of older people, and that they are more likely to be registered to vote and turn out to vote, has not been lost on politicians. Though the UK has yet to see the development of a Grey Power lobby, such as exists in the USA, there is no doubt that the economic and political strength of this group will grow and become a bigger part of the UK socio-economic and political scene.

7.4.3 Household and family change

Successive governments have made a great deal of 'the family' and the need to be supportive of family policies. Few, however, are prepared to pro-vide monetary support, particularly in the case of paid leave for fathers (MacErlean 1999b; Taylor 1999). The present Labour administration is divided in its perception of 'the family'. There are the traditionalists who want to support the 'Old Family', where the parents are married to each other. The pragmatists, on the other hand, recognise the reality of the 'New Family' (Sylvester 1999). This takes account of the variety of families that now exists – heterosexual/homosexual, married/cohabiting, siblings/step-siblings, and so on. This

demonstrates the blurring between the family and the household. Strictly speaking, the household is simply a domestic arrangement where one or more people share a common dwelling. The family, however, implies some type of relationship through blood, marriage or adoption (Benjamin 1989). People living alone are not considered to form a family (Gold 2000).

One aspect of household and family change that is worthy of note is the increase in cohabitation. The proportion of never-married women who were cohabiting in Great Britain increased from 9 per cent in 1981 to 27 per cent in 1997 (ONS 1999c). That figure is higher if divorced and widowed women who are cohabiting are included. East Anglia, the South West and the South East have the highest incidence of cohabitation and Northern Ireland, Scotland and Wales the least (ONS 1999a). There is no longer a social stigma associated with living together (Haskey 1999). Also the increase in formal female employment over the same period (see Chapter 8) means that women have more options and no longer have to rely on marriage for economic security. The increase in cohabitation has led to calls for it to be formally recognised, particularly from the Liberal Democrats, and there have been calls for it to be sanctioned by the Church of Scotland (Walker 1998). However, the Church of England still believes that the institution of marriage is the appropriate locus for family formation (Beaumont 1998).

Another issue that has been newsworthy in recent years, and has been the subject of much debate, is that of household formation (see Chapter 16). Since households and families are so closely intertwined they are both dealt with in this section. There is no doubt, however, that the number of households (but not the number of families) has been increasing in recent years and that this process is likely to continue for some time (Table 7.3). In 1901, for example there were 7.04 million households in England and Wales and by 1981 there were 19.49 million (Coleman and Salt 1992). By 2021, there are predicted to be 24 million (ONS 1999a). There are a number of reasons why households are increasing at a time when there is very little population growth. These relate to the decrease in household size. In

1901, the average household size in England and Wales was 4.6 persons; by 1989 it was 2.5 persons (Coleman and Salt 1992). One of the main reasons for this has been the increase in one-person households. In 1961, for example, only 10 per cent of households were occupied by one person. Now it is 30 per cent, and by 2010 it is expected to be 40 per cent (Rayner 2000). The main reasons for the increase in household numbers are as follows:

1 *Ageing*: As the population ages so the number of widowed people increases – mostly women. This is a major cause of single person household formation. Ageing also increases the number of people in non-private households – residential or nursing homes, and other communal arrangements.

2 *Marital breakdown*: The introduction of the Divorce Reform Act, 1969, in 1971 made the process of ending a marriage much easier in England and Wales. In fact, between 1971 and 1985 the number of divorces doubled (Ermisch 1989). Today, over one in three marriages fails (Beaumont 1998; see Chapter 15). In 1961, there were 27,000 divorces in the UK and by 1996 there were 171,000 (ONS 1999c). As well as an increase in marital breakdown, there has been a decline in the popularity of marriage. In 1996, there were half the number of first marriages than in 1970. Marriage is also being delayed. Between 1971 and 1996, the average age at first marriage rose from 25 to 29 for men and from 23 to 27 for women. This is due to a rise in pre-marital cohabitation and longer participation in further and higher education, especially among women (ONS 1999c).

3 *Growth of singletons*: Marriage has become less popular and increasing numbers of people are opting for singledom. These are the New Singletons, an increasing proportion of which are women (Rayner 2000). This is due to an increase in women's career choices and disposable incomes (see Chapter 8) and changes in lifestyle (see Chapter 9). There is also a strong link between migration and singledom. This is best exemplified in Inner London where young professional people are attracted by job prospects and live alone in the large private rented sector (Hall *et al.* 1999). Post-2000 is predicted to be the singles' century (Reeves 1999c).

Many of the above are inextricably linked to changes in the nature of families. The family is perhaps more fluid now than at any time in modern history (Stacey 1996, 1998). There are more people who are widowed; more who are divorced, re-married or separated; there are more stepchildren. Increased marital breakdown, together with an increasing number of women electing to rear children on their own, means that there has been an increase in lone parent families. One in five children now lives in such a family in the UK, and over 90 per cent of these are lone mother families (Baker 1999). Although there is likely to be an increase in the number of better-off and better-educated women opting for lone parenthood in the future, most lone mothers still tend to come from relatively deprived backgrounds and were themselves raised in a lone mother family (Harding *et al.* 1998; see Chapter 15). Most lone parent families with dependent children are in London, the North West and Northern Ireland (Table 7.3). However, most dependent children live in a family with two parents (ONS 1999a). Marriage was once seen as the starting point of a new household and a new family. But now that marital status is much less rigid than it once was it can no longer be used as a predictor of household or family status. There is a tendency for children to stay in the parental home longer, particularly males. One-third of men aged between 20 and 35 still live in the parental home, compared with only one in six women (Ellen 2000).

7.4.4 Immigration

A migrant to the UK is someone who has resided abroad for a year or more, and who states on arrival the *intention to stay* in the UK for a year or more. A migrant from the UK is someone who has resided in the UK for a year or more, and who states on departure *the intention to reside* abroad for a year or more (Vickers 1998).

The United Kingdom has been described as having a unique approach to immigration (Hollifield 1997; Layton-Henry 1994). It 'is *not* a nation of immigrants, and is emphatically not a "country of immigration"' (Cornelius *et al.* 1994: 21). As an early example of a modern supranational state the UK has long had free movement between its constituent countries. Even after the secession of what became Eire there was still free movement within the British Isles for UK and Irish citizens. Immigration restriction began in the UK with the Aliens Act of 1905, followed by the Aliens Restriction Act of 1914. These, with an amendment in 1919, and the Aliens Orders in 1920 and 1953, regulated foreign immigration into the country (Juss 1994; Rees 1993).

Because of the UK's imperial history, the Conservative government in the early 1960s reckoned that a quarter of the world's population was legally entitled to enter the country. During the 1950s, around 20,000 immigrants from the New Commonwealth entered the UK every year, mostly from the Caribbean. By 1960, the figure was 58,100, and by 1961 it had risen to 115,150 (Sked and Cook 1990). The government was keen to encourage this immigration because of the labour shortages in the post-war era of full employment. Some employers, particularly the NHS and London Transport, led recruitment drives in the Caribbean. Further, these immigrants were attracted to England because migration to the USA was denied to them as a result of the McCarren–Walter Immigration Act of 1952. However, the increase in non-white immigration alarmed some people and was highlighted locally because of the geographical concentration of immigrants in specific areas (D. Phillips 1998). This helped to fuel the 'racial' disturbances between blacks and whites in Nottingham and London's Notting Hill in 1958 (Marwick 1996) and 'indicated a new and unattractive aspect of British social life' (Sked and Cook 1990: 202).

Public alarm at the scale of non-white immigration resulted in the 1962 Commonwealth Immigration Act. This Act was blatantly racist as it set quotas from New Commonwealth countries but did nothing to control immigration from the mainly white Old Commonwealth or from Ireland. This was essentially a reaction to racism that was not exclusively expressed by the political right. It received its best expression by the infamous **'rivers of blood'** speech of Enoch Powell in 1968, prompted by the prospect of a significant influx of Asians who were being threatened with expulsion from Kenya. This was followed by the 1968 Commonwealth Immigrants Act. Shortly after came the 1971 Immigration Act, which no longer distinguished between Commonwealth and other immigrants, created a new offence of illegal entry, doubled the maximum fine for harbouring illegal immigrants and extended the time over which offenders remained liable to prosecution. British immigration policy is shaped by political and cultural considerations (Findlay 1994). Since the 1970s UK immigration policy has become more restrictive, particularly in the area of asylum seekers and refugees. Occasionally, threats to national health are employed in order to impose restrictions on access. These have ranged from smallpox and typhoid to HIV/AIDS and other sexually transmitted diseases (Graham and Poku 1998).

The publicity given to crime committed by immigrants and refugees has helped fuel the debate on immigration policy generally and asylum policy specifically. The right-wing press has made a great deal of the UK being 'swamped' by 'bogus' refugees and the government has promised to process asylum applications more rapidly. Allegedly high rates of petty crime committed by refugees from the Balkans has also attracted the attention of tabloid editors. More coverage, including some from the left-wing press, has been given to organised crime committed by immigrants such as the Jamaican Yardies (Vulliamy and Thompson 1999), Turkish heroin smugglers, West African fraudsters, Chinese illegal immigrant smugglers, Colombian cocaine dealers, Indian and Pakistani heroin and illegal immigrant smugglers and Russian 'mafioska' prostitution racketeers (Thompson 1999).

In the short to medium term, it is unlikely that any government will institute policies to open up immigration. One of the main worries about further

EU expansion eastwards is fear of an increase in immigration from eastern European countries. In the longer term, however, it is likely that the present restrictive policy will have to change. Below replacement level fertility means that population decline will become an economic and security issue. The simplest way of increasing population, and increasing it selectively, is by encouraging immigration (Browne and Reeves 1999). The low fertility levels and declining populations have given rise to calls for increased immigration quotas. The UK government has launched a campaign to attract skilled immigrants, mostly in the computer industry. In 1999, net immigration to the UK reached 185,000, an all-time record (Browne 2000a). Jean-Pierre Chevènement of France has claimed that in the EU between fifty to seventy-five million immigrants 'could sensibly be admitted over the next five decades' (*The Economist* 2000c: 46).

7.4.5 Ethnic minorities

The 1991 Population Censuses of England, Wales and Scotland asked a question on **ethnic group** for the first time. No such question was asked in Northern Ireland, where the main ethnic cleavage is along religious lines. A question on religion has been asked in what is now Northern Ireland since 1801. Unfortunately there was no attempt to disaggregate the white ethnic group, since the main source of interest from the government's point of view was secondary genetic characteristics (Compton 1996). The question is based on 'race' rather than 'ethnicity' and 'was aimed at identifying the size and distribution of the main *visible* ethnic minority groups in Britain' (Bulmer 1996: 35). So, while there is a mass of detailed information on the relatively small non-white ethnic groups, there are no data to differentiate the large, but heterogeneous, white population – although the Irish are treated as an 'honorary ethnic group by being included in many of the ethnic volumes tabulations through birthplace' (Coleman and Salt 1996: 13). The only way to analyse the various white ethnic groups is by country of birth data, an inexact way of determining the geography of ethnicity. That said, we can draw some

conclusions about the distribution of ethnic groups other than those recognised by the Office for National Statistics (ONS) and the government.

If we accept the definition of ethnic group provided in the glossary and drawn from Bulmer (1996) and *The Dictionary of Human Geography* (Johnston *et al.* 1983) then the biggest ethnic group in the UK is the Scots, constituting 8.8 per cent of the population. The Welsh form the second largest group, with 3.5 per cent. The Northern Irish account for 2.6 per cent. Some 2 per cent were born in the Irish Republic, but a much larger proportion than this can claim recent Irish ancestry. The percentage contribution other ethnic groups make to the UK population are: Indians (1.6), Pakistani (1.0), Black Caribbean (0.9), Jews (0.5), Black African (0.6), Bangladeshi (0.4), Chinese (0.3), other Asian (0.3), Black other (0.2), non-specified (0.2), Roma/gypsies (0.2) (Minority Rights Group International 1997; Schuman 1999).

Clearly defining an ethnic minority depends on geography, as well as the cultural and other traits that constitute ethnicity (see Chapter 11). Thus, while the Scots are the UK's largest ethnic minority, they do not constitute a minority in Scotland. Similarly, there are districts in London, Leicester and Birmingham where 'whites' will constitute a minority. In some parts of Wales, large numbers of English migrants mean that the Welsh are in a minority. The largest immigrant group in Britain is the English in Scotland and Wales (Dorling 1995), and the English form the largest ethnic minorities in both these countries. Thus minority status only occurs at a given level of spatial resolution.

From the latest estimates, the non-white ethnic minority population totals 3.6 million or 6.4 per cent of the British total. That is an increase from 5.8 per cent in 1992 (Schuman 1999). Given that immigration has been seriously curtailed, much of this growth is due to the higher natural growth among non-white ethnic groups. Their relatively young age structures predispose to lower mortality and higher fertility. And for cultural reasons, average family size is higher among some groups, particularly Pakistani and Bangladeshi. The South Asian group

– Indian, Pakistani and Bangladeshi – constitutes almost half of the non-white ethnic minority groups.

One of the most notable features of the ethnic minority populations is how concentrated they tend to be in specific areas. The indigenous ethnic minorities (see Chapter 11) are concentrated in their homelands (Conner 1986). Thus, as we might expect, the Scots are concentrated in Scotland, the Welsh in Wales and so on. The non-white ethnic minorities also demonstrate a high level of spatial concentration and are heavily over-represented in a small number of urban areas in England. Almost half of all the non-white ethnic minority population lives in London and it is predicted that by 2010 whites will be a minority in London (Browne 2000a). In Inner London, almost a third of the population is of non-white ethnic minority origin. It is also home to large numbers of white ethnic minorities from the UK and other countries, particularly Ireland. Together, the metropolitan counties of Greater London, Greater Manchester, West Yorkshire and West Midlands contain almost 75 per cent of the non-white ethnic minority population. In contrast, Scotland, Northern Ireland, East, North East and South West house very few of the UK's newer ethnic minorities.

Of all the groups, the Chinese are the most ubiquitous, and can be found widely dispersed throughout the country. This is due to their high involvement in the restaurant and fast-food trade. Other individual groups are more concentrated. London, for example, contains 85 per cent of the total Black African population and almost 65 per cent of the Black Caribbean. The Pakistani population is concentrated in the metropolitan counties of Greater Manchester, West Yorkshire and West Midlands. Even within the South Asian group there are clear preferences in area of residence. Over half of all Bangladeshis and 45 per cent of Indians live in Greater London, compared with less than 20 per cent of the Pakistani population (Schuman 1999).

This spatial concentration of the various non-white ethnic minority groups is often explained in terms of choice and constraint theories. The choice theory adopts the view that ethnic minorities will opt to live in specific areas where social, cultural and economic support is readily available. The constraint theory argues that ethnic minorities are constrained and limited by discrimination and exclusion. The reality for many members of the various ethnic minorities is likely to be a combination of the two.

7.5 SUMMARY

Population is at the heart of any study of human geography. Although the growth of population has slowed down in the UK some areas are experiencing population growth at the expense of other areas that are witnessing a decline in population. Although some of this difference is due to the balance of births over deaths, most is caused by differentials in migration. Fertility and mortality rates remain low and although there are more births than deaths the UK fertility rate is below replacement level and the population will begin to decline in terms of natural change. It will take a change in government policy over international immigration to prevent the population going into terminal decline. Even though some of the population is on the move within the country, the distribution remains very uneven over the whole country.

There are many aspects of population that impinge on policy at local and central government levels. The ageing of the population is of particular concern in terms of increasing dependency and financial support. The marked social changes in terms of family and household formation, marital breakdown and increase in extra-marital fertility also have policy implications. Between the early 1970s and the mid-1990s, the numbers of people marrying fell by 40 per cent, the annual number of divorces doubled, the number of lone parent families almost trebled and the proportion of births outside marriage quadrupled (Gold 2000). The growth of the non-white ethnic minority population went a long way in bringing about the draconian immigration legislation now in place. The concentration of these groups in specific areas also has policy implications, insofar as the provision of services and the implementation and monitoring of anti-discriminatory legislation are concerned. Normally, policies will derive from population issues – policies on immigration, 'race' relations, single mothers, pensions, nursing homes, health provision. Nevertheless, even though the UK has no 'population policy', governments can and do influence the dynamics of population through legal and fiscal means (Jackson 1998).

R E V I S I O N   Q U E S T I O N S

- What are the three main components of population dynamics?
- Why is the population of the UK so unevenly distributed?
- What is meant by an ageing population?
- Why is the number of households increasing yet the number of families decreasing?
- Why is geography so important in our understanding of ethnic minorities?

KEY TEXTS

Champion, T. (2000) 'Demography', in V. Gardiner and H. Matthews (eds) *The Changing Geography of the United Kingdom*, London: Routledge, pp. 169–89.

Champion, T. *et al.* (1996) *The Population of Britain in the 1990s: A Social and Economic Atlas*, Oxford: Clarendon Press.

Coleman, D. and Salt, J. (1992) *The British Population: Patterns, Trends and Processes*, Oxford: Oxford University Press.

Jackson, S. (1998) *Britain's Population: Demographic Issues in Contemporary Society*, London: Routledge.

WORK

8.1 INTRODUCTION

In Chapter 3 we noted that until the 1970s there were labour shortages in the UK, but since then near full employment has given way to slower growth and mass unemployment. Since the 1970s there have been insufficient jobs in the UK for the working population (see Chapter 3), and for increasing numbers in **paid work** it has become less secure and reliable (Hutton 1995). The income received from paid work is essential for maintaining well-being for most individuals and households in the UK (for a fuller discussion see Chapter 4). A distinctive geography of paid work has emerged, with regional divergence (a North–South divide), as well as growing inequalities at intra-regional and intra-urban levels. In this chapter we focus on the changing geography of paid work to see how changes in the economic environment, social trends and demographic developments have come to create a 'new landscape' of employment over the last few decades (Box 8.1). This introduction is followed by a discussion of paid and **unpaid work**, unemployment and economic inactivity. The third section examines paid work through the lens of sectoral change. Section four highlights demography, social change and paid work. In section five changing work practices are featured. But first we will briefly explore the relationship between home and work.

Since the Industrial Revolution, 'home life' (including unpaid work, such as childcare, cleaning, and the like) and 'work' (work for wages; that is, paid work) have taken place in different domains. These domains thus became separated spatially and deeply gendered, with 'home' a largely female domain and 'work' a male domain, with a stereotypical nuclear family comprising a male 'breadwinner' and a female 'homemaker' (Horrell and Humphries 1995; the notion of separate spheres is contested). A woman's participation in the labour

Box 8.1
The new landscape of employment

- the loss of over five million manufacturing jobs between 1960 and 1998
- the growth in the number of service sector jobs
- a reduced demand for traditional skilled manual labour – predominately men – alongside increasing participation and employment rates amongst women
- a greater premium being placed on higher level skills/qualifications, and a reduction in employment opportunities for those with no/few formal qualifications
- a growth in flexible working – notably part-time, contract and temporary working, and an enhanced prevalence of labour market insecurity
- increasingly those in full-time employment work more than their contracted hours
- the superseding of 'bureaucratic' organisational forms and career patterns by new 'flexible' and 'adaptive' forms
- male unemployment rates and those of some ethnic minority groups have remained persistently high
- a growth in the number of people disconnected from the labour market
- the distinctions between employment, unemployment and economic inactivity have become less clear cut
- the informal economy has continued to grow

Figure 8.1 Weaver's cottage, East Midlands; home and work are located in the same place

Source: Authors.

market during this period was one that was dependent on her social class and geographical location. Prior to the Industrial Revolution, and even during the **proto-industrial** phase, work and home would often be the same place, for example this weaver's cottage (Figure 8.1).

A distinct middle-class consciousness developed as a result of these economic changes. A man's status – in terms of class – became dependent upon his wife not going out to undertake paid work. The essentially Victorian and middle-class ideal of womanhood therefore stood for the sanctity of the family. At the heart of this family ideology was the belief that the male breadwinner went out to work to maintain his wife and children as dependants at home (Pennington and Westover 1989). But working-class women – single and married – were visible in the labour market, in factories, such as textiles (and on farms) and in the homes of the middle and upper classes as domestic servants (Massey and McDowell 1994). Some working-class women also undertook paid work but less visibly as homeworkers, by undertaking such tasks as taking in other people's washing (Pennington and

Westover 1989). A person's ability to participate in the labour market and therefore undertake paid work is dependent upon the amount of unpaid work they have to undertake. As women still largely undertake the bulk of unpaid work, men and women participate in the labour market on a very different basis.

8.2 THE CHANGING WORLD OF WORK

There have been a number of changes in the world of work in recent years. We now explore some of the most important of these.

8.2.1 What is work?

'Work' in its broadest sense includes unpaid work as well as paid work (Glucksmann 1995). The home is an important site of work, of domestic labour (done by household members, mainly women or off-loaded to others, becoming commodified). Glucksmann (1995) uses and extends Pahl's (1989)

example of a woman ironing a shirt. Such a woman may be working in a laundry, ironing for pay, or ironing for someone else (or in someone else's home) for the same reason. A woman ironing a shirt for her family would be carrying out unpaid domestic work, or ironing a shirt for a lover/partner, as an expression of her attachment and devotion. Is this work? Pahl suggests that it is not, rather an expression of interpersonal relationships (Crompton 1997).

Statistics on employment tell us how many people are employed, what sort of paid work they do, their working patterns, as well as the number of jobs and the industry they are in. It is important to make the distinction between the number of people with a job (employment) and the total number of jobs. One person can have more than one job. The International Labour Office (ILO) – an agency of the United Nations – sets out guidelines for the measurement of employment and unemployment. Under ILO guidelines anyone working for at least an hour a week is employed (Box 8.2).

Someone who works less than thirty hours a week is classified as part-time, and above those hours they are termed full-time. But within the category part-time the hours worked can vary from as few as two up to twenty-nine hours per week. Far more women than men hold part-time jobs. The reason most women worked part-time in the UK in 1997 was because they 'did not want to work full-time' (79.2 per cent). A further 9.6 per cent 'could not find full-time work' and 10.4 per cent were 'still at school or students'. This compares with 37.9 per cent, 24.1 per cent and 34.7 per cent for men in each of these categories. More women than men have a second job, 6 per cent compared with 3.8 per cent for men (ONS 1998a).

In the UK, as in the USA, full-time (and part-time) workers increasingly work in excess of their contracted hours (for some it is as paid overtime, for others – not just managers and professionals – no extra remuneration is received) (Hardill *et al.* 1997). In the last two decades in the UK working hours have not fallen; in 1997, the average weekly hours of all full-time employees was 44 hours compared with 43.7 in 1990 and 42.6 in 1984. In most other European Union (EU) countries during the same period the average weekly hours worked have declined. The greatest number of hours worked are in agriculture and fishing (48.2 hours) and the least in public administration, education and health (42.4 hours). On average men work longer in the formal economy (45.8 hours) than women (40.7 hours) (ONS 1998a). But remember, many women also go home to a 'second shift' of unpaid work. Research by the Institute of Management shows that over 50 per cent of women managers still take *sole* responsibility for organising the ironing, shopping, cleaning and cooking (Fieldman 1999).

8.2.2 Unemployment

The definition that is generally used for measuring unemployment is that proposed by the ILO (Box

Box 8.2 ILO classification of employed people

- employees (work for a company and have their National Insurance paid for directly from their wages)
- self-employed (work for themselves and generally pay their National Insurance themselves)
- unpaid family workers (these people do unpaid work for a business they own or for a business a relative owns)
- participants in government-supported training and employment programmes (all people aged 16 and over taking part in one of the government's employment and training programmes)

Source: ONS (1998c)

**Box 8.3 ILO
definition
of the
unemployed**

- out of work, want a job, have actively sought work in the last four weeks and are available to start work in the next two weeks or so
- out of work, have found a job and are waiting to start it in the next two weeks

Source: ONS (1998b)

8.3). The Statistical Office of the EU and the Organisation for Economic Cooperation and Development (OECD) and other countries use this definition (ONS 1998b). Those out of work and who do not meet the criteria of ILO unemployment are economically inactive. The ILO unemployment rate is the proportion of the economically active who are unemployed. The economically active are people who are either in employment or ILO unemployed (see pp. 86–90). The rate of unemployment in the UK has been falling since the most recent peak in 1992, and at the end of 1998 stood at 6.6 per cent (ILO rate). The highest unemployment is in Northern Ireland (8.3 per cent) and the lowest in England. At the regional level, unemployment ranged from 4.6 per cent in the South East and Eastern regions to 11.9 per cent in Merseyside (Pearce 1998).

Another way in which unemployment is measured is the count of claimants of unemployment-related benefits (called the claimant count). It is a by-product of administrative records of people claiming benefits. The claimant count records the number of people claiming unemployment-related benefits. These are currently the Jobseekers Allowance (JSA) and National Insurance (NI) credits. The claimant count is directly affected by changes to the rules governing entitlement to unemployment-related benefits. This means that comparisons over time are affected by changes to the benefit system (ONS 1998b). Over the last two decades who qualifies as 'unemployed' has changed many times, and those households with no earners are not spread evenly across the regions and localities of the UK (see Chapter 14).

As was noted above, the economically inactive are those who are neither in employment nor ILO unemployed. This group includes those who are 'homemakers', the retired, and also the growing number who have withdrawn from the labour market and who have become disconnected from the labour market and 'mainstream' opportunities. The region with the highest percentage of households where no one was officially in paid employment was Merseyside (26.1 per cent) and the lowest was the South East (12.4 per cent) (ONS 1998a).

A growing number of workers work 'informally' as undocumented workers. They do not appear in official statistics and are not protected by UK or EU legislation. Recent research has revealed that for many such work is undertaken as a 'survival strategy' (Williams and Windebank 1999). For documented UK workers their employment rights come from two sources, UK and EU legislation. UK legislation is augmented by EU directives, which have benefited women workers in particular, for example by improving maternity rights as well as addressing the rights of part-timers to name just two examples (Townsend 1997).

The distinctions between employment, unemployment and economic inactivity have become less clear cut. Beatty and Fothergill (1996: 638) in an examination of the UK coalfields conclude that, 'a large number of those who might normally seek employment have become "discouraged" workers, resigned to live off whatever state benefits and occupational pensions are available to them'.

8.3 SECTORS OF THE ECONOMY

Traditionally the economy has been divided into three sectors: primary, secondary (or manufacturing) and tertiary (or the service sector). During the twentieth century the proportion of the working population employed in each sector of the economy has changed with the proportion employed in primary and manufacturing industry declining, while the service sector has now grown to dominate the UK economy (see Chapter 3). Employment changes for each sector of the economy will now be reviewed.

8.3.1 Deindustrialisation and the decline in male employment

Between 1960 and 1998, over five million jobs have been shed by manufacturing industry, along with almost 700,000 in extractive industries such as coal mining. By 1998, manufacturing accounted for less than 18 per cent of employees in employment, compared with 42 per cent in 1955 (Pearce 1998). This job loss is one manifestation of **deindustrialisation** (see Chapter 3). No longer can the UK be described as the 'workshop of the world' (Hudson 1997). The UK is not the only advanced capitalist economy to experience such industrial job losses, most other older industrialised countries in north-west Europe passed their peak of factory jobs in the mid-1960s (Townsend 1997), but the UK was the first to deindustrialise. Jobs have been shed

in a whole range of industries such as textiles and clothing, motor vehicles and coal and steel (Beynon 1999). And this job loss has been largely of male – skilled, unskilled and semi-skilled – full-time jobs. In 1971, 98 per cent of males aged 45 to 54 years were economically active but by 1997 this had fallen to 91 per cent. It is estimated that only about 70 per cent of men will be economically active by 2011 (ONS 1999d). The proportion of men aged 50 to 64 neither in paid work nor looking for it has increased from 11 per cent to 27 per cent since 1976 (Rennell 1999).

The biggest regional loss in England was recorded in the West Midlands at 12.48 per cent and the smallest loss was in East Anglia at 7.31 per cent. At county level, in England, Staffordshire suffered the greatest loss at 15.10 per cent and Shropshire sustained the lowest loss at 4.13 per cent. In Wales, the biggest loss was in West Glamorgan (10.52 per cent) and the least in Powys (0.98 per cent). In Scotland, the highest was in the Central Region (13 per cent) and the least in Orkney Islands Area (1.36 per cent). Most of the heavy loss was in the old industrial heartlands (Figure 8.2). At this level of analysis only the Shetland Islands Area showed a gain in manufacturing employment at 1.54 per cent over the decade related to the oil industry and the revival of small craft industries.

8.3.2 Case studies of deindustrialisation

Coalfields

One of the best examples of deindustrialisation is the coal industry. In 1947 there were 958 mines in the UK employing 710,500 men. By 1998 there were only 18 mines open and only 12,600 men employed. Although a small number of women worked in the mining industry as caterers, secretaries and the like, women were forbidden by law to be miners by occupation. Coal production from traditional deep mines declined from 189 million tonnes in 1960 to 30.3 million tonnes in 1997. During the same period output from the cheaper, but more controversial, opencast coal pits has increased from 7.7 million tonnes to 16.7 million tonnes (Department of Energy various years). The biggest sectoral decline in jobs in recent years has been in energy and water, of which mining plays a considerable part. There was a decline of 50.2 per cent in jobs in that sector between 1988 and 1998 (Pearce 1998).

The decline in the coal industry was due to a variety of factors. Pits became exhausted or unworkable. Cheap subsidised foreign imports undercut the price of British coal despite investments in the more productive pits. And the long period of government by the Conservative Party, which was particularly antagonistic towards the mining unions, did not help the ailing industry and led to some of the

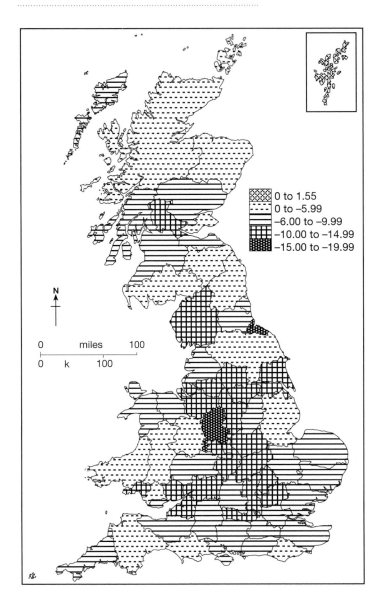

0 to 1.55
0 to –5.99
–6.00 to –9.99
–10.00 to –14.99
–15.00 to –19.99

N

| 0 | miles | 100 |
| 0 | k | 100 |

Figure 8.2 Percentage change in manufacturing employment, 1981 to 1991: Great Britain, counties England and Wales, regions and island authorities Scotland

Source: Data derived from OPCS, GRO(S) 1994, *1991 Census Key Statistics for Local Authority Areas Great Britain*, Table 14, London: HMSO.

most bitter industrial conflict ever seen in this country which led to violence, murder and unprecedented twentieth-century infringements on civil liberties. The government successfully undermined the traditional solidarity of the miners when the miners in the Nottinghamshire and Derbyshire coalfield formed their own union. The privatisation of British Coal in 1994 was the final nail in the coffin of the mining industry.

The massive decline of the coal mining industry has had devastating social and economic effects on several areas of the country. Many of the former coalfields suffer from high levels of long-term unemployment, particularly among men. The former coalfields also tend to have high levels of ill health, particularly limiting long-term illness. Depression, low self-esteem, alcoholism and suicide are also higher than average in former mining communities. These once tightly knit communities have seen other forms of social breakdown such as an increase in divorce and illegal drug use. Considerable effort and funding have gone into the former mining communities to ameliorate the worst effects of pit closure and economic decline. For example, former Nottinghamshire mining communities in Mansfield, Sutton-in-Ashfield, Newark, Chesterfield, Bolsover, Worksop and Retford have recently received £24 million of the government's Single Regeneration Budget to boost the local economies (Smith 1999). The government has set up a Coalfield Taskforce specifically to deal with the problems in former coalfield areas.

West Yorkshire wool textiles	In the nineteenth century the industrial base of West Yorkshire evolved with the wool textile industry. Moreover, it could be asserted that such towns as Bradford, Halifax and Huddersfield were built on wool textile money. In common with much manufacturing industry in the UK, the wool textile industry has undergone severe restructuring – technologically, of its markets, organisation and ownership and in its geographical location (Hardill 1990). The industry's operating environment was extremely turbulent during this period, and the forces affecting it included, firstly, demand changes, with the rise in casual wear, and secondly, a dramatic rise in the volume of imports of woollen cloth and garments, along with a loss of traditional export markets. These changes were compounded by the generally poor quality of management in the industry, along with a lack of restructuring and modernisation (Hardill 1990). Since the 1970s, the industry has simultaneously contracted and undergone a structural and technological revolution, but the decline in the number of jobs has continued almost unabated. In the early 1950s, for example, a weaver was responsible for up to four looms (according to the type of cloth woven). With the introduction of computer-controlled looms a weaver can be responsible for up to 120 looms (Hardill 1990).

8.3.3 The rise of the service sector

The post-war period has been characterised by a steady growth in the numbers of employees and the contribution of services to the UK economy, in terms of gross domestic product (GDP) and invisible exports. The proportion of employee jobs in the service sector has more than doubled in the last forty years to almost 76 per cent (Pearce 1998). The service sector can be examined by a number of binary divisions, such as producer and consumer services or public and private services, tradable or non-

tradable services, office-based services and the like. Services are very diverse, ranging, for example, from tourism to retailing, public services like hospitals, to computing and financial services. Services thus have an economic, social and cultural importance.

There is no one geography of services, but it is still possible to make some generalisations, because certain types of services are concentrated in very large metropolitan centres, others in provincial cities and others in small towns and rural areas. And we should not forget the global city service functions, especially financial, of the City of London

(McDowell 1997; Thrift and Leyshon 1999). Higher-order services such as banking and computing tend to employ male full-time workers, but a growing number are on fixed-term contracts; while those employed in retailing receive far less remuneration and tend to be employed part-time. Women and young adults (including students) dominate the retail workforce, but some retail outlets also employ older adults.

Dramatic changes have taken place in retailing in the last two or so decades. UK food retailers in the 1970s and 1980s developed logistically efficient stock control systems, quick response centrally controlled warehouse-to-store distribution networks and serve an increasingly large proportion of the population from the growing number of out-of-town/suburban supermarkets (Wrigley and Lowe 1997). And the trading hours of these stores have changed; we now have Sunday trading, twenty-four hour stores, late-night shopping and so on. There have also been some other important sociocultural changes in retail spaces, which are increasingly being considered as consumption/leisure spaces, such as shopping malls (Wrigley and Lowe 1999). In addition, most provincial UK towns and cities have leisure spaces where cafés, wine bars and clubs are clustered (Figure 8.3), some of which are in refurbished buildings such as banks, which are popular with young adults.

The 1990s have witnessed some shedding of labour in banking, insurance and finance. This began with the recession of the early 1990s, and has proceeded apace with company restructuring partly set in motion by the consequences of the deregulation of the financial sector in the 1980s. The distinction between banks and building societies has been blurred with deregulation. Moreover, some key functions have been relocated abroad, such as to India as in the case of NatWest bank to capitalise on labour cost differentials (Lakha 1999) or Ge

Capital relocating its call centre to India. In addition the increased penetration of information and communications technologies (ICTs) is transforming the structure and organisation of the banking sector. Nevertheless, the largest growth in the labour market has been in banking, finance and insurance, with 28.4 per cent increase in the number of employees between 1988 and 1998, compared with only 4.3 per cent for all jobs (Pearce 1998).

Moreover, with the use of Automatic Teller Machines (ATMs), in addition to those located at banks, a new geography of where one can undertake banking transactions has emerged. Cash can now be obtained from ATMs located at supermarkets, while debit cards can be used to obtain cash while also paying for produce at the checkouts of supermarkets. Also, increasing numbers of bank customers now complete transactions using the Internet or by telebanking, with the future of

Figure 8.3 Urban regeneration here in Nottingham's Lace Market provides employment

Source: Authors.

banking in investment terms being seen to be in Internet banking. Banking activities have thus become much more spatially dispersed thanks to the impact of ICTs. Thus, one is less likely now to undertake banking activities actually at the branch of a bank; that is, for the 78 per cent of the population with a bank/building society account. These technological changes have adversely affected the numbers employed in banking as bank customers are less tied to visiting banks to undertake banking transactions, as a result the number of bank outlets and bank employees has been pruned.

The number employed in tourism has continued to increase, as has the contribution of the sector to invisible earnings, through the fact that half of Britain's twenty-six million tourists come from overseas (Meikle 1999). While many jobs are seasonal and low paid in traditional tourist destinations, such as coastal resorts, which rely increasingly on day-trippers, tourism is growing in 'heritage sites' such as Britain's historic towns like Canterbury, Edinburgh and York (see also Chapter 9). Although more tourists visit London in the summer, it is a twelve-month tourist destination.

8.3.4 Agriculture

Although rural parts of the UK are primarily agricultural in terms of land use, farming is no longer the foundation of the rural economy today, nor is it the linchpin of rural society (MacFarlane 1998). For the UK as a whole, less than 2 per cent of the workforce is employed directly in agriculture. The last fifty years have seen an unprecedented growth in the productivity and output of UK agriculture. Agriculture has received public support – first through deficiency payments and then through the Common Agricultural Policy (CAP) – which gave continuous impetus to production (HMSO 1995). Despite very high levels of support and protection under CAP, farmers' incomes have generally been under pressure. Agricultural prices have fallen in real terms, while costs have risen substantially, in part because CAP itself has contributed to higher input costs.

Excluding crops that cannot be grown in the UK,

farmers and growers produce nearly three-quarters of the food and animal feed consumed (HMSO 1995). Farming also underpins associate industries such as contracting farm work (using casual labour), farm machinery and fertiliser and other ancillary industries and services, as well as processing and adding value to food products (DETR, MAFF 1999). Farm diversification is also important, and much farm-based work is now concerned with such activities as woodland management, running farm shops and equestrian businesses, the provision of sporting facilities, nature trails, holiday cottages and various agricultural services (HMSO 1995).

Finally, in the last decade or so the intensive nature of farming has seen some harmful consequences, which have resulted in the loss of consumer confidence, and demand for some products has declined. This is perhaps best illustrated by bovine spongiform encephalopathy (**BSE**) which causes new variant Creutzfeldt–Jakob disease (**nvCJD**), a degenerative brain disease which has killed around eighty, mostly young, people in the UK. The British beef industry has been decimated following the link between BSE and nvCJD. The EU banned British beef exports (from 1996–9) while UK beef consumption has declined dramatically. Incidents of BSE have largely occurred in England, and cases have been negligible in Northern Ireland and Scotland, yet Northern Irish and Scottish farmers have also been subject to the EU export ban, which was only lifted in July 1999.

8.4 DEMOGRAPHY AND SOCIAL CHANGE

In this section we focus on structural changes in the composition of the workforce under three themes: young adults and the labour market; the feminisation of the labour market, and ageing workers and the labour market. Structural change has a spatial dimension, with important regional divergence, as well as at intra-regional and intra-urban levels. In 1961, 25 per cent of the population were aged under 16 years, by 2011 the proportion will be 18 per cent. As a result, during the period under review the number entering the labour market has declined. At the same time the proportion of the population exiting the labour market has increased. For example, those aged 65 years and older numbered 12 per cent of the population in 1961 and will number 17 per cent by 2011.

8.4.1 Young adults and work

For most people, their late teens and early twenties are the most formative period in their lives as they conclude education and enter the labour market. How they go through this period has a very strong influence on the way in which they will spend the rest of their lives. There are clear regional and local differences in people's experiences during this crucial time (Champion *et al*. 1996). A general picture of people's progression through the ages of 16 to 24 is given in Figure 8.4, distinguishing men from women. The most obvious change is the fall in the proportion of those economically inactive. This largely reflects the end of the education process, but for women the proportion is also affected by some leaving the labour force from age 19 years for childbirth and child rearing.

The proportion of young people staying in education, after the minimum school leaving age of 16, has been growing for many years. But there is a large variation in the share of an area's 16- and 17-year-olds who are full-time at school or college

(Champion *et al*. 1996). Strong urban/rural contrasts are evident, but we should remember that Scotland has a very different education system (with different qualifications) than England, Wales and Northern Ireland.

The areas with low staying on rates include East London and the former Scottish and North East coalfield areas. And these low rates are not because young people easily find jobs there. In fact, it is the opposite, these areas have had high levels of unemployment and economic inactivity for several generations, and the young, especially white males, often never have the opportunity to become connected to the labour market. Moreover, they see little incentive to staying on at school because of bleak labour market prospects (Champion *et al*. 1996). Some ethnic groups, notably Bangladeshi and African-Caribbean males, also record high economic inactivity because of relatively low levels of educational attainment and discrimination in the labour market. In 1997, Labour announced a new policy initiative the New Deal for Welfare to Work. It especially targets young adults, aiming to reconnect them to the labour market and mainstream opportunities through training, counselling, and work placements (for a fuller account see Chapter 12).

The proportion of young adults in higher education has risen during the post-war period, and now almost one-third of each age cohort go to university to obtain a degree. Between 1951

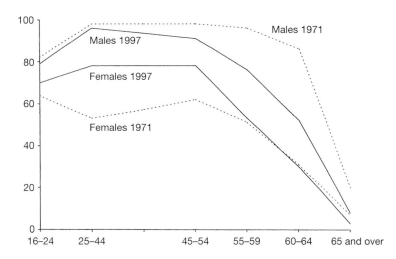

Figure 8.4 Percentage of males and females economically active by age, 1971 and 1997: UK

Source: Data derived from ONS 1999c, *Social Trends 29*, Figure 4.1.

and 1997 there was a 94 per cent expansion in full-time and part-time students in higher education. The proportion of part-time students has also increased from 17 per cent in 1951 to 35 per cent in 1997. Further, in 1951 only 22.5 per cent of higher education students was female, but by 1997 that had grown to 52.5 per cent (CSO 1961; ONS 1999d). Also student life is very different today, as many readers of this book know only too well, because of the decline in the value of the student grant, and many now hold part-time jobs to help them finance their studies. The proportion of 'graduate jobs' certainly has not kept pace with the expansion in the number of graduates, and some enter higher education with a burning desire to study, but for others it is more a reality of limited opportunities in today's labour market.

The Green Paper (consultation document) *The Learning Age* produced by the Department for Education and Employment (DfEE 1998) recognises that no longer can one set of skills and qualifications sustain a job for life. We are asked to think about systematic 'lifelong learning', which can help develop people's skills, orientations and confidences in a rapidly changing work environment. Educational achievement in the UK lags behind other EU countries, and it is being suggested that if

the country's skills base is improved then so would competitiveness.

8.4.2 Feminisation of the labour market

In 1971, just over half of women aged 25 to 44 years were economically active, now it is over three-quarters. This has occurred at a time when the number of men in formal employment has declined. It is estimated that by 2011 around 58 per cent of all women will be economically active compared with 70 per cent of all men (ONS 1999e). Figure 8.4 shows how male and female employment patterns have become more alike between 1971 and 1997. Since 1971, female economic activity has increased for women of all ages, except those aged 60 years and above. Note also how the decline in economic activity rates occurring at the childbearing years, which was clearly evident in the 1971 trend, had disappeared by 1997.

The types of jobs women undertake have changed dramatically during the century. Domestic service has been replaced by clerical work as the main occupation in England and Wales (Figure 8.5). The same pattern applies to the whole of the UK. Prior to the Second World War, many higher grade female

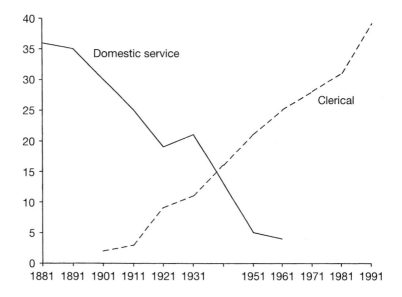

Figure 8.5 Percentage of females in domestic service, 1881 to 1961, and in clerical occupations, 1901 to 1991: England and Wales

Source: Data derived from Census of Population (data not available for 1941).

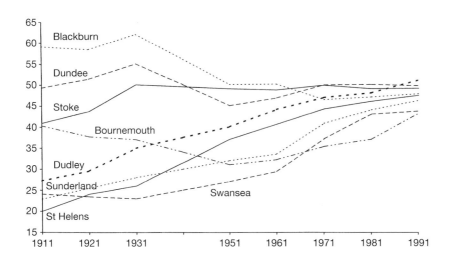

**Figure 8.6
Percentage of
females economically
active, 1911 to 1991:
eight British towns**

Source: Data derived from
Census of Population
(data not available for
1941).

occupations, particularly in teaching, the civil service, banking and local government, were closed to married middle-class women, who were expected to disengage from the formal labour market when they married.

Figure 8.6 demonstrates that the changes in female economic activity rates in the formal economy have not been even throughout the country. A selection of towns from different parts of Great Britain with very different industrial traditions have been chosen to underline that, while female economic activity rates have risen generally, in some places they have actually declined. In the former mining and heavy industrial areas – St Helens, Sunderland, Swansea and Dudley – where there was little local employ-ment for women, rates have increased from between 20 to 27 per cent in 1911 to around 44 to 50 per cent in 1991. Indeed, Dudley has the highest female participation rate of the eight towns. Bournemouth, once a major employer of female domestic ser-vants, has remained at around 40 per cent, despite a decline to 1951. The textile towns of Dundee and Blackburn, and the pottery town of Stoke-on-Trent, once employed large numbers of women and these show little in the way of real increase in female activity rates. In fact, Blackburn has declined from almost 60 per cent in 1911 to fewer than 48 per cent in 1991.

One of the most dramatic features of the labour market in the second half of this century has been the growth of married female participation. Figure 8.7 shows much the same pattern as Figure 8.6 except that there has been an increase in married female activity rates in all eight towns during the century. Even after the Second World War, when the marriage bar no longer existed and the scope for married female employment in factories, offices and the expanding retail trade widened employment opportunities for women to remain in employment after marriage, there were still wide discrepancies in married female activity rates in different parts of the country. Although there has been a marked convergence in married female activity rates during this century there is still a tendency for the areas which have traditionally had high levels of married female economic activity rates still to have high rates. Even after a century of change this traditional component still has a marked impact on the geog-raphy of married female employment rates (Graham 1988).

Also, although retail and sales occupations have been the fastest growing outlets for female (part-time) employment in the last two decades, around 40 per cent of female (full-time) occupations are still clerical or secretarial (EOC 1995). Even though few women will now withdraw from the labour market on marriage/cohabitation, childbearing remains the main reason why women do withdraw. Some 23 per

Figure 8.7 Percentage of married females economically active, 1911 to 1991: eight British towns

Source: Data derived from Census of Population (data not available for 1931 and 1941).

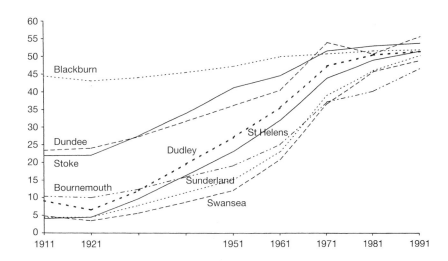

cent of women in their late twenties remain outside the labour market, and as women have their first child at a later age this is now affecting women in their early thirties. The number and age of children are the factors most likely to influence a woman's labour force activity.

Only 40 per cent of women with a dependent child below primary school age are likely to be in the formal labour force, compared with 65 per cent with a dependent child aged 5 to 10 years. Where the youngest child is aged 11 to 15 the economic activity rates are much the same as those for women with no dependent children, at around 70 per cent. Similarly, the more dependent children a woman has, the lower the likelihood that she will be in the labour market (EOC 1994). Something like 64 per cent of single mothers and 40 per cent of mothers in stable relationships with a dependent child aged under 5 years are economically inactive. For those with no dependent children the rates are 28 and 23 per cent respectively (ONS 1999c). Around half the married women in Great Britain with pre-school children now work at least part-time, compared with only a quarter in 1985. There are more dual earner households, in which the female partner is in paid work, in London and the South East than elsewhere (see Chapters 15 and 16).

8.4.3 An ageing population

Between 1991 and 2011, the 'Third Age' population is expected to increase by more than a quarter, representing a major demographic shift towards an ageing of the UK population as life expectancy has increased during the century for both men and women (ONS 1999e). However, the elderly do not constitute a homogeneous group, and exhibit increasing socioeconomic polarisation (see Chapter 15). The last two decades have seen the emergence of the 'wealthy pensioner' as an increasing social phenomenon as managerial and service class workers approached retirement with substantial financial assets (provided by occupational pensions and the accumulation of wealth from domestic property and other assets). Many were able to leave the labour market early because of their financial strength (Mallier and Shafto 1992) and can anticipate the 'Third Age' as a phase of their life, which offers new opportunities including leisure (see Chapter 9). This section of the elderly enjoy 'choice' relating to key lifestyle decisions (where and when to shop, place of residence(s), ownership of motor vehicle(s), holidays, and so on), and their lives are characterised by spatial mobility. This is having a considerable impact on the geography of age structure in the UK, with high concentrations of retirees in specific areas like the south coast of England, the Borders of Scotland, North Wales, and other attractive areas. This has a knock-on effect on medical and welfare provision in these areas.

But for the majority of people, retirement is still accompanied by financial insecurity and dependence upon state benefits (see Chapter 15). In contrast though, substantial numbers of manual and non-manual workers leave the labour market before statutory retirement age, this is more likely to be involuntary (see section 8.3.1), and consequent upon redundancy and subsequent unemployment/ economic inactivity (Collis and Mallier 1996). For this section of the population, their retirement years are increasingly being preceded by a period with a more tenuous attachment to the labour market or economic inactivity followed by retirement with limited material assets and lack of 'choice' relating to key lifestyle decisions (Beatty and Fothergill 1996). With changing work practices, growing numbers of people will be entering the 'Third Age' in the twenty-first century with limited control over their lives because of the growth of **non-standard employment** and labour market uncertainty, as members of the so-called 'risk' society (Beck 1992).

Linked with the ageing of the population is **ageism** in the job market. Legislation exists to prevent discrimination on the grounds of gender, disability and ethnicity (and religion in Northern Ireland), but no legislation outlaws ageism, which is a major problem in the UK and other developed countries. Only around 65 per cent of those aged 50 to 65 are in work, compared to over 80 per cent aged 35 to 49. This discrepancy is not fully accounted for by early retirement, having more to do with ageism and the discouraged worker effect. Many people over 50, once they become unemployed, simply give up applying for jobs. It is reckoned that ageism costs the UK £26,000 million a year, according to the Employers' Forum on Age (Rennell 1999).

This ageing population brings with it its own employment dimension as the growth in residential and non-residential care sectors demonstrates. These have greatly expanded the opportunities for skilled and unskilled work. Most of the jobs, however, have been in the unskilled, low-pay and long hours sector

in residential homes, many of which have earned considerable notoriety and the attention of the authorities, less for the conditions of the workforce than the treatment of residents.

8.5 WORK PRACTICES

The labour force has undergone several changes in recent years in the way that work is organised. Here we explore some of these.

8.5.1 Flexible work

There has been a proliferation in the number of part-time and temporary jobs (called non-standard employment) in the UK in the last few years, and of people holding more than one paid job. Part-time workers now total 6.7 million – 25 per cent of those in employment. About 44 per cent of women work part-time, compared with only 9 per cent of men (Pearce 1998). In 1991, 3.5 per cent of employees worked part-time. By 1997, it had increased to 12.7 per cent. The number of full-time workers has decreased accordingly. In 1998, 7.4 per cent of all jobs were temporary – 6.5 per cent of male jobs and 8.4 per cent of female. However, more women professed a preference for temporary work, while most men could not find permanent work. Nevertheless, the UK has a low level of temporary employment compared with most other EU countries (ONS 1999c). In 1997, 4.8 per cent of employees and self-employed, in the UK, had a second job – 3.8 per cent of males and 6 per cent of females (ONS 1998a).

Some petrol filling stations, supermarkets and other retail units remain open twenty-four hours a day in London and some provincial cities. This is part of the 'Americanisation' of society. Even where there are more restricted opening hours, employees work on a variety of shifts; they are nearly all part-time, and are students, women and some 'Third Agers'. Even in the better paid managerial and professional occupations (including academia), short-term contract employment and practices such as incentive payments or bonuses are replacing secure, life-time salaried conditions that previously marked managerial and professional employment in the 'core' economy (McDowell 1997).

The word 'flexibility' is currently used in a variety of different ways to describe a wide range of forms of work organisation which, though rarely new in themselves, have increasingly been introduced since the 1980s (see Chapter 3). In 1997, 18 per cent of male and 26 per cent of female employees in the UK were engaged in some type of flexible working arrangement. This consisted of those on **flexitime**, annualised hours, term-time working, job sharing, nine-day fortnight, four-and-a-half day week and zero hours contract. But this varied from place to place. Wales and the North East had the highest percentages of females working some type of flexible working pattern at 30 per cent and 29 per cent, respectively, while Merseyside had the lowest at 23 per cent. The North East had the highest percentage of males (22 per cent) in flexible work and Northern Ireland and Eastern region the lowest (15 per cent) (ONS 1998a).

New technologies coupled with increasing competitive pressures characterise new organisational structures (see Dex and McCulloch 1997). We can see 'internal flexibility' brought about by multiskilling, which enables workers to shift from one job to another as required; and the use of casual, on-call, temporary and part-time staff. The word is also used to describe the increasing use of subcontracting and of homework, both with and without the assistance of ICTs. The latter are often described as 'external flexibility'.

Flexible work covers a range of options (Box 8.4). Anglo-American capitalism today emphasises flexibility, workers are asked to be open to change at short notice, to become ever less dependent on regulations and formal procedures (Sennett 1998). Thus for many workers their connection to the labour market is characterised by uncertainty, for example in the total number of hours they will work that week, or for how long they will have a job contract for. And it has been claimed that flexibility gives people more freedom to shape their lives, but one has to question this for many flexible workers. Today it is no longer the norm to have a 'job for life'; that is, the deployment of a single set of skills through the course of a working life, a career with one organisation.

8.5.2 Self-employment

There has been a considerable growth in the number of self-employed men and women in the UK and it is

**Box 8.4
Flexible
work**

- self-employment, including subcontractors and freelancers but excluding the self-employed with employees
- part-time work, especially with very low hours which do not ensure eligibility to the National Insurance system and thus to many state benefits, to employment protection legislation
- temporary work
- fixed term contract work
- zero hours contract employment
- seasonal work
- annual hours work, shift work, job sharing, flexitime, Sunday working, overtime, term-time only work or compressed working weeks
- working at home
- teleworking

Source: Dex and McCulloch (1997)

predicted that 15 per cent of the UK workforce will be self-employed by 2006 (DfEE 1996). Indeed 30 per cent of businesses established in 1999 were by women. By 2009, the proportion is expected to be 50 per cent (Fieldman 1999). The new self-employed's actual work situation is far from that of the traditional small business owner (Corden and Eardley 1999). Entrepreneurial strategies of self-employment are emerging for managers and professionals (largely male and middle aged). Increased insecurity in job contracts and dissatisfaction with their terms and conditions of employment are resulting in some managers and professionals adopting an entrepreneurial strategy. They move from employee to self-employed status, often drawing on their social networks and business contacts using their accumulated skills and expertise (Sennett 1998).

Some careerists opt out of the 'rat race' and are attempting to 'take control' of their careers and workloads and therefore their lifestyles (Arthur *et al.* 1999). High-pressure careers and large incomes are traded in for a less frantic and more creative life by becoming self-employed in a different area of activity or by a spell of voluntary work such as VSO (Voluntary Service Overseas) (Knowsley 1999). A third group identified as pursuing an entrepreneurial strategy is international migrants. Self-employment is adopted as a strategy (following migration) to optimise the household's opportunities and advantages, often because of labour market discrimination (Phizacklea and Ram 1996), or because of the lack of harmonisation of qualifications.

8.5.3 Homeworking

As was noted earlier in this chapter, homeworking has long been part of working life in the UK (Pennington and Westover 1989). But the scope and nature of homeworking has changed (Box 8.5). Homeworking is therefore much more diverse than the image of a woman sweating over a sewing machine. There is no accurate count of the number of homeworkers; it could be 2.5 million (Phizacklea and Wolkowitz 1990). Homeworkers are found in clerical work, in manufacturing, and there are also professionals who homework as teleworkers as a self-employment strategy or telecommute (see below). While many homeworkers are women, they often undertake homework because they have dependent children/other caring responsibilities/other impediments to access to the labour market. They undertake homework to 'juggle' 'home' and 'work'. Men, especially those in mid-career, are amongst the professional and self-employed homeworkers and they often telework. It is estimated that almost 70 per cent of teleworkers are men aged 35–54 years (Pandga 1999). They use their computers to make a living. Telework encompasses many temporal and spatial patterns of work, class positions and forms of employment, and involves working *at* and *from* home. Big gender differences emerge between those who work at home or from home. Around 41 per cent of women work at home against 60 per cent of their male counterparts who operate from different locations using their home as a contact point. It is estimated that there are 1.1 million teleworkers, and their numbers are growing

Box 8.5 Homeworkers

These include:

- people working at home for a single employer
- people working at home for a variety of employers
- people running their own businesses at their own address (including live-in shopkeepers)
- people working at home for another family member
- live-in domestic servants

by 200,000 per annum (Pandga 1999). There has been a 13 per cent rise in the number of teleworkers between 1997 and 1998, and almost 5 per cent of the working population are engaged in some form of **teleworking** (Bibby 1999).

Teleworkers include: *the self-employed*, such as freelance editors, proof-readers and translators (Pandga 1999); *employed home-based teleworkers* (also called *telecommuters*) – those home-based for part of their working week (Baines 1999). They tend to be male, with employee status and relatively highly skilled, with high trust relationships and some degree of greater autonomy as to when and where they work. The home is the *extension* of the workplace, much like in pre-industrial times. These numbers are now increasing with **hot-desking** and **hotelling**; part of their home becomes the office, and diaries are 'monitored' by email. For these the home *replaces* the workplace; and finally there is *mobile telework* which is practised by sales representatives, for whom the car is their office.

8.5.4 Unpaid work

Overall, women spend more time than men doing unpaid work. For example, women spent an average of just over an hour per day cooking (in May 1995) compared with around half an hour for men. Women spent five times as long as men on cleaning the house and over eight times as long doing the laundry. Working age working women spent 3 hours 15 minutes per day, those not working spent 5 hours 3 minutes on unpaid work (cooking, routine housework, shopping and care of children and adults) compared with working age working men (1 hour and 2 minutes per day) and those not working (1 hour and 50 minutes) (ONS, EOC 1998). Many women also 'care at a distance' for their elderly relatives, and growing numbers of women over 50 years are withdrawing from the labour market to care for elderly relatives. These are the 'sandwich' generation, who often withdrew from the labour market to care for their dependent children, and are now facing elder care.

The post-war period has been characterised by the possibility of some unpaid work being undertaken on the basis of purchased commodities (ready-made clothing and factory-prepared foods) and on a less-intensive basis (domestic appliances such as electric irons, vacuum cleaners and so on) (Glucksmann 1995). As a result, the time spent on some of these unpaid tasks has declined. There is also a long tradition of some unpaid work being commodified (paying for someone else to do the work).

At the end of the nineteenth century over one-third of all working women were employed in domestic service (see Figure 8.5). These working-class women were employed as either daily or live-in domestic servants in middle- and upper-class homes. Domestic service declined dramatically after the Second World War; indeed, after 1961 the category 'domestic servant' disappeared from Census of Population definitions. But domestic service has not altogether disappeared; the nature of the work has changed with the use of full-time and part-time workers, often employed informally in a wider range of homes. Several agencies exist that provide domestic help. Recent research shows that the number engaged in domestic service, such as cleaning, ironing clothes, gardening, childcare and so on, has actually increased in the last few decades, especially in the homes of the expanding middle classes, managers and professionals (Gregson and Lowe 1994).

As more women are now undertaking paid work, caring work for the young and elderly is becoming increasingly 'marketised'; that is, carried out for money rather than for love, duty or social obligation. While childcare is now commodified by the use of childminders, nannies, nurseries, crèches, nursery school or playgroups which represent 'professional' childcare, the care of children by employed mothers is still largely achieved with help from partners, parents, and friends (that is, 'informal' childcare). But the reality is that in the UK state-funded childcare is very limited when compared with other EU countries, but the government is extending state support through childcare vouchers and the encouragement of 'family-friendly' working environments. As a result, the costs of childcare are too great for many women workers, other than those who earn managerial and professional salaries; so many women use informal childcare arrangements or juggle a part-time job with childcare. We should also remember that private sector caring jobs are insecure and poorly paid. State-funded childcare has a distinctive geography, with Scotland having better provision than elsewhere.

8.5.5 Workaholism and stress

One consequence of the work practice changes described above is that increasingly workers – whatever the nature of their job contract – may be concerned about their job security. In order to demonstrate their commitment to the job and organisation, the pressure is on not to be the first to leave for home; this has been described as 'presentism' (Crompton 1997; Dean 1998). Workloads for a whole range of employees have increased, not least because there are often fewer people around to do the work. In some organisations an individual may under-record the hours taken to complete the required assignments, in order to achieve output targets for a shift, so as not to jeopardise their remuneration.

Many of those in paid work today have lifestyles that are characterised by workaholism (Dean 1998). The working week is getting longer and many people work in excess of their recorded hours (Hardill *et al.* 1997). This 'tyranny of time' hinders combining paid work and domestic life. Sennett (1998: 26) acknowledges the conflict between family and work, asking 'how can long-term purposes be pursued in a short-term society . . . which feeds on experience which drifts in time, from place to place and from job to job'. Moreover, stress and other health-related problems (ONS 1998c; Scase *et al.* 1998) often accompany such lifestyles. Stress has been linked to coronary heart disease (see Chapter 10) and recently to a substantial proportion of sickness absence (MIND 1992), increased staff turnover, low morale and poor job performance (Sutherland and Cooper 1990).

Scase *et al.* (1998) have highlighted the growth in recorded stress by managerial and professional women, and expressed concerns about the impact on children of the 'invisibility' of parents in the family home, with worries about children becoming among other things 'mall rats' (Sennett 1998). A similar concern was raised in the 1950s and 1960s with 'latch key kids'. One-third of British women aged 16 to 74 reported a large amount of stress compared to one-quarter of men. For both British men and women, reported levels of stress were highest amongst those in the professional and intermediate social classes (ONS, EOC 1998). Indeed, recent press reports have highlighted the new reality of the home–work interface for managerial and professional dual career households (Freely 1999). A growing minority of women professionals now withdraw from the labour market following childbirth (often the birth of their second child) because of the problems of having a meaningful home life and working excessive hours. Different work patterns can lead to widely diverging lifestyles (Case study 8.6.1).

8.6   SUMMARY

The past two decades have seen a much changing socioeconomic landscape of employment, characterised by uncertainty, instability, flexibility, and job losses – especially in the manufacturing sector and certain service sectors. Female participation rates have continued to rise. Will Hutton, in his book *The State We're In*, has captured the human impact of the new world of work. He (1995: 14) has described the UK as 'a 30–30–40 society' comprising 30 per cent 'disadvantaged' (note the broad definition used, incorporating those on government schemes and the economically inactive as well as the unemployed). A further 30 per cent are 'marginalised and insecure' (part-timers, casual workers, those on fixed-term contracts, the self-employed and fully employed for less than two years who thus don't qualify for employment protection), and the 40 per cent 'advantaged'. These are full-time employed and self-employed who have held their jobs for more than two years and part-timers who have held their jobs for more than five years.

8.6.1 Case study

'Workaholic Britain'

Workaholic Britain is turning into the 'grab and go' society

BY CHERRY NORTON
Social Affairs Correspondent

Weekends are the only time the Deans can get together as a family *John Voos*

'I SEE SO LITTLE OF THEM'

RUTH AND Hugh Deans live with their two children, Alice, three, and Harry, seven months, in a three-bedroom semi-detached house in Beckenham, Kent. They have a combined annual salary of between £45,000 and £50,000. Hugh, a landscape gardener, works part-time. Ruth, 34, is a marketing consultant in London. She works 12-hour days, getting up at 6.45am and commuting for an hour each way to work. "Time is so precious," she said. "At the weekends we always do something special together as a family because I see so little of them in the week." The couple eat take-outs at least twice a week. "We must spend about £30 a week on take-aways," she said. They pay a cleaner and someone to do the ironing. The family's clothes are bought by mail order.

Sophie Chalmers and Andrew James live well on £30,000 *Chris Jon*

'WE SEE THEM ALL THE TIME'

ANDREW JAMES and Sophie Chalmers live in a six-bedroom converted mill in Wales, with 20 acres, with their three children. They have a combined income of £30,000. Eight years ago the couple left their well-paid jobs, and moved out of London.

They now run a magazine called *Better Business* working at home. They still have a nanny but overall spend much less on maintaining a high-octane lifestyle. "We spend much less on clothes, eating out and new cars. We no longer have to spend money to compensate for having a stressful life," said Andrew.

The couple have two holidays a year, one somewhere warm without the children. "We don't have to spend every minute with the children because we see them all the time," said Andrew.

FAMILIES ARE spending more than ever before on "convenience living" to cope with the demands of modern life, according to a survey published yesterday.

Three-quarters of working parents said that lack of time forced them to pay somebody else to do their cleaning, ironing, cooking or DIY, and nearly half were too busy to spend quality time with family or friends.

Experts believe that the nationwide survey of 1,000 adults, conducted by NOP, the polling organisation, confirms that the "Grab and Go" society is here to stay with families paying financially and emotionally to balance the pressures of work and home.

"Today's working families are richer in material terms than their counterparts of just 30 years ago but are becoming increasingly time poor. It seems that we're all working a lot harder just to stand still and we have become so busy that we have to spend more and more money to try to keep our households running smoothly," said

8.6.1
Case study
continued

Bridget Walsh, the group marketing manager of Abbey National, who commissioned the research.

Nearly three-quarters of the families said they bought take-away food at least once a week because they could not find the time to cook.

People in London, with or without children, were the most likely to buy take-aways, 49 per cent overall, followed by those in Yorkshire at 45 per cent and Scotland at 43 per cent. Nearly two-thirds of people who worked full-time, 63 per cent, found it difficult to find time to visit the supermarket and shopped instead at local convenience stores, adding more than £20 a week to their food bill.

The increasing pace of life has made it more difficult for nearly 70 per cent of the population to manage their time. The survey showed that 37 per cent of those questioned had forgotten an important anniversary or birthday in the last 12 months. People with children found that they were too busy to go to the cinema, 49 per cent, to organise holidays, 36 per cent, or do the gardening, 37 per cent.

However, it is not just household chores that can be farmed out to other people. For an annual fee of £500 organisations such as TEN, Time Energy Network, will do just about anything for you.

"Time is a luxury," said Alex Cheatle, founder of TEN. "We do lots of birthdays, buying cards and presents and reminding people of anniversaries."

When it came to planning their finances, a third of families with children said they did not have sufficient time to manage their day-to-day finances, compared with 23 per cent of households without children.

"Friends, hobbies and relaxation are all increasingly sidelined in the relentless pursuit of wealth and the need to be super-parents," the survey concluded.

R E V I S I O N   Q U E S T I O N S

- What do you think are the key reasons for disconnection from the labour market? To what extent does this vary geographically?
- What has been the impact of EU integration on the UK labour market? How strong has this been in the different sectors of the economy?
- Agriculture employs few workers but has a greater economic significance. Discuss.
- What have been the major changes in the world of work for women?
- Far more women than men work part-time, and 79.2 per cent say they 'do not want to work full-time'. Why do you think this figure is so high?
- More women than men have a second job. What strains and stresses must these people face in trying to cope with juggling more than one paid job as well as other responsibilities?

KEY TEXTS

Beynon, H. (1999) 'The end of the industrial worker?', in J. Bryson, N. Henry, D. Keeble and R. Martin (eds) *The Economic Geography Reader*, Chichester: Wiley, pp. 357–60.

Crompton, R. (1997) *Women and Work in Modern Britain*, Oxford: Oxford University Press.

Hudson, R. (1997) 'The end of mass production and the mass production worker? Experimenting with production and employment', in R. Lee and J. Wills (eds) *Geographies of Economies*, London: Arnold, pp. 302–21.

Thrift, N. and Leyshon, A. (1999) 'In the wake of money: the City of London and the accumulation of value', in J. Bryson, N. Henry, D. Keeble and R. Martin (eds) *The Economic Geography Reader*, Chichester: Wiley, pp. 333–40.

Wrigley, N. and Lowe, M. (1999) 'New landscapes of consumption', in J. Bryson, N. Henry, D. Keeble and R. Martin (eds) *The Economic Geography Reader*, Chichester: Wiley, pp. 311–14.

CONSUMPTION AND LEISURE

9.1 INTRODUCTION

Until the post-war period, the pursuit of **leisure** activities and conspicuous consumption in the UK was largely the preserve of the upper classes because they had the time and financial resources (Cannadine 1998; Veblen 1953). The period after the Second World War heralded the era of mass consumption, and of the 'affluent worker' (Bocock 1999). However, since the 1980s there has been an upsurge of new forms of consumption, and particular attention paid to it during the 1990s, essentially due to intellectual and political developments within the academy as much as to the changes in consumption itself. There has been a decline in Marxist theorisations that focus on production, with economic and social geographers embracing the cultural turn within the discipline (Bryson *et al.* 1999: Lee and Wills 1997; see Chapter 3). In addition, the number of jobs in the manufacturing sector has declined while service sector jobs allied to consumption have grown (see Chapter 8). Over the last two decades, consumption patterns and lifestyles in the UK have become more differentiated and less easily defined by social status. **Disposable income** has increasingly become just as important a determinant of leisure and consumption patterns as social class and background.

The notion of 'lifestyle' emerged as part of the attempt to capture this set of changes in consumers' patterns of purchasing. 'One's body, clothes, speech, leisure pastimes, eating and drinking preferences, home, car, choice of holidays and so on are to be regarded as indicators of the individuality of taste and sense of style of the owner/consumer' (Featherstone 1987: 55). Through leisure and consumption one acquires an 'identity' (Clammer 1999). Moreover, the whole process of consumption itself has become more variable, for example, with the emergence of round the clock consumption and the advent of new technologies (Thrift and Leyshon 1999; Wrigley and Lowe 1999). In this chapter we focus on these changes. But first we will explore the meaning of leisure and consumption.

First consumption. This literally means the use of commodities for the satisfaction of needs and desires. Consumption includes not only the purchase of a range of material goods, from cars to television sets, but also the consumption of services such as travel and of a variety of social experiences (Bocock 1999). Consumption, together with production, distribution and exchange, are key features of a capitalist economy – these are not separate processes. It is only through consumption that commodities become real objects. Consumption also creates the need for new commodities. In the 1970s and 1980s most interest in consumption focused on a form of collective consumption on one commodity, housing. In the mid-1980s this narrow focus began to broaden out through work on the consumption sectors and in the late 1980s social and cultural geographers undertook work on the service class. Currently the geography of consumption is chiefly fixed on social and cultural issues to do with the way in which commodities and their meanings have become intertwined. Advanced capitalist societies, like the UK, have developed the process of consumption into a major social activity that uses up large amounts of time, money, energy, creativity and technological innovation to sustain it. Witness the traffic queues before Christmas at such out-of-town shopping centres as Meadowhall, Sheffield or the Metro Centre, Gateshead.

Turning to leisure, which is commonly defined as either:

- an 'attitude' of a feeling of freedom. It has been suggested that leisure is a 'lived experience' rather than a simple state of mind;
- a kind of 'social' activity, that is chosen and separate from activities we are obligated to do, such as work tasks or required household commitments;
- a specific 'time' period. As was noted in Chapter 8, the Industrial Revolution established an industrial lifestyle for urban societies like the UK, a lifestyle that encouraged the 'work ethic' and recreation in the time left over after 'work' (Stokowski 1994: 3).

Leisure and consumption have become blurred activities, but with a distinctive geography, for some segments of the UK population, who increasingly express themselves through lifestyle statements. As is explored in Chapters 14 and 15, not all the UK population has the necessary disposable income to be a member of this consumer society. Bearing this in mind, we focus on leisure and consumption in the 1980s and 1990s in the remaining part of this chapter, answering the following key questions:

- How is a consuming lifestyle possible in the UK?
- How have leisure and consumption patterns changed?
- What are the key features of a consuming lifestyle in the UK today? What differences are there in terms of age, gender, social class and ethnicity?
- In what ways do leisure and consumption patterns vary within the UK?
- What is the relationship between place and consumption?

The definition of leisure as 'time outside work' creates several difficulties, as the activities of paid work, unpaid work in the home and leisure are somewhat blurred. For example, do-it-yourself (DIY) activities, such as improving the house, could be a hobby but could also be work activities designed to increase the value of a person's home. But DIY is done in non-work hours, and the activities do not normally result in monetary compensation. Some workers undertake certain work tasks outside typical working hours (see Chapter 8) and they think of these tasks as being creative (academics and writing books, but not teaching; composers and writing music but not conducting orchestras).

9.2 LEISURE AND CONSUMPTION IN THE UK TODAY

Leisure and consumption have undergone a number of changes in the last few years and we now look at some of these in more detail.

9.2.1 The development of the consuming lifestyle

The separation of 'work' and 'leisure' emerged with the Industrial Revolution (see Chapters 3, 4 and 8). In this chapter we focus on more recent changes, but before that we summarise the general post-war trends that have shaped work and leisure:

- the number of hours worked per week (for many people) has been reduced and time off with pay for holidays and vacations for most workers is guaranteed;
- improved life expectancy and per capita income;
- the growth of recreation, leisure and tourist industries around the world;
- improved transportation, especially car ownership and mass air travel;
- labour-saving devices in the home;
- over the last two decades social polarisation has resulted in increasing affluence amongst some population segments, along with decreasing wealth for many others, and groups of leisure consumers defined by age – such as young adult, middle-aged, the retired – have emerged (see Chapters 14 and 15).

Some of the above developments have made life easier for the majority of the British population. They have also made the tasks of social reproduction less time-consuming. In the process, leisure and consumption have assumed a time and space separate from work, but are often planned, scheduled and coordinated in the same way as work. These patterns are strongly gendered in the same way that work is (see Chapter 8). Such activities as recreation, play, sport and tourism represent visible aspects of leisure in society. For those with disposable income leisure time is used to express 'lifestyle', and disposable income is used to purchase goods and services and properties that also make such lifestyle statements.

9.2.2 Changes in leisure and consumption patterns

There has been a number of ways in which leisure and consumption have changed in recent decades. These include an increase in disposable income as a result of fiscal policy and the growth in real wages (for most, though not all), the development of 'consumerism as a way of life', a growth in individualism, rapid developments in technology, changing holiday patterns, the commercialisation and professionalisation of sport, and a switch away from religion and voluntary work.

Growth in disposable income

Between 1991 and 1997, real household disposable income per head almost doubled (ONS 1999c). This has had a discernible impact on both leisure and consumption. Consumer spending has increased and lifestyles have changed. The increase in car ownership is perhaps the most obvious manifestation of increased income. Even in the UK, where cars are supposed to be the most expensive in Europe, they are still relatively cheap, since the car is now becoming a personal, rather than a household item. The percentage of households in Great Britain with one car has remained the same since 1971, but during the same period the percentage with two or more cars has trebled (Figure 9.1). Increased car ownership and car dependence have obvious environmental, social and health consequences (see Chapters 10 and 16). When motorists combine the consumption of the car with their leisure pursuits mayhem can ensue. For example, residents of West Bridgford, where Nottingham Forest Football Club's ground and Trent Bridge cricket stadium are situated in close proximity to each other, find that they are often unable to park their own cars because of the preference of spectators to arrive by car, even though the area is well served by public transport.

Another aspect of disposable income is the increase in recent years of children's disposable income. Children derive income from a number of sources. During the 1990s there was a real increase in average weekly pocket money. Income from jobs has also increased, particularly among the 14 to 16 age group. Handouts from friends and relatives have also increased. While retailers and advertisers are increasingly targeting children, over half of their own money goes on sweets, ice cream and the like (NTC 1999a). It appears that they pressurise their parents via 'pester power' into buying the more

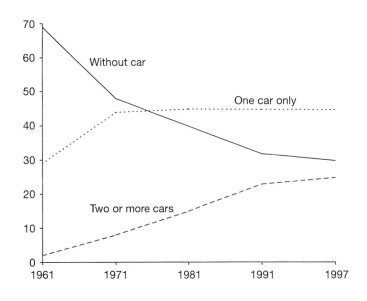

Figure 9.1 Percentage of households with regular use of a car, 1961 to 1997: Great Britain

Source: Data derived from ONS 1999c, *Social Trends 29*, Table 12.7.

expensive items – designer clothes, computer hardware and software, and so on.

Another way that disposable income has increased is through the extension of credit. Almost everything can now be bought on credit. Although companies spend considerable amounts of money and time checking creditworthiness and chasing up bad debts, that is nothing compared to the money and effort expended trying to persuade people to use credit. Acquiring things 'on tick' no longer attracts the stigma it once did. Consumers, including university students, are actively encouraged to accept quite substantial debts as the norm. Up until recently it was common, particularly among the upwardly mobile working class moving to more middle-class areas, to forgo certain commodities in order to keep up appearances and pay the mortgage (itself 'tick'). West Bridgford, in Nottingham, is known locally as 'Bread and Lard Island'. In Scotland, such areas attract the epithet 'Spam valley'.

The growth of individualism

There is a number of ways that this has affected leisure and consumption patterns. The rapid growth in the number of single person households, decline in fertility, growth of divorce and rising levels of income for some women are some of the more obvious manifestations of this phenomenon. The growth in individual training regimes via cycling, swimming and gymnasia, at the expense of team and competitive sport, is a further reflection of this. There has been an increase in the obsession with health and fitness, our bodies and performance. For many working age Britons success (expressed through the workplace), identity and lifestyle are inextricably linked with keeping slim and youthful, and how one looks is part and parcel of the image one wishes to emphasise in the workplace in an effort to enhance one's career (McDowell 1997). The health and fitness industry is one of the fastest growing in advanced capitalist countries, including the UK. Gymnasia, once the preserve of body builders and competitive athletes, situated in YMCA buildings or country clubs, have become one of the

most visible images of individual leisure in the built landscape, from large-scale, multimillion pound endeavours on **greenfield** sites to more modest arrangements in former factories and warehouses in inner city areas. Most British towns and cities have seen a remarkable growth in the number of these health and fitness centres, but membership of these centres is costly and requires a monthly fee that is beyond the means of many households. In 1998, for example, working out/aerobics was the second most popular leisure or sporting activity in the country after walking/rambling (which is of course done in the *name* of leisure: most of these walkers and ramblers would never dream of walking to work, the shops, etc., and, of course, most of the trips to the gym are made by car). While 21.8 per cent claimed to take an active part in the latter, 20.3 per cent claimed to take an active part in the former. The percentage of people taking regular exercise by walking doubled between 1980 and 1996, but the percentage of people working out increased sixfold (NTC 1999a). There is a paradox here, which has not been lost on manufacturers and advertisers. Although there has been a growth in individualism, there has been a simultaneous increase in uniformity of appearance. The trend at present is for everybody between the ages of 16 and 36 to wear black – as if participating in a giant wake. It is becoming increasingly difficult to buy apparel that does not bear a logo or designer name. We are urged to be ourselves in a constant stream of advertisements, but we can only be ourselves by buying product x or wearing something with the logo of company y. Brand loyalty is everything. Advertisers are increasingly looking for new ways to target and get through to consumers. There is an 'insatiable search for virgin, unbranded space' (Leith 2000: 13).

Changing holiday patterns

Social and economic changes, including rising disposable income, increased car ownership but, crucially, time available for holidays with increased holiday entitlement, three-day weekends and various flexible working arrangements, have combined to boost demand for both domestic and international tourism and recreation since the war. Around 60 per cent of people in the UK take a holiday in any one year (Boniface and Cooper 1994), but the poor and elderly largely do not participate. The domestic holiday market has seen some fundamental changes over the last two decades or so and is no longer characterised by a week or two spent at a traditional resort such as Blackpool or Clacton. There has been a continued decline in the length of stay, a growth in the shorter holiday market, growth of business and conference tourism, a shift away from traditional coastal destinations to towns and countryside, and an increased volume of trips to friends and relatives (Boniface and Cooper 1994). Many short holidays are taken as 'additional' holidays to complement the main holiday (often taken abroad). The greatest market growth in tourism has been in overseas travel. In 1998, UK residents took 50.9 million trips abroad and spent £19,500 million (Cope 2000). Most overseas trips are inclusive tourism to mainly Mediterranean destinations, but travel to the Americas, Africa, Asia and Australia/New Zealand is increasing. While some long-haul travel is geared towards package holidays for growing numbers, especially young adults, this form of tourism revolves around longer-term backpacking, gap-year-type working holidays.

Professionalisation/ commercialisation of sport

More and more sports (for example, golf, rugby football, tennis) and associated activities (for example, motor racing, snooker, darts) have come under the influence of big business, particularly in the form of corporate sponsorship (Bale 1982, 1989). Many sports would find it hard to exist without the sponsors who underwrite many famous tournaments and venues. Sponsorship, however, does not derive from altruism, and sport has paid a price for the **Faustian bargain**. Nowhere is this better illustrated than in the changes that have taken place in the UK's major sporting activity, association football. Until relatively recently, football was essentially a low-cost, working-class male leisure activity, but for

some premier league clubs the cost of match attendance is becoming costly. Football stadia are no longer open terraces filled with spectators who stand for the entire game. Increasingly they are covered, with the crowd seated, and they offer other facilities such as leisure clubs, conference facilities and restaurants. Merchandising of the team is big business. Some of the leading clubs are now quoted on the stock market. In 1998, English Premier League clubs alone spent £150 million on transfer fees (Kuper 1999). In 1999, the Office for Fair Trading challenged a £743 million deal between the English Premier League and BSkyB and the BBC (Chaudhary 1999). Such is the strength of the 'soccer mafia' that England's new football stadium at Wembley has been designed almost exclusively for football when it was intended as a multi-sport venue (Campbell 1999). The government was so dismayed at the commercialisation of the UK's national sport that it set up a Football Task Force designed to divert some of the large sums earned by the big clubs back into grassroots football (Gregoriadis 1999). However, it would appear that the game is becoming more exclusionary and the gap between the bigger and smaller clubs is increasing.

Developments in technology

Technological developments have done a lot to change leisure and consumption patterns. Improvements in air transport have widened horizons in terms of holiday destinations, while the relative decline in car prices has made this ubiquitous mode of transport, and most obvious of lifestyle statements, a very common commodity. Indeed, the increase in car ownership and use, and the concomitant decline in public transport (except flying), has had the biggest single impact on leisure and consumption generally. Those excluded from the car culture are often more circumscribed in their leisure activities (access to venues and the like) and consumption patterns (access to shopping, garden, DIY centres, and so on). Although the government is committed to curtailing out-of-town retail developments in favour of inner city regeneration, that policy appears to be not working. The recent Bluewater development in north Kent, for example, has a car park for 13,000 vehicles (Glancey 1999).

The mobile phone is perhaps one of the most obvious of recent technological developments. Although most mobile phone ownership is still concentrated amongst adults, the fastest growth is among teenagers and children (ONS 1999c) with all that entails for future health problems (see Chapter 10). This is a development that has spatial implications, in that the user no longer needs to access a fixed-line unit, unless they happen to be in the Highlands of Scotland or some other sparsely populated area which is poorly served with trans-mitters. Rapid progress in the Internet, particularly the development of the World Wide Web, encryp-tion, faster download times and efficient search engines, means that this is increasing as a leisure pursuit – simply browsing – or as a means to develop existing or new interests, from the most basic and innocent to the more obscure and prurient. This is another development that has compressed space and reduced geographical confines, people become accessible everywhere and private tele-phone conversations occur in very public spaces. However, it appears that many will be excluded from this form of communication. The primary piece of equipment for Internet access is a personal computer. Ownership of these increased between 1986 and 1998 from 16 per cent of all households in the UK to 28 per cent of all households (ONS 1999a). However, as Figure 9.2 shows, even at a crude, regional level, home computer usage varies from place to place, reflecting relative disposable incomes, lifestyles and class.

Thirty years ago, the death of the cinema was predicted. Indeed, the post-war period saw a decline of city centre cinema due to growth in television ownership and video rentals. Cinema attendance declined, particularly between 1952, when there were 1,300 million admissions in Great Britain, and 1984, when a low of 53 million occurred. But, due to better cinema design, sound and production, this increased to 124 million in 1997. In 1997, 54 per cent of adults attended the cinema regularly,

compared with only 31 per cent in 1987 (ONS 1999c). This has led to a resurrection of cinemas in the form of multi-screen venues both in suburban commercial parks, with huge car parks, and in city centres.

Consumerism as a way of life

Up until recently, a major focus of social identity was the family unit. With the demise of the nuclear family as the main unit of social organisation increasing numbers of people derive their identity through their lifestyle statements which are increas-ingly stemming from leisure and consumption. Even children are increasingly viewed as fashion accessories. As long ago as the 1960s, American demographers were debating whether babies were consumer durables (Blake 1968). Now childlessness is also viewed as a lifestyle statement (see Chapter 10) and the terms 'child-free' or 'unburdened' are replacing the term 'childless', which carries a barren stigma. The child-free feel that it is better to be 'Thinkers' (two healthy incomes, no kids, early retirement) than 'Sitcoms' (single income, two child-ren, oppressive mortgage) (Summerskill 2000b). The leisure and consumption industries are targeting these people with 18-and-over holidays, hotels, restaurants and so on. But advertisers have not been slow to target children as consumers in their own right; no longer are they simply the consumers of tomorrow. For example, children no longer dress as children but as miniature adults (Hitchens 1999). Fashion is a critical area in which once-rigid age barriers have become blurred. More and more television commercials are being geared towards children, and supermarket layouts have long been designed with children in mind. Consumerism as a way of life starts as soon as a child can work the TV remote control, if not earlier. Conversely, there are those Peter Pan characters who never grow up. These 'kidults' are actively courted by retailers and manufacturers. These are the 'Middlescent' – adults with adolescent tastes. Sony Playstations, mini-scooters, trainers, Harry Potter novels are just some of the adolescent items being bought by adults – mostly men (Summerskill 2000c).

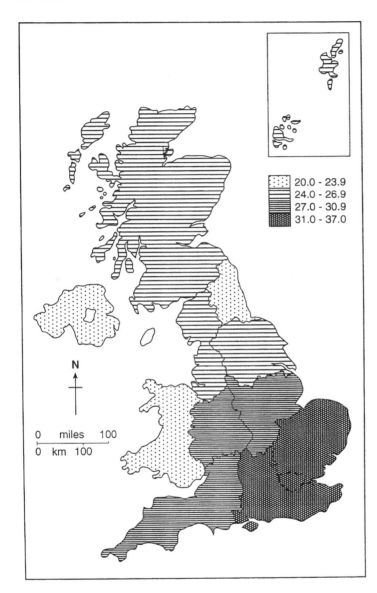

20.0 - 23.9
24.0 - 26.9
27.0 - 30.9
31.0 - 37.0

N

0   miles   100
0   km   100

Figure 9.2 Percentage of households with a home computer, 1998: UK

Source: Data derived from ONS 1999a, *Regional Trends 34*, Table 8.13.

Increasing loneliness, either voluntary or involuntary, has helped drive the consumerist machine. Many people, especially women, succumb to compulsive shopping sprees. They binge on shopping the way increasing numbers of young people binge on alcohol or bulimics binge on food (see Chapter 10). The 'shopaholic' is now recognised as a person in need of remedial treatment, much like the alcoholic or compulsive gambler. Shopping, according to Goss (1999) is the second most important leisure activity in North America, and watching television the first – much of programming actually promotes shopping, both through advertising and the depiction of model consumer lifestyles. Increasingly, 'surfing' the Internet is becoming not only a leisure activity but also a means of consumption as e-shopping, e-advertising and e-commerce.

Decline of religion and voluntary work

There has been a marked decline in formal Christian religious activity in post-war years (see Chapter 11). In 1970, just over nine million people in the UK were members of mainstream Christian churches. In 1995, this had reduced to 6,361,000. Of these, members' regular attendance has also dropped. That said, non-Christian religious membership increased from 451,000 to 1,295,000 over the same period (ONS 1999c). This is a result of growth in non-white ethnic minority population and a growing increase in non-Christian religions among the 'indigenous' population. Although religious observance has been in decline for a number of decades, up until relatively recently Sunday was seen by many as a special day. Entertainment and consumption were strictly controlled. Now with increasing shopping and leisure facilities opening on a Sunday, and an increase in Sunday working as a result, the Christian Sabbath has become much like any other day.

Perhaps related to this decline and the growth in individualism and more women juggling paid and unpaid work, fewer people are becoming involved in voluntary work. One recent survey suggested that half of the adults in England had no interest in doing voluntary work at all (ONS 1999a). This has led to a call from politicians, including the Prime Minister, Tony Blair, for more people to give freely of their time to 'good causes'.

9.2.3 Consuming lifestyles in the UK

Time and money are the crucial commodities as far as leisure and consumption are concerned, (Green *et al.* 1990). Being in paid employment in many ways enables leisure and, particularly, consumption. Yet as we noted in Chapter 8, many people in paid employment in the UK now work over 40 hours a week. There is thus an imbalance, for many, between time and money for leisure and consumption. As with paid employment patterns, so with leisure and consumption patterns. These are also influenced by age, class, education, gender and ethnicity.

One of the most distinguishable variables in leisure and consumption patterns is age. Although most age groups will engage in similar types of leisure activities and consume many of the same types of commodities the precise nature of leisure and consumption will vary with age. Thus, while most people enjoy foreign holidays, the '18–30' type venues and programmes will not attract children, the middle aged and the elderly. We all shop, but much shopping is not only gendered it is geared towards age groups and life cycle stages. Then there are activities which are more clearly determined by age. Garden centres, for example, rarely attract teenagers and twentysomethings. Few pensioners go clubbing, and so on.

Similarly, social class, and all that entails, still has an effect on leisure and consumption patterns, even though, as was noted above, disposable income has as big an influence as class *per se*. Besides, class is still very highly correlated with income, and this combines to affect the geography of aspects of leisure and consumption. Another component of social class is education. This can influence leisure and consumption – for example, Serengeti safari versus Skegness; opera versus rock concerts; Arundhati Roy versus Barbara Cartland; and so on. Stereotypes aside, pigeon racing and whippet breeding are still more common in urban northern parts of England than in southern and rural parts, while the reverse pattern tends to prevail in the case of polo matches and gymkhanas. Even in our relatively egalitarian and 'Americanised' society a person's social class can often be determined by the way they spend their time and money. This balance between time and money, which is at the heart of any analysis of leisure and consumption, is particularly problematic for those who are money poor and time rich – many pensioners and unemployed – and those who are money rich but time poor – many professional and managerial people caught up in the cycle of career development and job insecurity (see Chapter 8). Although much leisure activity can be undertaken without recourse to heavy monetary expense, even relatively minor expenses such as public transport to 'free' venues – parks, libraries, shopping malls and the like – can impact on a low fixed budget. Besides, most leisure and consumption activities in modern society are geared towards financial exchange.

Just as paid work is gendered, so too are patterns of leisure and consumption. Working women, for example, spend more time socialising than men do, but less time eating out or drinking out (ONS, EOC 1998). That said, it is common now for women to have nights out in pubs and restaurants. Gone are the days when public bars displayed signs saying 'men only', and when 'respectable' women could only get a drink in the company of a member of the opposite sex. According to Katharine Viner (1994: 14), 'Girls "letting their hair" down is very much a feature of the post-modern city.' That may be true in London and some other cities, but in textile towns of the north it has been common practice since the nineteenth century for women. In Dundee, for example, relatively high earnings in the jute and flax trades meant that women's nights out have been a regular feature of the social scene since the 1840s (Close 1992; Graham 1988). And according to J.B. Priestley (1994: 135), writing in 1933, in Nottingham 'the enormous numbers of girls employed in the lace trade [were] more independent and fonder of pleasure than most provincial young women'.

For women, especially, leisure and consumption can be 'blurred' with work, as well as with unpaid work in the home. The blurring of work and leisure and consumption spaces has been emphasised in a study of merchant bankers in the City of London (McDowell 1997). For many women a lived reality is ironing whilst watching television or combining childcare with a child-centred leisure activity, such as visiting a park or a fast food outlet (Gilroy 1999). Moreover, in a recent study of the leisure activities of women academics, four main forms of leisure were identified (Todd, cited in Utley 1998). These include: *leisure as an extension of work*, such as conferences, drinking and eating with colleagues and networking; *partner leisure*, such as going out for a meal, or simply synchronising diaries to relax together; the most significant category is *family leisure* which consists of being with children, going to parks, and the like (here there is a blurring of leisure and the

tasks of social reproduction); and *personal leisure*, which includes health and fitness, arts events, and so on.

Ethnicity can also affect leisure and consumption. Gaelic games, such as hurling and Gaelic football, as played in Northern Ireland, or shinty, as played in Scotland, are rarely played in England (and are thus rarely televised). Conversely, England's national sport of cricket is televised to the other countries of the UK where the game is rarely played (see Chapter 11). For some non-white ethnic minorities certain leisure and consumption patterns can be discerned. The social lives of women and young people outside the home are constrained among some groups, for some Muslim women dress codes prevent participation in such activities as swimming at public swimming pools. Conspicuous consumption is associated with others and has been parodied by the Jewish comedian 'Ali G'. Yet other groups devote considerable amounts of their time to religious and voluntary endeavour (see p. 101). As the distribution of income, social classes, ethnicities and age groups is not even, at both regional and local level, this is reflected in the distribution of leisure and consumption.

9.3 NEW LANDSCAPES OF CONSUMPTION

The space of consumption is as important as the physical activity of consumption. During the 1990s the leisure and consumption and cultural practices of the British week, including the weekend, have undergone a fundamental transition. The working day for some has a 24-hour dimension with, for example, extended trading hours for retailing, with some supermarkets open 24 hours a day, ATM machines allow us access to cash all day, and the Internet is reshaping lives (see Chapter 8). And as already shown, the British Sunday is no longer the 'day of rest' reserved for church, Sunday lunch and a snooze (Gregson and Crewe 1994). The spaces, places and scenes of leisure and consumption in British town centres vary radically at different parts of the day.

9.3.1
Case study

Nottingham's
Lace Market

This is well illustrated by Crewe and Beaverstock's (1998) description of the Lace Market of Nottingham. During the day it is a space for work and retailing, but it is an area with a dynamic evening economy (see below). But the spaces, places and scenes of these streets can be spaces of fear, especially for women and members of ethnic minorities at some parts of the day (Valentine 1989). The age structure also changes dramatically as the day progresses. Time and space can combine to alter the nature of leisure and consumption.

We can identify an 'evening economy' of the spaces, places and scenes where people come together to eat, drink, dance or listen to music. These are important markers of self-identification. Moreover, the night-time economy is becoming an increasingly important marker of the post-industrial city. It is estimated that the evening economy generates between 5 and 15 per cent of local GNP, and the nightclub industry alone is worth an estimated £2,000 million a year, with over a million people spending an average of £35 per week clubbing (Crewe and Beaverstock 1998). Focusing on the Lace Market area of the city of Nottingham, they describe an area busy with street activity and life, of bars, cafés, clubs, restaurants and gymnasia. The area has developed a reputation for a particular brand of nightlife favoured by the young (Figure 9.3). The evening economy also has a multicultural dimension. For example, for how many young people does an evening of clubbing include an Indian or Chinese meal? For young Asians in the UK one important cultural expression is through Bhangra music and gigs (see Chapter 11). But the evening economy does have a distinctive regional dimension.

Figure 9.3 Nightlife in Nottingham's Lace Market

Source: David T. Graham.

There are strong differences between the weekend city scene in England and Wales compared with the scene in Scotland and Ireland. While the 'circuit' is a major feature of English and Welsh cities it is a very rare occurrence in central parts of the towns and cities of Scotland and Northern Ireland. From preliminary investigation, it seems that this is due to more effective policing in the latter two countries – Northern Ireland is one of the world's most policed democracies, and Scottish forces practise zero tolerance more effectively than those south of the border. The much more liberal, almost continental, licensing laws in Scotland mean that binge drinking and associated behavioural problems are less frequent. In both Scotland and Northern Ireland, the majority of pubs are owned by individuals or small businesses, and even city centre pubs are more like 'locals' – patrons simply do not feel the need to roam around town. By contrast, most pubs in England and Wales are owned by big companies (a Japanese bank is the biggest landlord in England) or breweries. Thus, people have no loyalty to a particular pub and tend to be attracted to theme pubs – so-called Irish pubs being the most ubiquitous and best example. This process of trying to make everything the same has been termed **'McDonaldisation'** or **'Disney-fication'**, which links globalisation to economic and cultural processes (Crouch 2000). However, it would seem that the theme pub's days are numbered, with new policies being put into place to 'prompt the rebirth of the traditional, unthemed, city-centre pub' (Brandwood 1999: 10).

Geographers have highlighted several consumption spaces in the UK today:

The shopping mall

There has been a wealth of research on shopping malls, but as Wrigley and Lowe's (1999) recent work emphasises, shopping malls are very diverse. For example, virtually every local shopping centre has within it a version of a mall; there are also huge out-of-town malls (Figure 9.4); and there are malls constructed as part of tourist strategies. Shopping malls have been constructed in city centres, where they are at least accessible to the thirteen million people with no access to a car. But the biggest and most alarming are out-of-town shopping centres – what Goss (1999) has called 'cathedrals of consumption'.

**Figure 9.4
Bluewater
shopping mall:
Kent**

Source: *The Guardian, Arts*, 27 September 1999, pp. 12–13.

Figure 9.5 The street provides a free venue for leisure activities, as this rather unconventional cyclist demonstrates

Source: Darren Regnier.

The street

The street provides a free venue for leisure activities (Figure 9.5). In some locations, such as the Lace Market in Nottingham or Little Clarendon Street in Oxford, there are spaces where it is trendy to 'hang out' (Crewe and Beaverstock 1998; Wrigley and Lowe 1999). But there are also streets, such as Belgrave Road, Leicester or Green Road, Newham, which have been transformed by the entrepreneurial activities of the members of the Asian diaspora. These streets form the nucleus of flourishing manufacturing, wholesale and retail business districts where a whole range of Asian goods and services can be purchased (Hardill 1999; Nash and Reeder 1993).

Car boot sales

Gregson and Crewe (1994) note that car boot sales are an alternative consumption space, in both working-class and middle-class neighbourhoods, but that they exist for different reasons. There are also significant differences between 'neighbourhood specific' boot fairs and their massive 'out of town' counterparts. Car boot fairs are places for buying

and selling and a site of pleasure and curiosity. Car boot sales are one manifestation of the massive 'second hand' market, of interest in 'collecting' second-hand style and retro-chic (McRobbie 1989). They are also one of the most popular ways of off-loading the proceeds of thefts and burglaries.

The captured market

This is where individuals, confined in places for other reasons, are induced to consume. These consumption sites include airports, hospitals, cruise ships or cross-channel ferries. At such locations the blurring of leisure and retail takes place.

Home: the new consumption landscape

The home through 'home shopping' is now part of the new consumption landscape, through mail order catalogues, lifestyle magazines and through Internet shopping (Wrigley and Lowe 1999). In addition to being an investment, for some people home ownership is also an important element in social display and a focus for the expression of cultural values and identity (Hamnett 1999). The neighbourhood/locality chosen – whether for rented or bought properties – often reflects the desire for a particular kind of living space, a space which extends beyond the house itself out into the surrounding environment, 'a civilised retreat' and sites for anchoring 'middle-class' identities (see Chapter 15).

9.4 SUMMARY

Leisure and consumption now form a large part of most people's lives. Globalisation and expansion of the free market, together with newer and more effective means of mass communication, make it more difficult to escape from the consumerism that is at the heart of the capitalist mode of political economy. But leisure and consumption mean different things to different people. Changes in patterns of leisure and consumption have been rapid over a relatively short space of time. And these patterns are affected by age, class, gender and ethnicity. There is thus a marked geography of leisure and consumption. While technological change (cable and satellite TV, Internet, mobile phone) has allowed people to access leisure and consume without changing location, other features of leisure and consumption are very geographically specific – shopping malls, ice rinks, swimming pools, national parks, Alton Towers (Staffs), Holme Pierrepont (Notts), the Millennium Dome, and so on. There is, then, a variety of ways, often complex, in which space and place and leisure and consumption combine.

R E V I S I O N   Q U E S T I O N S

- Define leisure and consumption. In what ways do the two interact?
- In what ways have patterns of leisure and consumption changed in the UK in recent years? Compare your own experiences with those of your parents at the same age.
- What is meant by the consuming lifestyle? What factors have caused society to become more consumerist?
- In what ways does geography influence leisure and consumption and vice versa?

KEY TEXTS

Bocock, R. (1999) 'Consumption and lifestyles', in J. Bryson, N. Henry, D. Keeble and R. Martin (eds) *The Economic Geography Reader*, Chichester: Wiley, pp. 279–90.

Crouch, D. (2000) 'Leisure and consumption', in V. Gardiner and H. Matthews (eds) *The Changing Geography of the United Kingdom* (3rd edition), London: Routledge, pp. 261–75.

Hamnett, C. (1999) *Winners and Losers: Home Ownership in Modern Britain*, London: UCL Press.

McDowell, L. (1997) *Capital Culture*, Oxford: Blackwell.

Wrigley, N. and Lowe, M. (1999) 'New landscapes of consumption', in J. Bryson, N. Henry, D. Keeble and R. Martin (eds) *The Economic Geography Reader*, Chichester: Wiley, pp. 311–14.

HEALTH AND WELL-BEING

10.1 INTRODUCTION

One feature of life in the UK of perennial interest is health and well-being. Health provision is a constant topic of debate for politicians and policy-makers, a perennial source of reporting and misreporting in the media (Adam 2000; Moeller 1999) and an unremitting source of concern for patients. The influenza epidemic of the winter of 1999/2000 demonstrated how vulnerable the **NHS** is and how fascinated the media and public are with health care provision. Relatively recent health scares, such as BSE and nvCJD, multidrug resistant **TB** (MDRTB), **HIV** and **AIDS**, indicate how susceptible we are to the vagaries of microbes. The conviction of Harold Shipman, as the most prolific mass murderer in UK history, showed how one of the most trusted symbols of the community – the general practitioner – could prove fallible and diabolical. It also highlighted the lack of rigorous professional control among health care practitioners.

As we saw in Chapter 9, an increasing number of people are spending more time and money in pursuit of a healthier lifestyle. At the same time, large segments of the population are pursuing unhealthy and dangerous practices. Attitudes to health are closely linked to social class, education, ethnicity and gender. This manifests itself in a geography of health with some areas being much less healthy than others (see Chapter 14). However, unlike many of the other issues dealt with in this book, the geography of the phenomenon is not simply a result of human geography. Some aspects of health are a direct result of the physical geography of where people live (and die). In this chapter we explore some of the main themes in health and well-being in the UK.

10.2 MORTALITY

One way of looking at health is to look at mortality, since it is, as far as clinicians are concerned, at one end of the health spectrum and thus indicative of health and health care. This has already been covered in Chapter 7 but here we look at some specific health-related topics.

The nature of mortality in advanced capitalist countries is such that most death is due to **chronic disease** rather than **infectious disease**. Heart disease, for example, is the UK's biggest cause of death – killing 400 people daily (Browne 1999b). As societies develop, and diet, education, infrastructure and health care improve, so the nature of disease and mortality changes. This is known as the **epidemiological transition**. This 'focuses on the complex change in patterns of health and disease and on the interactions between these patterns and their demographic, economic and sociologic determinants and consequences' (Omran 1971: 510). That said, there are a number of infectious diseases that are causing concern among health authorities and public alike. The rapid speed of social, technological, environmental and population change is such that the emergence of new diseases and re-emergence of old diseases is a real threat (*Population Trends* 1997). Increased international population movement, through tourism, business travel, migration, refugees and educational travel, widens the scope for the spread of infectious disease (Graham and Poku 1998). Many of the new cases of MDRTB in London are a result of people visiting relatives in the Indian subcontinent, as well as poverty. Even the seemingly antiseptic conditions in hospitals can be potentially fatal with the methicillin resistant *Staphylococcus aureus* (MRSA) causing death in postoperative and intensive care wards. The following observation chillingly reinforces how vulnerable human populations are, even in advanced capitalist countries:

> October 1977 saw the global eradication of one of mankind's great plagues when the last naturally occurring case of smallpox was tracked down in Somalia. Then, within two years, in the summer of 1979, the first official diagnosis in the United States of America of a patient with acquired immunodeficiency syndrome (AIDS) signalled the start of a new global scourge.
>
> (Cliff and Haggett 1990: 93)

Because chronic disease has replaced infectious disease as the major cause of mortality the biggest

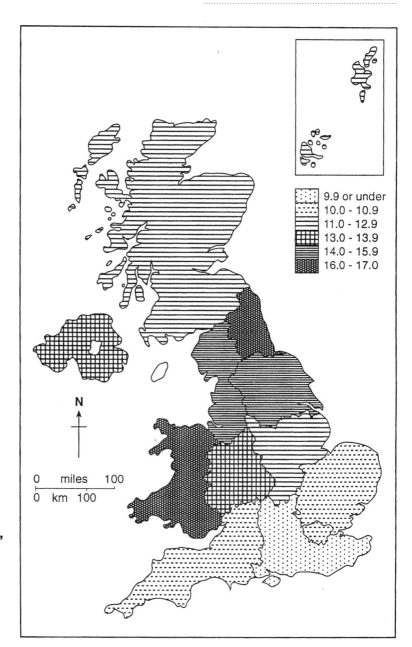

9.9 or under
10.0 - 10.9
11.0 - 12.9
13.0 - 13.9
14.0 - 15.9
16.0 - 17.0

N

0   miles   100
0   km   100

Figure 10.1 Percentage of people reporting general poor health: Scotland, Wales, Northern Ireland, Government Office Regions, 1997

Source: Data derived from ONS 1999a, *Regional Trends 34*, Table 7.3.

cause of death in the UK is circulatory disease, accounting for some 45 per cent of total deaths (*Population Trends* 1997). And as we have seen (Chapter 7) mortality varies from place to place. The overall pattern reflects the familiar North–South divide in socioeconomic and cultural phenomena (see Chapter 14) and this spatial pattern is long standing (Britton 1990; Dorling and Tomaney 1995). Figure 10.1 shows the regional pattern of self-reported poor health. We now explore some of the reasons for this geography of mortality and health in more detail.

10.3 THE GEOGRAPHY OF HEALTH AND WELL-BEING: SOME CAUSAL FACTORS

One of the most notable features of health and well-being is the marked variation from place to place. For example, people in Wales, the North West and North are 40 per cent more likely to die of stomach cancer than residents in the South East. Bootle, Hartlepool, Stoke-on-Trent and Salford have 50 per cent worse rates for the same disease compared to the rest of the country (Coleman and Salt 1992). Table 10.1 shows how crude mortality varies within the UK even at regional level. Even by this measure, which takes no account of age or sex structure (so that the youthful population of Northern Ireland shows a relatively low mortality for the first two causes, whereas the opposite holds true for the older population of the South West), it is clear that mortality varies by cause over different areas.

Health differentials become more marked at finer spatial resolutions. For example, people in Springburn, Glasgow, suffer a rate of chronic illness of 155 per thousand, compared to a rate of 36 per thousand in Wokingham, Berkshire. In terms of infant mortality, there are 6.79 deaths of infants under age one per thousand live births in Springburn, but only 5.32 in Wokingham (McKie 2000). Glasgow and Belfast have the unenviable record of having some of the worst heart attack rates in the world. In a ten-year-long study of thirty-seven cities, Glasgow had the highest rate for women and second amongst men. Belfast had the second highest rate for women and the fourth for men. 'Glasgow's problems come from high levels of smoking and poor diet causing high cholesterol and blood pressure. The intake of fruit and vegetables, which protect against cancer and heart disease, is low and the population generally does not take enough exercise' (Boseley 1999b: 16).

Where people live also affects their state of mind and mental health. Judging by the number of anti-depressants prescribed, Hillingdon, on the outskirts of London, is the happiest place in England. The suburb has low unemployment and low crime rates and a balanced age structure. The least happy, by the same criteria, is Morecambe, in Lancashire. This is a deprived and declining resort with high unemployment and high crime rates (Thomas 1999).

There is also an urban–rural dimension to health and well-being (Baggott 1998; Curtis 1995). While the countryside is often viewed as a healthier

Table 10.1 Causes of death (rates per 100,000 population), 1998, UK and statistical regions

Area	Circulatory system	Cancer (all neoplasms)	All accidents	Motor vehicle accidents
UK	472	275	21	7
North	523	306	22	7
North West	511	287	21	6
Yorkshire and Humberside	478	279	19	7
East Midlands	459	268	23	8
West Midlands	465	271	22	7
East Anglia	463	274	25	9
South East	418	254	17	5
South West	514	287	20	6
Wales	524	302	23	5
Scotland	529	300	28	7
Northern Ireland	427	223	26	10

Source: ONS 1999a, *Regional Trends 34*, Table 2.2

environment than the city, there is still a high risk associated with modern agricultural methods. Pollutants, in the form of pesticides and inorganic fertilisers, can be dangerous. A higher incidence of psychiatric ill health is to be found in towns and cities than in rural areas (Shepherd 1984), though this does not explain the very high suicide rates among farmers. Atmospheric pollution is higher in towns and they tend to be more dangerous in terms of violence and road traffic accidents.

Such spatial variations occur because the underlying factors affecting health vary between places. These factors are many and varied. They are often closely interrelated and it is difficult to determine individual causes. Some causes of ill health are long established and well known. Public health observers have long recognised that poverty, poor housing and inadequate diet have adverse effects on health (e.g. Booth 1889–1902; Chadwick 1842; Mayhew 1851–62). Other ailments and diseases are the result of more recent practices, behaviours or technologies. For example, the development of a new variant of CJD because of greed and bad practice in the farming industry causing BSE; the development of multidrug resistant TB (and whatever else may develop) as a result of antibiotic misuse; the spread of HIV/AIDS due to sexual promiscuity and failure to follow safe sex advice; a possible epidemic of brain diseases as a result of radiation from mobile telephones; and so on.

10.3.1 Environmental factors

The UK has a very varied set of physical environmental conditions. Climate and weather vary and these can affect health and well-being. Lack of daylight causes seasonal affective disorder (SAD), which can result in anxiety and depression – Aberdeen has been called the saddest place in the UK because of the high incidence of SAD due to the lack of daylight in winter (O'Sullivan 1999). Geological conditions are also seen to be important in determining health spatially. Background radiation is higher in the igneous rocks of upland Britain than in the sedimentary rocks of the lowland areas of the south and east. Although the health implications

of this are not fully understood there was considerable concern in 1988 about dangerously high levels of Radon-222 (a natural radioactive gas derived from the decay of underground uranium) discovered in many houses in Cornwall (Jones 1990). Trace elements derived from the soil are also important in stimulating, as well as preventing, certain ailments. Water hardness is also implicated in health (Foster 1992). Hardness is normally a function of suspension of calcium and magnesium in the water. Most hard water is found in the south and east of the UK and although there are drawbacks to this, in terms of 'furry' kettles and latherless soap, most adverse health conditions are associated with soft water. There is a strong correlation between water softness and cardiovascular disease (Jones 1990). This might go some way in explaining the higher incidence of this disease in the north.

Air quality is also known to affect health. Up until relatively recently, most atmospheric pollution derived from the burning of coal in factories and homes. This resulted in severe local pollution, especially when certain weather conditions prevailed. Perhaps the best-known public health catastrophe related to atmospheric pollution that occurred in London in December 1952. A temperature inversion caused impurities to be trapped in the lower air, resulting in the deaths of 4,000 from chronic bronchitis (Jones 1990). Some areas, such as Northern Ireland, are still dependent on coal or peat fires for domestic fuel and the local pollution in many towns there can be particularly bad.

The biggest source of local atmospheric pollution is motor vehicles (Meacher 1997). Road transport accounts for 20 per cent of the UK's carbon dioxide emissions, 75 per cent of carbon monoxide emissions and 50 per cent of the nitrous oxide emissions (Walters 1999). This is the major cause of the dramatic increase in childhood asthma (Boseley 1999c). This and other respiratory diseases, as well as brain damage from lead in petrol, have been sources of concern in many urban areas close to busy roads. Often, these are in inner city areas where locals own few cars but suffer the highest pollution level. As well as being a major cause of illness among children, asthma is the most commonly reported cause of long-term illness for them. Asthma increased dramatically between 1976 and 1997, from around 10 per 100,000 to 40 per 100,000 (ONS 1999c). Some 1.5 million children suffer from the disease; fifty a year die from it; fifty a day are hospitalised; the cost to the NHS is £6,000 million a year. Air pollution accounts for some 25,000 premature deaths a year (Walters 1999), and is a major cause of distress for the UK's 3.4 million asthmatics (Brown 1998). One reason for the increase in asthma cases, and indeed other diseases, is increased awareness of personal health and an increase in self-reported illness.

Air pollution is worst in the large conurbations. For example, out of fifty-six cities surveyed by the government, London, Liverpool, Manchester, Birmingham, Leeds, Leicester, Nottingham and Bristol had the worst pollution. Some traditional industrial cities, like Newcastle, Sunderland and Glasgow, are less badly affected. This is due to relative traffic congestion. Nitrogen dioxide causes asthma attacks, heart problems, induces breathing problems and lowers immunity to allergic reaction (Arlidge 1999a).

Although EU legislation has forced manufacturers to improve engine performance in order to reduce pollution, and petrol is now unleaded, the sheer increase in motor vehicle numbers has led to no real decline in pollutants derived from car exhausts. In 1987 there were 17.9 million licensed motor cars in Great Britain. By 1997 that figure had increased to 22.8 million. Added to this are buses, coaches and lorries, with over 80 per cent of goods being transported by road. The total number of motor vehicles is predicted to reach 35 million by 2025 (Walters 1999). The fact that more than half of drivers do not believe that they should pay higher taxes for the sake of the environment (ONS 1999c) suggests that motor vehicle pollution will continue to be a major health problem. Car ownership in the UK is lower than in Sweden, Portugal, Luxembourg, Italy, Germany, France, Belgium and Austria, but patterns of use are different. UK motorists are tied to their vehicles more than their European counterparts and use them more for travel to work and short walkable journeys (see Chapter 16). Within the UK, car ownership is lowest in the North, Scotland, Northern Ireland and London. Although 70 per cent of the workforce drives to work, that figure is lowest in London, where less than half do so. Scotland also has a lower than average rate (ONS 1999a). The road lobby has gone to great lengths to try to convince the public that the link between increased respiratory disease and increased car use is spurious. Industrial polluters have also had some success in creating public confusion over the health and pollution association (Phillimore and Moffat 2000).

10.3.2 Behaviour and lifestyle

A range of factors associated with behaviour and lifestyle have been implicated in determining health and well-being. And, as we saw in Chapter 9, lifestyle and many of the behaviours examined below vary according to age and gender, as well as varying geographically.

Occupation

There is a long established relationship between occupation and health (Marmot and Feeney 1996). Occupation has a strong influence on wealth, education and other socioeconomic and behavioural factors that affect health. Also, the working environment may have a direct influence on health. Fishermen, construction workers, miners, foundrymen, publicans, stevedores and printers have traditionally had higher mortality rates than university teachers, office managers, politicians, social workers, secretaries and physiotherapists (Benjamin 1989). Suicide is notably high among farmers. Some jobs are especially prone to accidents and exposure to dangerous substances. The growth of tertiary occupations at the expense of primary and secondary occupations, as well as an increase in health and safety legislation, has seen many of the formerly dangerous jobs decline or disappear, or become less dangerous and unhealthy. Although accidents are most common among manual workers, increasing levels of occupational ill health are appearing among professional and other white-collar workers. For example, in 1996, 24 per cent of men with manual occupations were involved in a major accident, compared with 17 per cent of males in non-manual occupations. The rate was the same for women in each group – 15 per cent. In the same year, 71 per cent of males and 76 per cent of females in social classes I and II reported suffering stress. This compares with 52 per cent of males and 64 per cent of females in social classes IV and V (ONS 1999c). This is linked with the longer hours that higher status white-collar workers are working generally (see Chapter 8). Other white-collar ailments increasingly reported include posture-related back injuries, VDU-induced eyesight defects and repetitive strain injuries.

This stress and the time constraints this imposes on increasing numbers of people has been termed 'hurry sickness' (Reeves 1999d). There has been a complete reversal between status and time over the last two or three decades. The better-off now have very little spare time, whereas they used to have a lot of time (see Chapter 9). White-collar workers are also likely to suffer from sick building syndrome. This occurs when air condition systems do not work properly. In these circumstances, in buildings in Tower Hamlets, London, 85 per cent of staff suffered extreme tiredness, 80 per cent had regular headaches and sore throats and 60 per cent complained of more serious maladies (Gillies 1999). Faulty air conditioning can also provoke asthma attacks and spread legionnaires' disease. It seems that the link between occupation and health is increasingly operating through social class. That is, social class is determined by occupation and it is the factors associated with class that are responsible for health differentials rather than the nature of the job itself.

There is also a strong correlation between unemployment and ill health. This holds true for both men and women and for psychological and physical health (Bartley *et al.* 1996). The association is not straightforward and there are a number of ways it may be operating. Being unemployed can lead to stress and material hardship. Unemployment might be the result of low self-esteem or other

characteristics that can also cause ill health. Poor health can result in a poor work record and this makes unemployment more likely.

Diet and exercise

One variable that has long been known to influence health is diet. Lack of food will result in poor health and eventually death, as will too much food. Balance is also important. Lack of some foods leads to ill health just as an excess of certain foods predisposes to poor health. A deficiency of fresh food, vegetables and salads is associated with increased risk from stomach cancer (Britton 1990). Consumption of fruit and vegetables helps reduce the impact of cholesterol in the diet and thereby reduces the incidence of heart disease (Boseley 1999b). Excess salt in the diet can lead to high blood pressure (Marmot 1984).

Lifestyles affect diet. The growth of single person households, especially among busy professionals, has meant increased consumption of processed and ready-made meals. There is a geography to diet (see Chapter 14), which has a North–South dimension: 'Generous Northern breakfasts are a revelation to Southerners increasingly used to insipid muesli, and the traditional "Ulster Fry" looks like a particularly defiant rejection of nutritional rationality' (Coleman and Salt 1992: 364).

As noted above, the British are more likely to use the car for short journeys rather than walk (see Chapter 9). Almost 13 per cent of all car journeys are less than 1,000 yards (Brown 1998). This not only has a negative health impact on the motorist, but the excessive number of short journeys adds to atmospheric pollution. There are also issues linked to the general social health of an area. In communities where more people walk, a sense of community is fostered and crime rates are lower. Over-reliance on cars also leads to obesity, especially among children. This also links with the reduction in street play, walking generally and walking to school specifically. In 1971, around 70 per cent of 7-year-olds walked to school, now it is about 3 per cent (Hugill 1998). The 'school run' is the biggest contributor to local congestion, and outside schools are among the

"What shall we watch — Delia, Floyd, Rick Stein, Rhodes or Ready Steady Cook?"

Figure 10.2 Obesity usually reflects a certain pattern of leisure and con-sumption and tends to be incompatible with good health

Source: *Private Eye*, 7 March 1997, p. 20. Reproduced by permission of *Private Eye*.

worst areas for accidents as views are obscured by the ranks of parked cars.

A badly balanced diet and lack of exercise can lead to obesity. Sedentary lifestyles are very much a feature of 'consuming' societies (Figure 10.2; see Chapter 9). In 1980, 6 per cent of men and 8 per cent of women in Great Britain were technically obese. By 1997, the respective percentages were 17 and 20. Obesity related health problems account for between 6 and 8 per cent of the NHS budget (Reeves 1999e). It is obesity among children, however, which is the most worrying. A combination of junk food and lack of exercise has increased the number of obese children dramatically in the last ten years, leading to an explosion in US-style 'fat camps' for youngsters officially classed as obese. A decrease in physical education in the school curriculum is also to blame for the crop of mini 'couch potatoes' now being reared (Campbell 2000a). The government's continued sale of school playing fields and other recreational sites has also been criticised (Campbell 2000b). The result is poor health. In Scotland, children as young as 11 are developing heart disease due to obesity (Thompson 2000).

As we saw in Chapter 9, British society is becom-ing increasingly obsessed with body weight. The 'dieting industry' is one of the fastest growing in the country. At the beginning of the 1990s, as much as 90 per cent of women were worried about their weight (Crewe 1992). Dieting is not confined to females, but teenage girls who diet seriously are eighteen times more likely to develop eating disorders. Anorexia nervosa and bulimia are the most common and best known. Anorexia is a less common but longer recognised disease than bulimia. The former only affects one in a thousand, compared with 3 per cent of females affected by bulimia (Crewe 1992). Like all psychiatric illnesses there are complex underlying causes, generally related to self-image and self-esteem. It has also been linked to a decline in religion and rise in materialism (Crewe 1992; see Chapter 9). If not cured, these disorders can cause long-term health problems and even death. Some well publicised cases of possible eating disorders include Victoria Beckham, Calista Flockhart, Elizabeth Hurley and Kate Winslet (Paton Walsh 1999). Lena Zavaroni, the entertainer, died as an indirect result of anorexia.

Sexual behaviour

Levels of unprotected sexual intercourse, par-ticularly among young people, is a major cause of concern among health professionals (Anning 1999; Arlidge and McVeigh 1999). British teenagers have the worst sexual health in Europe (Boseley 1999d). Sexually transmitted diseases are now the most common infectious diseases in the UK (Coleman and

Salt 1992). Increasingly these are becoming drug resistant, and the most dangerous, HIV, is untreatable. Around 30,000 people in the UK have contracted HIV, of whom a third do not know they are infected – the figure for 1999 shows that the incidence of infection is rising (PHLS 2000). However, expensive 'combination therapies' can prolong the life of someone who develops AIDS by 20 years, though these have side effects. However, because of the high cost of these drugs, health authorities are increasingly rationing them (Browne 1999c). In 1999, for the first time in the UK, there were more diagnoses of heterosexually acquired HIV (1,070 cases) than homosexually acquired cases (989) (PHLS 1999a). It appears that, as in Africa and Asia, the disease will soon be as common among women as men; this is particularly likely since, like other sexually transmitted diseases, teenage girls are particularly vulnerable to infection for physiological reasons (Mbanje and Koch 1999).

Chlamydia infection is also growing at alarming rates and this can result in serious long-term gynaecological and obstetrical problems, such as pelvic inflammatory disease, infertility and ectopic pregnancy. In 1996, 6,236 males aged 16 to 24 in England and Wales were diagnosed with chlamydia, compared with 12,754 females; by 1998, these figures were 8,721 and 17,717 respectively (PHLS 1999b). Put another way, infection rates for young males were 2.9 per thousand and for females 6.4 per thousand in 1998. These diseases are not distributed evenly throughout the country. In Nottingham, for example, the incidence of some sexually transmitted diseases is triple the national average. The city is one of the country's three worst places for gonorrhoea and also has one of the highest teenage pregnancy and abortion rates (Hunter 1999).

Risks of ovarian cancer increase with sexual promiscuity and early age at commencement of sexual activity (Coleman and Salt 1992). On the other hand, it seems that sex, or at least giving birth at a relatively young age, protects women against breast cancer. Single women, women who have never given birth, and women who first give birth after age 35 have higher risks (Foster 1992). The high level of unprotected sex has led to the UK having the highest teenage pregnancy rate in Europe (see Chapter 7). Most of these end in birth. But 'a pregnant teenager is considered a high risk obstetric patient because she has a higher risk than normal of developing anaemia and pre-eclampsia. She also has a higher risk of maternal mortality. Her baby has an increased risk of infant mortality and of being low birth weight' (Botting 1998: 24). In 1996, 37 per cent of all teenage pregnancies ended in abortion, compared with 50 per cent of conceptions to girls under 16 (ONS 1999c). Following the introduction of the 1967 Abortion Act in 1968, clinical abortion made this common but dangerous method of family planning legal and safe. The procedure is still not entirely without risk, and death and sterility can result. The Act does not apply to Northern Ireland and women there travel to Great Britain in order to undergo the abortion. In 1997, 191,670 abortions were performed in Great Britain (Pearce 1998) and the abortion rate increased by 20 per cent in the wake of the 1999/2000 New Year celebrations.

Violence and accidents

Although many formerly dangerous jobs have disappeared or are subject to health and safety regimes, violent and accidental injury and mortality are still important among a section of the population. Young men, in particular, are the victims of road traffic accidents as the accident and emergency units of hospitals are well aware. Insurance companies can also testify to the predilection of young males for dangerous driving, and levy policies accordingly. Young men also tend to suffer most from sports injuries, and the popularity of so-called extreme sports adds to the injury and death toll among this specific group. They are also the group most likely to be involved in street violence and the victims of violent attack. Much of this violence is concentrated spatially and temporally – in city centres and at weekends. Very often alcohol helps fuel the violence. The most violent parts of the country are Gwent, London, Greater Manchester, the West Midlands and Nottingham (Willey 2000); the least violent part is Scotland (Montgomery 1999), where liberal licensing laws preclude the mass 'chucking out' time at 11 p.m. seen in the rest of the country.

However, more young men take their own lives than die in road accidents. Some 600 boys commit suicide every year. Much of this is associated with 'laddism' (see Chapter 11). As a recent study showed, while young women's confidence is growing, new lads are sad, suicidal and refuse to get help. Although females mimicking bad male behaviour is evident in the media and in city centres any weekend, it would seem that the culture of the 'new ladette' is something of a myth (Reeves 1999c). Young men are three times more likely to kill themselves than young women (Spencer 1999). Males, aged between 15 to 24, are twice as likely to suffer depression as those of that age a decade ago. Suicide among males has increased from 10 per 100,000 between 1976 and 1981 to almost 20. This is a much higher rate than for females. As the chief executive of the Samaritans noted, 'Depressed and suicidal men abuse drugs, commit crime, use and suffer from serious violence. The impact on the community and families is huge' (quoted in Arlidge 1999b). Also, men, macho to the last, opt for more violent forms of suicide – hanging, wrist slashing, drowning, shooting, and so on. Females opt for less drastic methods – drug overdose in particular. Relatives and friends can often summon medical help when a slower acting and reversible method is employed. Suicide tends to be higher in the North than the South of the country (ONS 1999c). Since unemployment, deprivation, poor education and lack of skills are positively correlated with suicide this further manifestation of the North–South divide is not surprising (see Chapter 14).

This rise in depression among young men has been partly explained by the increase in singledom as a way of life (Gerrard 1999a; see Chapter 7). While women appear to thrive on the single lifestyle, many men are suffering from it. Women develop diverse social networks and watch their health – mental and physical. Men, on the other hand, tend to lapse into degenerate lifestyles – junk food, excess cigarettes and alcohol, and lack of exercise – which lead to degenerative diseases and depression (Reeves 1999c).

It is not only young men who die in road traffic accidents. Although road traffic deaths have declined from 6,000 per year in 1984 to 3,400 in 1997, motor vehicles still account for a large amount of death and injury. Around 43,000 people are seriously injured on UK roads every year (Walters 1999). Of these, some 5,000 are children, of which 150 are killed. The motorist has now replaced infectious disease as the UK's biggest child killer. Most of these deaths are in urban areas. Every year, 244 people are killed or seriously injured in Manchester alone (BBC1 2000a). The fear of having their children being killed or injured on the roads leads many parents to keep them indoors. These 'battery children' are less healthy mentally and physically than 'free-range children' who are encouraged to play outside (Brooks 1999).

Excess speed is the major reason for accidents. On both motorways and dual carriageways over 50 per cent of motorists exceed the speed limit (ONS 1999c). It is estimated that a reduction in limits, from thirty to twenty miles per hour, could save 2,500 children from death or injury, yet the government has instructed police forces to ignore drivers who exceed limits by five to six miles per hour (BBC1 2000a). Motor manufacturers refuse to build slower vehicles and speeds in excess of national limits are seen as a major selling point and extolled in advertising campaigns. The paradox is that drivers and passengers have never been safer as more safety features, such as airbags, crumple zones and seat belts, are incorporated into cars. But this gives drivers a false sense of security and more pedestrians are being hit as a result. The car industry has known since 1985 how to make cars safer for pedestrians but has failed to produce such vehicles for financial reasons (BBC1 2000a).

Although there are plans to make more pedestrian friendly 'home zones' in residential areas, campaigners believe that motorists who kill should face much stiffer penalties (Harrabin 2000). The government is committed to reducing child deaths and serious injuries on the roads by 50 per cent by 2010 (Nicolson 2000). However, politicians are reluctant to tackle any of these problems for fear of 'losing the vote of Mondeo man' (BBC1 2000a). Pedestrians, represented by such organisations as the Pedestrians' Association, Brake and Roadpeace cannot hope to compete in terms of lobbying clout or financial backing with the vast motoring lobby, consisting of organisations such as the Road Haulage Association, Automobile Association, Royal Automobile Club, Society of Motor Manufacturers and Traders, Tarmac, and so on (see Chapter 6). The recent government volte-face on the road building programme, and the abject failure of its transport policy, is testimony to the power of the motoring lobby (see Chapter 12).

Addiction

While cigarette smoking has been declining, it continues to be the biggest single cause of preventable illness and loss of life in the UK, and is linked to about 120,000 deaths a year – almost a fifth of all deaths (Pearce 1998). Lung cancer is not the only risk incurred by smokers. They may develop bronchitis, TB and other cancers – mouth, stomach, breast, cervix (Clarke 1999). Also, the risk of developing ischaemic heart disease is as much a risk as lung cancer (Benjamin 1989). It can also cause blood clots, which might lead to amputation of affected limbs. Traditionally many more men than women have smoked, but now the proportion is much the same, with 29 per cent of men and 28 per cent of women smoking. However, among children, more girls (15 per cent) smoke than boys (11 per cent) (ONS 1999c). Smoking has declined among the higher social classes but has increased among the young, especially females. The increase in smoking among women, coupled with a decrease among men, means that lung cancer is becoming more common among women. Between 1987 and 1996, for example, the number of females diagnosed with lung cancer rose by over 20 per cent, during which time the number of cases among men fell by 11 per cent (Trueland 2000). Figure 10.3 shows how the habit varies from region to region.

Women are also drinking more. The group with the fastest growing levels of alcohol consumption is young women, particularly those in professional occupations (McVeigh 2000; Trueland 2000). The proportion of men exceeding the safe limit has remained the same at around 24 per cent, but the proportion of women exceeding the limit has increased from 9 per cent in 1984 to 16 per cent in 1999 (Reeves 1999f). The recommended limit for women is lower than that for men because of relative size and different physiologies, but it seems women are particularly vulnerable due to different biochemistry. There is the additional problem of smoking and drinking during pregnancy. These can cause foetal damage, yet almost half of all teenage mothers smoked during pregnancy, although average levels of alcohol consumption were low (Botting 1998). This underlines the major difference between these two powerful drugs. Most alcohol consumers are users of the drug who do not become addicted, whereas most tobacco consumers are abusers and do become addicted. This demonstrates that nicotine is one of the most addictive drugs known to pharmacology. In moderate doses, alcohol has beneficial effects on health (Baggott 1998). That said, heavy alcohol consumption can lead to chronic diseases, such as cirrhosis of the liver, and contribute to heart disease as well as to a range of psychiatric conditions. Binge drinking, common in the 'lad' culture, is especially dangerous. Huge quantities of alcohol once a week are much worse for health than moderate drinking more often (Reeves 1999f). Alcohol consumption in the UK has increased by 10 per cent since 1975. There is also a geography of alcohol consumption. Blackpool has the highest consumption for both men and women, Newcastle and Sunderland have the highest consumption rates for men, and Leeds, Manchester and Bradford the highest rates for women (McVeigh 2000).

The growth in consumption of illegal drugs is a cause of concern in terms of health, as well as crime and other anti-social activities. And, as Porter (1999: 318) has observed, 'health is compromised in areas . . . where people are intimidated by high levels of crime and disorder'. The link between drugs and crime is two way, in that some drug dependence leads to crime and some of the most predictable supplies are to be had within the prison system. The outcomes of illegal drug use can range from job loss to death. Some well publicised heroin and ecstasy

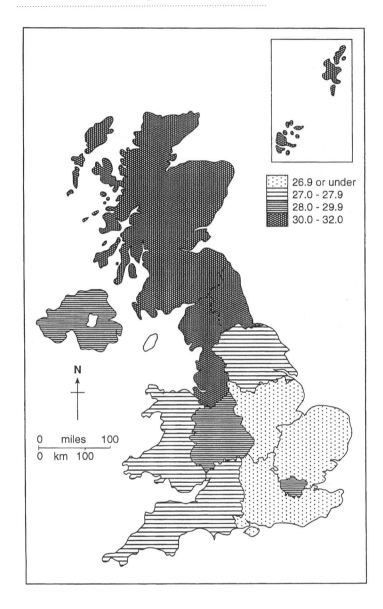

26.9 or under
27.0 - 27.9
28.0 - 29.9
30.0 - 32.0

Figure 10.3 Percentage of adults who smoke: Scotland, Wales, Northern Ireland, Government Office Regions, 1997

Source: Data derived from ONS 1999a, *Regional Trends 34*, Table 7.6.

deaths have shown how the lack of control of the quality of some substances can lead to fatal dosages entering this large but unregulated market. Methadone, heroin, and temazepam account for most drug related deaths. Casual drug use in Scotland rose from 18 per cent of all 16- to 59-year-olds, in 1993, to 23 per cent in 1996. In 1996, over 40 per cent of 16- to 29-year-olds in England and Wales had used an illegal drug (Bradley and Baker 1999). As well as tackling issues of source and supply, the government spends money on education, treatment and rehabilitation.

We are witnessing a new range of addictions that have an impact on health. We saw in Chapter 9 that there is an increasing trend in keeping fit. While we would expect this to have a positive impact on health and well-being, there is a minority (the most obsessed) who put themselves, and occasionally others, at risk. The dangers of steroids and other performance enhancing substances are well documented. The growth of so-called extreme sports puts the lives of participants and rescuers in danger.

'Shopaholics' have already been mentioned in Chapter 9. Although this might seem like a trivial disorder and worthy of jocularity, like all addictions it is the result of some underlying condition or set of conditions. Apart from the economic and social results of this – debt, guilt, stealing, lying and family breakdown – the causes of the disorder can lead to health impairment. It is difficult to tell how common this is, but one estimate is that there are 2.5 million compulsive shoppers. Like most disorders, this is gendered. Only 7 per cent of men are compulsives, compared with 16 per cent of women (Insley 1999). In many ways this is much like gambling addiction.

In Chapter 8, we saw how many people in the UK work longer hours than the EU average. There are obvious health issues associated with this – stress, fatigue, skipped meals, increased accidents, and so on. But there is evidence that the new 'flexible' labour market – little job security, poor working conditions and low incomes for the majority – is producing more 'workaholics'. In the USA, Workaholics Anonymous attempts to break the cycle of addiction by means of a staged programme, but it is failing to retain and recruit members because addicts are too busy to attend sessions (McLean 2000).

One of the most common, widespread and least studied of the modern addictions is car dependency (Newman 1999). This has all the symptoms of other forms of addiction – denial, dependence, withdrawal, attachment, behavioural changes and so on. There are clear health implications, as outlined above. There is the added stress associated with car dependence. This often manifests itself in violent behaviour, the best known being 'road rage', which often leads to injury and sometimes death. That said, car owners are healthier than those who do not own a car (Marmot 1996), largely because car owners are relatively better-off.

10.4 HEALTH INEQUALITIES

There are a number of reasons why good or bad health is not distributed evenly. The structure of the population demographically, socially and economically are among the main reasons for variation and inequalities in health.

10.4.1 Age

As we have seen, the most important predicator of health and well-being is age. This is explained in terms of the life cycle during which people become prone to certain illness at different stages in their lives. For example, of the 20,000 people who die each year from accidents or violent incidents in Great Britain, most are under 25 years old (Baggott 1998). The older a person is the more likely they are to suffer from or succumb to a chronic disease related to degeneration of mind or body. For example, 62 per cent of people aged 65 and over report long-term illness or disability, compared with only 27 per cent aged 16 to 44 (ONS 1999c). An awareness of the underlying age structure of a population in a given area is crucial before the patterns and trends of health and well-being can be understood. This is why as many health, well-being and mortality indices as possible are standardised to take account of the age distribution of the population. Because age structures differ from place to place (see Chapter 7) we would expect there to be different health needs from place to place. Thus there are more facilities for children and young people in relatively youthful places like Milton Keynes than in Bournemouth where the relatively greater number of elderly have different health care needs

10.4.2 Gender

The gender differences towards attitude to health are quite marked:

- men are half as likely to visit their general practitioner as are women;
- men are a third less likely to go for regular dental inspections;

- women are a third more likely to be on prescribed medicines than men;
- men are half as likely to look at the fat or sugar content of the food they are buying;
- men are twice as likely to die from lung cancer as women;
- the incidence of testicular cancer has doubled since the 1970s;
- men have almost four times the risk of heart disease as women, below age 65;
- the average lifespan of a man is still around five years less than that of a woman.

(Browne 2000b)

However, as was noted in Chapter 7, there has been a closing of the gender differential in mortality, in recent years, as women increasingly adopt the unhealthy lifestyles and behaviours of men. The health of men appears to be improving. Fewer men than women are taking up smoking and more men are giving up. There remain, however, important sex-specific differences in disease patterns. Cancer mortality differs markedly between men and women. The highest incidence rates are of lung cancer among men and breast cancer among women. Deaths among men aged under 65 from the most common causes – lung cancer, heart disease and strokes – have fallen to less than half the level of 1972 (Pearce 1998). Although men are more prone to suicide, women still suffer more from depression. Some 40 per cent will report symptoms at some point, compared with fewer than 20 per cent of men. 'Today's twentysomething is three times as likely to suffer serious depression as her Fifties counterpart was' (McVeigh 1999: 5). There are likely to be clinical reasons for this, but since women are more likely to seek medical advice than men, more depression is recorded and treatment is given. Men fail to seek advice and much depression ends in suicide. Also the psychiatric disorders manifest themselves differently by gender. Compulsive disorders among women include anorexia and other eating disorders, and compulsive shopping. Among men, excess alcohol consumption is the most likely result of depression. One innovative way to try to improve male health and increase men's awareness

of health is the establishment of 'pub clinics', in the Midlands, where health professionals set up basic clinics in pubs and offer men a range of medical tests and advice (Browne 2000b).

10.4.3 Social class

One of the longest standing and strongest predictors of health status is social class (Jones 1990; Jones 1994). Although the class stratification most widely used in **morbidity** and mortality studies is not without its critics (Marmot 1996), not least because of the very general nature of each stratum, it does demonstrate very clearly the strong health divide which exists in the UK. This is strongly linked with poverty and background, which is why the social class of the father is such an important predictor of an individual's health (Curtis 1995; Dorling and Shaw 2000). All the attributes of social class impact on health. Generally speaking, higher income allows those in social classes I and II to afford better housing, choose a healthier diet, improved access to sporting facilities and a healthier lifestyle in general. Being better educated, they are more likely to heed warnings and advice about their health, as well as being better able to deal with fellow professionals in the health services. They are also more likely to have access to lawyers and private health insurance. There is a very close link between class and ability to get a decent diet. This is not just to do with income. Often, those in the lowest social classes live in areas where food choice is poor and they lack private transport to get to areas where a better and cheaper range is available (Davison 1998).

Of the sixty-six major causes of death among men, sixty-two are more common among social classes IV and V than other social classes. Of the seventy major causes of death among women, sixty-four are more common in women married to men in social classes IV and V. Someone born into social class I can expect to live seven years longer than someone born into social class V (Baggott 1998). Five times more working-class people die from lung cancer than middle-class people, and three times as many die from heart disease. Figure 10.4 shows how the class gradient persists across the major causes of death.

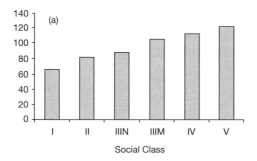
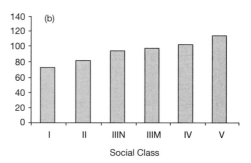
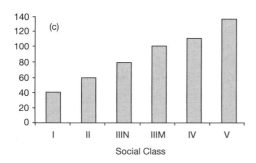
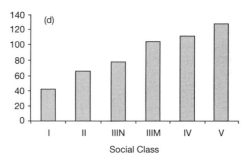

Figure 10.4 Standardised mortality ratios of males aged 15 and over, by social class, England and Wales, 1971: (a) malignant neoplasms, (b) cerebrovascular disease, (c) respiratory disease, (d) lung cancer

Source: Data derived from S. Harding (1995) 'Social class differences in mortality of men: recent evidence from OPCS Longitudinal Study', *Population Trends* 80: 30–7, Table 4.

There is evidence that the provision of health care still varies according to class and that deprived areas are less well served than prosperous areas. It has been estimated that around '2,500 people die every year who would be alive if they had the same access and standard of care as the middle class' (McSmith 1999a: 8). 'Birthweights vary by social class and low birthweight has a particularly strong association with infant mortality' (ONS 1999c: 120) and gives the child a poor start in terms of health generally. Since the late 1970s, there has been a widening of health inequalities between the highest and lowest social classes (Baggott 1998).

Although social class is an important determinant of health and well-being a lack of money *per se* is not the problem (Baggott 1998; Le Fanu 1999), since the poorest spend more on alcohol, tobacco, processed foods and the like. Poverty comes in other forms such as lack of self-esteem, lack of hope, lack of knowledge, lack of control, lack of housing, lack of paid work, lack of skills, lack of aspiration and so on. One of the strongest links between deprivation and health is education (Blane *et al.* 1996). Also, as Wilkinson (1996: 75) points out, it is not the overall income that is important but the distribution of that income – 'the most egalitarian rather than the richest developed countries . . . have the best health'. Thus, as income disparity becomes greater in the UK, so will the health divide. There are clear spatial implications here, with higher morbidity in poorer working-class areas, such as inner cities, while the wealthier suburbs tend to have better health.

10.4.4 Ethnicity

There are a number of ways in which ethnicity can impact on health. There are certain genetic conditions that predispose to illness and disease. Perhaps the best example is sickle cell anaemia (Baggott 1998). This hereditary disease affects Black people and, although it confers some immunity to the worst symptoms of malaria, can lead to anaemia and premature death. There is strong evidence for a relationship between blood groups and disease, and a map of blood types strongly reflects the general pattern of mortality in Great Britain (Jones 1990). However, as Wilkinson (1996) argues, it is very difficult to disentangle genetic and environmental factors in disease **aetiology**.

TB infection rates are highest among non-white ethnic groups, particularly among Black Africans, among whom the rate is over fifty times the rate in the white ethnic groups (PHLS 1999c). People of Indian origin have lower than average mortality from certain cancers – stomach, intestine, rectum, 'trachea, bronchus and lung', breast, ovary, cervix and skin (Adelstein and Marmot 1984). However, once the behavioural traits of the host population are widely adopted these differentials tend to disappear. There are cultural considerations to account for. Diet varies among ethnic groups and this affects health, most notably the prevalence of rickets in the children of some Asian communities. Attitudes to health also vary according to cultural practice (Curtis and Taket 1996). In some groups, women are reluctant to consult a male doctor, though many health authorities now accommodate this. Language problems can lead to misdiagnosis, as well as misunderstanding (Jones 1994).

Much of the ethnic differential in health is due to the concentration of some groups in specific areas, especially in the relatively deprived inner city areas where unemployment is high, housing conditions are generally poor, levels of violence are high and medical services are poor. Levels of poverty are higher among non-white ethnic minorities (Blackburn 1999). That said, the ethnic group with the greatest health disadvantage in the UK is the Scots (Dorling and Shaw 2000). This is due to long-term relative poverty, some of the worst housing conditions in Europe and cultural factors – smoking, drinking, high fat intake. This is in spite of Scotland being at the forefront of medical innovation and provision.

10.5 HEALTH CARE PROVISION

The NHS was set up in 1948, in the face of fierce opposition from most of the medical profession, as a result of the National Health Service Act of 1946. There were a number of characteristics associated with the new service:

- it would be free at the point of delivery;
- it was to be a universal service;
- it was intended to be comprehensive.

The service would be free to the user, but the general population would pay through taxation and National Insurance contributions. Everyone would be able to use the service whenever the need arose. The aim of the legislation was that the service would cover all aspects of health care provision – medical, nursing and other specialist services; mental health and physical health problems; personal and public health services; curative and preventive services (Kendall 1995).

Despite the manifest criticisms of the NHS, the UK enjoys a relatively high standard of universal, free at the point of delivery and well distributed health service. That said, there are still variations in the delivery and quality of health care from place to place. This manifested itself fairly early on in the history of the NHS (Jones 1994) and was perhaps best articulated by a GP working in a deprived area of Wales. He postulated the inverse care law after observing that:

> In areas with most sickness and death, general practitioners have more work, longer lists, less hospital support and inherit more traditions of clinically ineffective consultation than in the healthiest areas; and hospital doctors shoulder heavier caseloads with less staff and equipment, more obsolete buildings and suffer recurrent

crises in the availability of beds and replacement staff. These trends can be summed up as the inverse care law: that the availability of good medical care tends to vary inversely with the need of the population served.

(Tudor Hart 1971: 412)

Perhaps more notably (and pertinently in the current political climate) he summarises:

The availability of good medical care tends to vary inversely with the need for it in the population served. This inverse care law operates more completely where medical care is most exposed to market forces, and less so where such exposure is reduced. The market distribution of medical care is a primitive and historically outdated social form, and any return to it would further exaggerate the maldistribution of medical resources.

(Tudor Hart 1971: 405)

The opposite situation to inverse care is territorial justice, where expenditure and other resources more closely correspond with need. The areas with the poorest health would get better health care through targeted funding and resource allocation (Jones and Moon 1992). This is the aim of a strategy newly announced by the government.

The UK lags behind other advanced capitalist countries in the treatment of heart disease. In 1995, 314 men per 100,000 died from heart disease, compared with 233 in the USA, 92 in France and just 60 in Japan. Much of this is due to poor diet and lack of exercise, but there are major differences in treatment of the disease. Fewer pacemakers are fitted than in most other countries and waiting lists for treatment are longer. The UK has fewer heart surgeons than most other countries – 16 per 100,000 population, compared with 30 in France, 42 in Germany and 75 in Italy (Browne 1999b). Cardiac arrest survival rates in Seattle are fifteen times that of London. Around 180,000 people in the UK suffer cardiac arrest every year. In the UK, only 3 per cent survive, compared with 11 per cent in the USA. The provision of more trained personnel and more

defibrillators, up to the US level, could save 20,000 lives a year (Browne 1999d).

The survival rate in the UK, once cancer has been diagnosed, is one of the lowest in Europe. The UK survival rate for breast cancer is only 60 per cent compared with 70 per cent in Germany and 80 per cent in France (Dillner 1999). This suggests that there are flaws in the health care delivery systems. These include lack of adequate drugs, a shortage of specialist oncologists, non-implementation of agreed reforms, a conservative medical culture, excessive waits for treatment and a lack of investment. Within the country, some areas are worse than others. Birmingham has a better record than Oxford, for example. In 1998, North West health authority had 40 per cent of its cancer patients waiting unacceptably long for treatment, followed by Trent (37 per cent) and Northern and Yorkshire (36 per cent). The best figure was in the West Midlands, with 7 per cent (Bower and Boseley 1999). The government is aware of the UK's poor treatment of both heart disease and cancer and has made a commitment to reduce waiting lists and improve services (McSmith 1999a). Although 'new' Labour has abandoned the 'old' Labour aim of egalitarianism through wealth redistribution, the appointment of a Minister for Public Health and the 'New Contract for Health' suggest a seriousness of purpose in its stance on health generally (Porter 1999). Further, unlike the previous Major and Thatcher administrations, which denied a direct causal link between deprivation and ill health, the Blair government accepts the correlation and is committed to tackling the problem (Baggott 1998). In 1998, the government published a consultation paper – *Our Healthier Nation* – with proposals for a health strategy to tackle the causes of health problems in England. Similar documents have been produced for the other countries (Pearce 1998). The two main aims of the programme are:

To improve the health of the population as a whole by increasing the length of people's lives and the number of years people spend free from illness. To improve the health of the worst off in society and to narrow the health gap.

(Department of Health 1998: 5)

The government has announced a plan to save 148,000 deaths from stroke and coronary heart disease by 2010. Funding will go on training more cardiac surgeons, creating a number of 'chest pain clinics' and 'stop smoking clinics', increasing ambulance response times, the purchase of extra defibrillators and an extensive education campaign (Murray 2000). The programme is to be directed by a 'heart tsar' who will boost the campaign to reduce preventable deaths already started by the 'cancer tsar'.

The Labour Party's stance has been criticised for following the approach of previous administrations insofar as it adopts the strategies of 'lifestyle management' as a solution to premature death and ill health, rather than tackling the root causes of inequalities. This sits well with the individual responsibility advocated by 'new' Labour and its rejection of collectivism as an approach to public health problems (Purdy 1999). However, the Labour-inspired Health Action Zone programme, which emphasises partnerships and public/ private/voluntary coalitions in the delivery of health care, suggests that there is still a communitarian side to health care under 'new' Labour. But despite increased investment waiting lists are still too long (Figure 10.5) and treatment within the NHS has become a postcode lottery, with the quality of treatment very much dependent on the patient's address, rather than condition.

Only a small proportion of the illnesses and dysfunctions that manifest themselves in the population actually lead to formal care. This is known as the 'illness iceberg' (Hannay 1980). That is, much of the ill health in the population generally is never made known to the health services and practitioners. People have coping mechanisms whereby they learn to tolerate ailments and the symptoms of illness. If all ill health in the population was brought to the

attention of the health care services at the same time they would be overwhelmed (Gillespie 1995). As the influenza epidemic of the winter of 1999/2000 demonstrated, it does not take much to over-stretch the NHS. This coping mechanism is especially strong among men who make much less use of the NHS than women. Screening programmes for women's diseases are well established, if not without problems. 'Well Women clinics' are commonplace. Similar facilities for men are scarce. Indeed, according to the chairman of the Men's Health Forum, 'The money going into men's health is pitiful. There is far more money going into breast cancer research – £38 million – compared to just £47,000 for prostate cancer research' (quoted in Browne 2000b: 16).

There has always been an alternative to the NHS, indeed the medical profession insisted that doctors should be able to practise privately if they wished (Jones 1994). Private health care has traditionally been used by the wealthy, but is increasingly being used by people with private health insurance. However, private insurance and private medicine are not without hidden costs in the form of

"There's a two year waiting list"

Figure 10.5 Waiting lists are no laughing matter for patients

Source: *Private Eye*, 7 March 1997, p. 25. Reproduced by permission of *Private Eye*.

corruption and fraud (Palast 1999) or death of a partner (Ryle 1999). The use of private health care is not distributed evenly. There is, for example, a high concentration of private long-stay beds in the coastal locations of southern England, a reflection of the high proportion of elderly there. Most private health care in the UK is concentrated in the wealthy South East, especially Berkshire, Buckinghamshire and Surrey (Mohan 1999).

10.6 SUMMARY

As we have seen, health and well-being are highly complex multidimensional phenomena. According to ONS, the state of the population's health has greatly improved since the middle of the last century. But this improvement has not been across

the board. Some things are better, some worse and some have stayed the same (Box 10.1).

Although the dental health of the general population has improved, according to the list, that of children has deteriorated. This is due to the lack of dental practices willing to undertake NHS work. In parts of Devon and Cornwall, dental decay and problems doubled between 1993 and 1999 (Rowlat 1999). Also, it must be borne in mind that these are general observations. One thing clear from this chapter is that when analysing such a multi-dimensional issue as health and well-being, certain contradictory evidence emerges. Thus men are becoming healthier generally, but some sub-groups are not. Similarly with women.

The paradox of modern society is that while the majority have never been better-off and have never had as much to do in terms of leisure and

Box 10.1 Health conditions: continuity and change

Little sign of change in health:

- all-cause mortality for men aged 35–9
- self-reported acute and chronic morbidity
- healthy life expectancy
- risk factors for heart diseases
- physical activity
- smoking in young adults

Signs of improved health:

- all-cause mortality for all women and men under 30 and over 40
- certain cancers
- life expectancy at all ages
- dental health
- reductions in smoking in older adults

Signs of worsening health:

- all-cause mortality for men aged 30–4
- certain cancers
- proportion overweight and obese

Source: Population Trends (1997: 46)

consumption, there has never been as much unhappiness, stress, depression, suicide, antisocial behaviour and other deviant pathologies (James 1998). Much of the general malaise in society has been accounted for by the decline in the size and strength of the family, increased mobility (see Chapter 7), decrease in formal religious observance (Chapters 9 and 11), and rapid economic and social change in general.

REVISION QUESTIONS

- What is the epidemiological transition?
- Is the UK healthier now than in the past?
- Account for geographical variations in health and well-being.

KEY TEXTS

Baggott, R. (1998) *Health and Health Care in Britain* (2nd edition), Basingstoke: Macmillan.

Dorling, D. and Shaw, M. (2000) 'Life chances and lifestyles', in V. Gardiner and H. Matthews (eds) *The Changing Geography of the United Kingdom*, London: Routledge, pp. 230–60.

Jones, H. (1994) *Health and Society in Twentieth-Century Britain*, Harlow: Longman.

Purdy, M. and Banks, D. (eds) (1999) *Health and Exclusion*, London: Routledge.

Wilkinson, R.G. (1996) *Unhealthy Societies: The Afflictions of Inequality*, London: Routledge.

CULTURE AND IDENTITY

11.1 INTRODUCTION

By definition the UK, as a multinational state, is both multicultural and multiethnic – a place of multiple identities. The Acts of Union in 1536, 1707 and 1800 created a state with six languages and a wide range of customs and traditions. However, while the UK was a successful economic and political Union, the customs, traditions and identities of the constituent nations survived. Scotland, in particular, was adamant that it retain its unique education and legal systems. The Welsh and Irish, through language, literature and music, retained a strong sense of identity. Within England, some areas retained a stronger sense of identity than others, particularly the West Country and the North East (Critchley 1986). England was the hegemon within the Union, with 54 per cent of the new state's population in 1801, though 200 years later that has increased to around 83 per cent (see Chapter 7). The new country may have been the first modern state, but it was not and never has been a nation-state. Indeed, the UK has 'the rare distinction of refusing nationality in its naming' (Anderson 1991: 2). Its official name is the least used, and no other country suffers such confusion over nomenclature – even official documents fail to get it right (Crick 1991; Nairn 1988).

According to Brockliss and Eastwood (1997: 2–3), 'Within the new United Kingdom the space for regional, ethnic, national, linguistic and religious identities was constantly contested.' Yet 'a viable British State was created out of a myriad of ethnic, religious, economic and spatial loyalties' (pp. 3–4). This is because the various Acts of Union, from 1536 to 1800 were designed 'for political uniformity rather than a true unity of cultures' (Jones 1998: 19).

The countries of the UK have a long and, until recently, relatively honourable tradition of accepting and absorbing immigrants from a variety of backgrounds. Successive waves made specific impacts and local areas benefited and were enriched by the Italians, Poles, Jews, Ukrainians, Greeks, Turks, Portuguese and so on who settled in the last 300 years. It is, however, the more culturally diverse immigrants from further afield who arrived in the post-war years who have attracted the most media, academic and political interest in terms of culture and identity within the UK. Indeed, much of the metropolitan literature on multiculturalism takes as its starting point the *Empire Windrush* landing Jamaican immigrants in London in 1948, rather than the 1707 Act of Union, or even before. As Blake (1999: 109) has noted, 'England has been culturally and ethnically hybrid for the last 2,000 years.' This has led to a complexity of British culture based on the English culture, 'which comes in a dozen major regional variations', and a 'vibrant group of non-English counter-cultures' based on Celtic identity (Davies 1999: 963). Added to this are a variety of other collective identities linked to overseas colonisation and empire building, as well as more recent developments in Europe, which Cohen (1994) has called the 'fuzzy frontiers of identity'. The problematic relationship between the UK state and 'its ethnic and regional minorities has become the subject of one of the most important debates in modern British life' (Smyth 1997: 244–5). In this chapter we explore this debate and these 'fuzzy identities', mostly in terms of place and identity, and in so doing we encounter other aspects of culture and identity, such as gender, ethnicity, religion, class, which make this subject so complex. We also heed Storry and Childs (1997) who point to the difference between UK cultural identities and cultural identities in the UK.

11.2 UNDERSTANDING CULTURE AND IDENTITY

As we saw in Chapter 5, culture and identity are extremely complex phenomena. Identity can stem from a number of sources and manifest itself in a number of ways. A person can feel a strong sense of identity as an individual – as a member of a common-interest group, or a social class, or an ethnic group, or the same gender – through community ties or through a common national heritage. 'Identities', as Linda Colley (1992: 6) advises, 'are not like hats.' Most people can and do wear several identities at once. Identity is often defined through reference to others (Bennie 1995). As Cohen (1995: 36) states, 'You know who you are, only by knowing who you are not.' A sense of identity stems from and is reinforced through cultural attributes (food, language, music and so on) that form 'the raw material for making identity emblems' (D.-C. Martin 1995: 13). For Mackenzie (1978) identity is based on difference and can stem from state, nation, religion or class. Culture and identity can be informed by history shaping ancestral 'roots' and by geography promoting a sense of 'belonging' (Jones 1998; D.-C. Martin 1995). According to Smith (2000a), a would-be Englishman needs two things as identifiers, a football team and an accent. People, then, become aware of their culture through their identities (Cohen 1982), while identity is shaped by culture.

Identity can also be linked to politics (Brand and Mitchell 1997). With little to distinguish the mainstream political parties now, much of this is channelled into nationalist movements. Similarly, politics can shape culture. Witness how close 'new' Labour wants to be identified with modern cultural icons from sport, popular music, theatre and so on. This was exemplified in the establishment of the Department for Culture, Media and Sport in England and Wales, out of the former, stuffier-sounding Department of National Heritage.

Identity and culture are often defined in terms of imagery, or even stereotypes. One image of Britain (that is, England) conveyed to an American audience consisted of country churches, pub signs, swans on the Avon, sheep grazing in green fields, Oxford quads, Windsor Castle, Buckingham Palace, Nelson's Column, the Palace of Westminster, Big Ben, red buses and black taxis. In this selective imagery there was 'no football, no London Underground, no coal mines or steel works or ports, no M25' (Marqusee 1994: 24).

However, Americans, like other foreigners, no longer swallow the old imagery whole. The lager lout in Union Jack shorts, the millionaire pop star, the sleazy tabloid journo on the make, the City slicker and the fascist skinhead are replacing the stiff-upper-lips. When Americans come here and take a good look they find a small country, stripped of empire and world status, a land of low pay, skinflint benefits, social division, economic and political stagnation.

(Marqusee 1994: 24)

Figure 11.1 Mural on a Belfast gable supporting Unionist terrorism

Source: Peter Shirlow.

**Figure 11.2
A Nationalist mural
demonstrates lack
of support for the
Protestant-
dominated Royal
Ulster
Constabulary**

Source: Peter Shirlow.

**Figure 11.3 Youth
gangs mark their
'turf': The
Meadows,
Nottingham**

Source: David T.
Graham.

A 'sense of place' is often crucial in determining identity and culture (Emmett 1982). As Mackenzie (1978: 130) puts it, 'Those who share a place share an identity.' National identity will be determined by geography, as will local and regional identity. Thus scale is important. Local identity can take many forms – from the territoriality that affects many aspects of life in Northern Ireland to a pride in

**Figure 11.4
Corporate identity:
McDonald's,
'Everywhere and
nowhere'**

Source: David T.
Graham.

a local accent. People even physically mark their territory to help reinforce that identity. For example, Protestants paint kerb stones in 'Loyalist' areas in the UK colours of red, white and blue, as well as flying the Union Flag, while in 'Nationalist' areas the Irish tricolour is flown and the white, green and gold of Ireland pick out the kerb stones (see Figure 13.1). Ethnic murals, of sometimes astonishing artistry, adorn the walls of buildings throughout the 'six counties' (Figures 11.1 and 11.2). Youth gangs also mark their 'turf' with graffiti, and corporate identity increasingly makes its mark on the landscape in the form of logos, colour coding and other symbols (Figures 11.3 and 11.4). This symbolism is a form of what Larsen (1982: 136) calls 'cultural shorthand'. Locality is often linked with ethnicity so that a sense of place and a sense of belonging become intertwined. As Cohen (1982: 3) points out, 'ethnicity and locality are both expressions of culture'.

Although definitions of culture and identity are difficult, there is no doubt that these impact on many of the other topics in this book, from leisure through to politics and health. Identities, in particular, may be intangible but their effects, as Macdonald (1993: 7) makes clear, can be every bit as important as 'more readily grasped economic and social phenomena'.

11.3 CHALLENGES TO CULTURE AND IDENTITY: WHITHER ENGLAND?

There are a number of challenges to identity that have come to the fore in recent years. The 'break up of Britain' has caused a rethink of identity, as has the development of the EU. As Marr (1995: 1) points out, one of the major problems of politics is 'how to preserve small national identities within larger structures'. The post-war immigration of non-white peoples from a variety of cultures, most of whom settled in a handful of English cities, has also proved to be as much a challenge to identity as an extension of the UK's deep-seated multiculturalism. Identity, then, is not unproblematic. For example, British and English identity are often conflated (Hall 2000; Seton-Watson 1977), largely because, for most English, Great Britain is simply another term for Greater England (Osmond 1988). But even something as seemingly clear-cut as Scottish identity is difficult to define (Dickson 1999; Nairn 1997). Myth also plays a large part in identity (Anderson 1991; Gruffudd 1999), and exploding or creating myths can impact on identity.

One of the most talked about, vexing, and biggest of identity crises, which has potentially far-reaching consequences for the future of the UK, is that of the UK hegemon – England – and the *emerging question of Englishness'* (Nairn 2000: 215). Witness the plethora of books and articles on Englishness (e.g. Barnes 1998; Davey 1999; Paxman 1998) and the difficulty most writers have defining 'Englishness', or, as Neal Ascherson (1999) calls it, 'Anglitude' (Nairn 1997; Taylor 1991). This is a long-standing problem (Dodd 1986). For example, H.V. Morton's, *In Search of England*, first published in 1927, is now in its *fortieth edition*. One problem is that English identity has never been seriously challenged (McCreadie 1991). Another is the relative heterogeneity in such a small country. The question constantly asked is 'which England?' (Marr 1999a; Wright 2000). It is the England of the Home Counties that is often invoked as being quintessentially English and symptomatic of English identity (Taylor 1991). Yet this exclusionary, white, middle-class realm, fails to represent the diversity in the country. It is England, after all, or at least some urban parts of it, which is the locus of the 'new post-*Windrush* multiculturalism' – peopled, as Cohn (1999: 3) puts it, by those who 'live in England, but not by Englishness'. However, the first race riots and the first recorded mixed-marriage in Britain occurred in Wales (Younge 2000a).

The Union was a particularly unsettling event for English identity, since, according to Elton (1992: 233), 'the English experienced the largest and most traumatic change when they turned into the British'. It is not clear, though, as we argue below, that they

ever did. Now, with the 'abolition of Britain' the English must reinvent themselves (Hitchens 1999) as they 're-emerge from their British phase' (Elton 1992: 234). As Nairn (2000: 215) points out, 'a far deeper uncertainty attaches to post-British England than to any dilemmas currently experienced in Scotland, Wales or Ireland'. This is because in England the state and nation were much more closely linked than in the other parts of the UK. Thus, while the Scots can feel British and Scottish and the Welsh British and Welsh, the 'idea of dual citizenship is extraordinarily difficult for many English people to grasp' (Jones 1998: 254). No one has tried or succeeded in creating a plural English identity (Marquand 1995a).

This question of Englishness and English identity is crucial for the future of the UK (Marr 1995). Many of the images of English **nationalism** are negative and associated with football hooligans, xenophobia, jingoism and militarism. The image is rather backward looking (Bragg 1999; Poulton 1999) and is bedevilled with nostalgia and decline (Bryson 1997). It is also masculine (Crolley 1999; Mackay and Thane 1986) and white. Because far right groups, such as the British National Party, Combat-18 and the National Front, have appropriated some of the symbols of English (as well as British) nationalism, there is a reluctance among the English to display their sense of identity in the same way other groups might. Thus, while the English openly make claims to patriotism at home and abroad, many are reluctant to identify with English nationalism for fear of being tainted with racism or Fascism. As the Brimson brothers observed, 'the Scots have the advantage of being able to revel in their Scottishness without being branded racist, and that helps to bring them together in a way that is impossible for the English' (Brimson and Brimson 1996: 20).

That is, the 'Ingerland' factor epitomised by the number-one-cropped, tattooed, St George Cross (or Union Jack) waving, xenophobic lager lout. This is the icon of Englishness for many foreign observers (Armstrong 1999). As Ford (1999: 32) put it, 'Englishness has become the race that dare not speak its name.'

**Box 11.1
How to toast
St George**

EAT: Beef on the bone, bacon and eggs, steak and kidney pie, mushy peas. 'Some of the boiled puddings take a lot of beating on a cold day.' No organic food or curries.
DRINK: Real ale or cider from a Heart of Oak tankard. 'None of that lager stuff, please. English beer and English pubs are unique. Breweries play an essential role in our history.'
GO TO: The local tavern on the village green, or organise your own medieval banquet or dinner dance. Visit a place of historic interest.

LISTEN TO: Elgar, Elizabethan harpsichord music or the Last Night of the Proms. Sing sea shanties.
RECITE: Shakespeare. His birthday is St George's day.
PLAY: Darts, bowls, cricket and bridge. Wave your handkerchief, strap bells to your ankles and dance around the maypole. Parade the town on horseback and ceremonially 'beat the bounds' to celebrate local freedoms.

Source: Royal Society of St George Rules, 1999 (cited in Arlidge 1999c)

Nevertheless there has been an increasing re-awareness of English symbols in recent years – St George's Day and the English flag, for example. The patriotic Royal Society of St George, whose Internet home page features a thatched cottage and a village cricket scene, quadrupled its membership between 1998 and 1999 to over 20,000 members (Arlidge 1999c), and offers guidance to patriots on how to celebrate the day (Box 11.1). It seems the English people are trying to claim these icons back from the lager louts, though recent events in Europe, at the Euro 2000 football competition, demonstrate that the louts still claim these emblems as their own (BBC1 2000b).

One problem of Englishness is that the majority of English people never learned to distinguish 'British-ness' from 'Englishness' (Anderson 1992). England's hegemonic position and relative heterogeneity have meant that while the terms – derogatory or affectionate depending on your point of view – 'Jock or Mac', 'Paddy or Mick' and 'Taffy' are used to describe Scots, Irish and Welsh, respectively, no such term has attached itself to the English (Paxman 1998), although the terms 'Sassenach', in Ireland and Scotland, and 'Saxon', in Wales (Emmett 1982), are used as a collective pejorative for English people.

Moreover, most 'British history' is in fact 'English history' (Cannadine 1995; McCreadie 1991) and there has been a failure 'to bring into the main-stream of British culture an awareness of both the pluralism and the shared identity of the British past – a past which, once shorn of the distorting excesses of English national feeling, can be the better appreciated as a complex collective experience of considerable consequence not only in the shaping of the curious cultural constellation of British identity but also in the making of the modern world' (Wrightson 1989: 260).

A 'British sense of identity is difficult to grasp' (Crick 1995: 170) and can only be defined in multi-national terms (Kearney 1991; Marquand 1995b). In purely linguistic terms, the Welsh are the only Britons (Seton-Watson 1977). Thus we have refer-ences to British and English identity from politicians, which demonstrate this confusion. For Earl Baldwin,

England (Britain) was evoked by sounds and sights and smells of a rural idyllic past:

> The sounds of England, the tinkle of the hammer on the anvil in the country smithy, the corncrake on a dewy morning, the sound of the scythe against the whetstone, and the sight of a plough team coming over the brow of a hill, the sight that has been seen in England since England was a land, and may be seen in England long after the Empire has perished and every works in England has ceased to function, for centuries the one eternal sight of England.
>
> (Baldwin 1937: 16)

This is not far short of the imagery invoked by a more recent 'English' Prime Minister, John Major, who described the essence of 'Britishness' as 'long shadows on county grounds, warm beer, invincible green suburbs, dog lovers and pool fillers and old maids cycling to holy communion through the morning mist' (quoted in Carrington 1999: 75). A clearer example of confusing England with Britain would be hard to find.

More recently, there was the view of the leader of the Conservative Party, William Hague (1999), for whom the British are 'a mature, tolerant, entrepreneurial, multi-ethnic, charitable, law abiding and private people'. British identity, he argued, was more than the sum of its parts, which is Welsh, Scottish and a somehow separate English consciousness. He falls, though, into the British equals English trap.

And then the Prime Minister, Tony Blair (2000), jumped on the identity bandwagon by arguing in much the same vein, that 'Britain is stronger together, than separated apart.' British 'identity is not some remote and abstract issue' but 'lies in our shared values not in unchanging institutions'. We must not, he argued, 'retreat from an inclusive British identity to more exclusive identities'. The qualities that contribute to British identity, according to the Prime Minister, are 'creativity built on tolerance, openness and adaptability, work and self-improvement, strong communities and families and fair play, rights and responsibilities and an outward looking approach to the world that all flow from our unique island geography and history'.

Even the legal nature of Britishness is 'fuzzy', 'vague' and 'malleable'; the concepts of nationality and citizenship have been constantly confused and the British Nationality Act (1981) did little to make things clearer (Davies 1999). According to Linda Colley, a Britishness based on citizenship would be one 'with no necessary ethnic or cultural overtones' (quoted in Young 2000: 22). This would also help include non-white ethnic minorities:

> Britain in the final analysis is made up of the peoples who inhabit it. Once they were Celts, Romans, Angles, Normans, and Saxons. Now they include many people of African and Caribbean descent as well as Bengalis, Kurds, Sikhs, Indians, Turks and Greeks. This multi-cultural diversity has developed without any help from and despite politicians . . . Britain and 'Britishness' are, as always, a work in progress.
>
> (Tisdall 2000: 15)

Another challenge to identity is the EU. For the Nationalist community in Northern Ireland this is not problematic, since Ireland has a very long history as a European nation with long and strong links that have been strengthened through membership of the EU. For the 'Loyalist' community there is more ambivalence. It can see the undoubted benefits that have accrued to the people of the Republic of Ireland, but having such close ties to the Union they are swayed by the isolationist sentiments of the English majority. Scotland has also had long-term links with France and the Baltic nations. For example, after Edinburgh and Glasgow, the largest Burns Night celebrations are held in Moscow. Wales, too, has had attachments to European nations, mainly through literature and music. England, however, has long been antipathetic towards European integration. As Bragg (1999) points out, there is greater support for the EU in Scotland and Wales, where the pro-European parties are the pro-independence parties. But in England, the most successful independence party is the UK Independence Party, which is fiercely anti-Europe.

'The myths, the iconography, the symbols of English nationhood are . . . non-, or even anti-, European (Marquand 1995a: 290). The English see little in common with Europe and tend to look towards the Old Commonwealth and the USA for allegiances (see Chapter 13). Yet, as Jones (1998: 251) notes, 'It is a great myth that England owes nothing and borrowed nothing of importance from mainland peoples and was separate from them and superior in its splendid isolation.' The English people, he reckons (p. 256), 'will need to develop a sense of historical and cultural identity with their European neighbours'.

One other challenge to culture and identity arises in assuming homogeneity. This is a major problem in English identity and has always been a problem with British identity. Even with smaller units heterogeneity makes the situation more difficult. There is a major cultural cleavage between the Highlands and the Lowlands of Scotland, for example, and the 'east and west coasts of Scotland are separate countries' (Campbell 1984: 102). The bifurcation of cultures and identities is the root cause of the problems in Northern Ireland where identity can be referenced by 'Irish', 'British' or 'Ulster' allegiances (Moxon-Browne 1991). The divisions between 'Welsh-speaking' Wales, 'non-Welsh speaking but Welsh-identified Wales' and 'British' Wales are plain and, if anything, growing deeper (Bowie 1993).

Similarly, there is a tendency to subsume the more recent cultures, within the generalised 'Black and Asian community' thereby 'forgetting the primacy of culture' as a source of identity (Davies 1999: 984).

Black evangelical Protestants who march with the Orange Parades in Liverpool may have more in common with other British Protestants than with Rastafarians from Jamaica or African immigrants from Nigeria or Ghana. Pakistani Muslims, Punjabi Sikhs, and Indian Hindus do not necessarily have much in common beyond skin colour, whilst other Asians, like the Chinese, belong to a totally different world from that of the Subcontinent.

(Davies 1999: 984)

Most individuals have multiple identities, which take on a variety of forms according to context (Smith 1991). A person from Merseyside might describe themselves as a Scouser, or Liverpool Irish or English, depending on where they were and with whom they are talking. Our identities are fabricated and moulded by many cultural attributes, each of which in turn can become emblems and symbols selected, consciously or unconsciously, as reinforcers of identity (see Chapter 1).

11.4 FABRICATORS AND SYMBOLS OF CULTURE AND IDENTITY

Culture and identity are shaped and fostered via a myriad of forces, such as place, education, media, gender, class and so on. The link between some of these has not been lost on the government, which established the Department for Culture, Media and Sport in England and Wales. The Secretaries of State in Scotland and Northern Ireland oversee cultural affairs in their respective countries. This devolution of cultural responsibility is mirrored in the existence of separate Arts Councils for the four countries. It has been argued that modern technology has displaced symbolism and thus weakened sense of identity and homogenised culture. Fabricators and symbols weaken with time and may become, as Stalker (1994: 72) notes, 'relegated to ceremonial or recreational functions – in parades of national costume or the gastronomic appreciation of regional cuisine'. But symbols are still potent forces in shaping identity and culture (Cohen 1986). In fact, some technologies, particularly the Internet, have helped stimulate cultural forms and strengthen identity by allowing communication between like-minded people. Of course, being part of a multinational and multicultural state can lead to confusion. In the UK, many symbols are shared. There is the Union Flag, a common monarchy, the BBC and so on. But many are unique to the component nationalities. This confusion is demonstrated by the ultra-nationalist chant of the less savoury element among the England football supporters

– 'Two World Wars and one World Cup, doo-dah doo-dah'. England may have won the World Cup, but it did not win any world wars.

11.4.1 Place

One feature that is central to cultural identity is a 'sense of place' (Carter *et al.* 1993; Friedman 1994). This can be a real sense of place or, like many symbols of culture and identity, mythical or imagined. For example, the popular and outsider image of Scotland is one of Walter Scott romanticism and the rural idyll, which conjures up images of pipers, glens and so on, compared with the reality of a modern urban country suffering the problems of postindustrialisation (Graham 1993). As Campbell (1984: 5) put it, 'praising haggis and tartan is rather like telling a black American how much you enjoy nigger minstrel shows'.

Senses of English identity are often formed by images and ideas of landscape (Driver 1999). J.B. Priestley (1994: 47), for example, had the Cotswolds as 'the most English . . . of all our countrysides'. This sentiment is echoed by Jones (1998: v), for whom 'there is no more quintessentially English area than the Cotswolds'. However, Heffer (1999: 39) is antipathetic towards the 'phoney chocolate-box representations of the nation', which includes the Cotswolds.

The links between sense of place and identity vary with spatial scale according to context (Smith 1991). Massey (1994) has talked of a 'global sense of place' and uses the multicultural Kilburn area of London as an example. Many people can identify with a region. In the North East of England, for example, according to Townsend and Taylor (1975: 385), the 'Geordie can be seen as possessing a common cultural and linguistic identity'. Writing in the interwar period, Mess (1928: 24–5) noted that, 'Tyneside is one of the districts of England with the most marked characteristics in custom, character, manners and speech.' After the war, Spence *et al.* (1954: 18) observed that, 'In the north of England, though possibly much less so than a few years ago, there is a sense of local independence and almost a detachment from the rest of the country.' There is,

then, a strong regional identity, yet people also have a county identity, often through support of a sports club. At a more local level different forms of identity can be seen. Local identity is linked to local attachment, which is 'a positive evaluation by the individual that the locality, however broadly defined, provides certain qualities which are not provided in other areas' (Townsend and Taylor 1974: 1).

Townsend and Taylor's interesting work on identity in the North East of England (1974) showed that local identity was stronger than regional identity. People tended to identify themselves as Wearsiders, Teessiders and Tynesiders rather than the ubiquitous Geordie, though people living on the Tyne were most likely to identify themselves as Geordie. Most people in the North East believe that the Geordie possesses a distinct identity (Colls 1992). This is based on a tough, hard-working, drinking image and generally working-class characteristics and a distinct accent and dialect of which four have been identified. This highlights the 'well-known local antipathies between the three estuarine nuclei of the region' (Townsend and Taylor 1974: 31). This is borne out by the claim of Beynon *et al.* (1994) that Teesside has a unique identity, it may be in the north but is not of the North.

Thus, like many other fabricators of identity, sense of place is somewhat malleable and multifarious in terms as to how it moulds identity. Myth can be important and individuals and groups can be very selective in how place impacts on their identity. Many identities and cultures, of course, are not spatially defined, and are less dependent on spatial metaphors than others – for example, diasporic identities (see Chapter 5). Also, many youth cultures are based on a globalised, transatlantic set of images. Even youth cultures, though, are affected by gender, class and ethnicity, however much the youths themselves might try to deny this.

11.4.2 Education

Education is often seen as a way of fostering identity – as a means of retaining a language, teaching a certain perspective of history, reinforcing social class or whatever. Schools, according to McCreadie (1991: 40) are the 'forcing-houses of national identity'. Scots were understandably keen to keep their quite separate educational system as a condition of the Union. This was much more open and egalitarian than that of England at all levels, and to a large extent has led to a more egalitarian society (Brett 1976; Harvie 1998). For example, for many years Scotland had five universities compared with only two south of the border. Indeed, it was a proud boast of Aberdonians that for well over two hundred years the town possessed as many universities as the whole of England (Graham 1972; Simpson 1963; Wyness 1971). In Wales, the school system is seen as the prime guardian and promoter of the Welsh language and culture. In England, however, education has never been valued, except for a few. According to one education expert, Dr Martin Stephen (quoted in Bright 2000: 1), English culture 'encourages young people to worship bodies not brains. It is the culture of the celebrity. It is not cool to be clever.'

Generally speaking, schools in the UK 'should take account of the ethnic and cultural backgrounds of pupils, and curricula should reflect ethnic and cultural diversity' (Pearce 1998: 132). The main division in education is the provision of separate schools for Roman Catholics. There are a number of state-funded schools for Islamic pupils, and schools throughout the UK are funded privately to cater for specific groups which seek to promote or foster the group culture and identity.

Another aspect of education is the way it reinforces class identities. This can range from the career and social circles an individual acquires, to the way language is used as a common identifier in terms of vocabulary and accent. Even in a supposedly egalitarian society some people identify and rely on the old school/university connections more than others.

11.4.3 Language

One fundamental fabricator and symbol of culture and identity is language. English was the official language of the new UK state and bilingualism was not yet common in much of Ireland and Wales.

There were wide linguistic differences between the Lowlands and Highlands of Scotland. From the outset, in linguistic terms, the UK was exclusionary. Educational reforms, economic integration and increased migration would diminish and weaken cultural diversity within the UK but 'the displacement of older identities could create new and equally potent differences' (Brockliss and Eastwood 1997: 1). The minority languages of the UK have sometimes been repressed by explicit government policy, such as the banning of Gaelic (Erse) after the 1745 Jacobite Rebellion (Evans 1995), or by biased employment policies, or have been treated as second class against English and subjected to wilful neglect and lack of funding. As Hutchinson (1996: 8) cogently reminds us, at the end of the nineteenth century 'more British civil servants spoke Urdu' than Scottish Gaelic.

Although forms of Scots (Lallans) are still spoken in the Lowlands and parts of Northern Ireland, the language declined after the Union (McCreadie 1991; Smyth 1997). Aspiring Scots switched to English as their main language in order to prosper in the new state (Gibbon 1934), and by the mid-eighteenth century books began to appear to allow translation from Scots to English (Hechter 1975). The Society for Propagating Christian Knowledge in Scotland, which was founded in 1708, equated the spread of English with the progress of civilisation (Ferguson 1978). Scots has only recently been revived through recognition and funding from the EU, United Nations Educational, Scientific and Cultural Organisation (UNESCO) and the Westminster government, and the work of the Scots Language Society. The ban on teaching Gaelic in schools was not lifted till 1918. In 1958, Gaelic became the medium of instruction in primary schools in the *Gàidhealtachd* (Minority Rights Group International 1997), but as recently as the 1960s pupils were punished for speaking Gaelic in preference to English in some schools and the Post Office refused to deliver mail addressed in Gaelic (Davies 1999).

Although self-expression through Welsh was, until recently, seen as a sign of backwardness and inferiority (Evans 1995), this language has fared better than those north of the border. The translation of the Bible into Welsh and the recognition that Welsh was the official language of the non-conformist chapel meant that even by 1901 just over half of the population of Wales were Welsh-speakers (Minority Rights Group International 1997), compared with universal use a century before (Williams 1989). Indeed, the chapels have been seen as 'important carriers of Welsh cultural identity' (Davie 1994: 95).

Of the UK's other indigenous languages, the last monolingual Cornish speaker died in 1799, though bilingual Cornish speakers survived for another century (Casey 1977). But the language has been revived by various Cornish societies, although widespread use is hindered by there being three competing versions (Anthony 1999). Nevertheless, there are more people speaking the language now than during the last 250 years, thanks to the efforts of Kesva an Tavas Kernewek (The Cornish Language Board) (Casey 1977). Manx is another Celtic language, which became extinct in 1974, although technically speaking the Isle of Man is not part of the UK. Again there have been successful attempts to revive this native language.

Today, the Westminster government is keen to promote both Welsh and Gaelic through education, broadcasting and legislation. Welsh is taught in primary schools as a first or second language and English is not a statutory language for 5- to 7-year-olds in Welsh-speaking schools. Some Welsh language skills are now compulsory for all pupils up to age sixteen. The work of Welsh language activists (Cymdeithas yr Iaith Cymraeg) resulted in bilingual signs in most parts of Wales (Jones 1997). Recently, there has been a resurgence in Welsh language printing and broadcasting. The Welsh language broadcasting of the television channel SC4 (Sianel Pedwar Cymru) proved particularly successful (Osmond 1988; Thomas 1991). The Welsh Language Act, 1993, established the principle that in public business and the administration of justice in Wales, Welsh and English should receive equal treatment. In parts of Wales, according to Bowie (1993: 168), 'identity is framed very much within the politics of the Welsh language'. The Gaelic broadcasting committee ensures that Gaelic language radio and television programmes are broadcast in Scotland. Gaelic is taught in schools in the *Gàidhealtachd* and some other areas. Only around 70,000 (1.5 per cent) people speak Gaelic in Scotland compared with 508,000 (19 per cent) of the population of Wales that claim to speak Welsh. Gaelic has been used in the new Edinburgh Parliament and Welsh is used extensively in the new devolved government in Wales. There are moves to encourage Gaelic among the young people of the unionist community in Northern Ireland in an attempt to encourage them to embrace a Celtic and British identity at the same time (Hardie 2000). For, example, in 1992 the Northern Ireland Office spent £1.2 million promoting Gaelic projects (Smyth 1997). The government policy for Northern Ireland is trilingual – English, Gaelic and Scots (Anon. 1999).

Other ethnic minority schoolchildren can learn their community language at school and increasingly there are radio and television programmes aimed at non-English cultures. Bilingualism is common among some of these groups but tends to be most common among younger groups and males. For example, only 10 per cent of Bangladeshi females aged 50 to 74 can speak English, compared with over 50 per cent of males in the same age group. Between ages 16 and 29, 68 per cent of Bangladeshi women speak English, compared with over 90 per cent of males in the same age group (ONS 1996). Bilingualism, and the hybridisation of identity that results from this, tends not to produce identity confusion (Marr 1999b) but, according to Emmett (1982: 219), allows 'entry to and membership of two cultures'. Yet 'Britain is one of the few countries in the world where being monolingual is frequently considered preferable to being bilingual in any of the country's other native languages' (Kay 1986: 18). The problem for the monoglot English is that English is now a universal language and the dominant form is that of the USA. Thus, despite a great literary tradition, 'English is becoming simply too successful to be useful to English national identity' (Marr 2000: 97). That said, within England it is estimated that over one hundred languages are spoken (Marr 1999a) and within Britain at least twelve languages can claim over 100,000 speakers (Smyth 1997).

Accents and dialects of English are also seen to be symptomatic of culture and identity (Mackenzie 1978). Various societies attempt to keep dialects alive in these globalised times. The age of some of these, for example, the Yorkshire Dialect Society (founded in Bradford in 1897), is testament to the threat received pronunciation (RP) was seen to be after the introduction of universal schooling. Until comparatively recently, using anything less than BBC RP was seen to be uncultured, even though fewer than 3 per cent of the population use it (Nairn 1988). Now regional accents are in vogue and so sought after that young people whose upbringing is RP affect a banal, pseudo-working-class 'Estuary English' or 'Mockney' (Brook 1997). Speaking with a Jamaican inflection is now seen as cool among some white working-class youth – the wiggers, or white wannabes (Jacques 1997). Increasingly, people, especially in the arts and media, are selecting accents the way they select items of clothing – to suit the occasion (Smith 2000a). That said, accent can be a major career barrier. Of all the regional accents, Birmingham and the West Country cause by far the most barriers in the workplace. Welsh, Scots and Irish accented English is still considered 'a cut above regional accents when it comes to credibility and authority' (Hilpern 1999: 6). For Priestley (1994: 136, 253, 290), who admitted to liking regional accents, the Nottingham accent was 'ugly speech', the speech of Liverpool was 'a thick, adenoidy, cold-in-the-head accent, very unpleasant to hear', the Newcastle accent was 'a most barbarous, monotonous and irritating twang'. Even the names given to meal times (dinner/tea, lunch/dinner) and to items of furniture (settee/couch/sofa) can mark people by region, as well as by class.

11.4.4 Media

The media are important moulders and emblems of identity, whether it be ethnic, national, regional, gender, political or lifestyle. They also foster cultural norms, especially language (Thomas 1991). The indigenous minorities are catered for in terms of radio and television programmes, as we have seen. There are also newspapers and specialist magazines in the non-English indigenous languages. Many newspapers and magazines are produced for the non-indigenous ethnic minorities, in English – *The Voice* being the leader – and native languages. Some are popular enough to be published daily, such as the Chinese *Sing Tao*, the Urdu *Daily Jang* and the Arabic *Al-Arab*. There are also many radio and some television programmes targeted at non-white ethnic minorities. In the UK, national newspapers mean English newspapers. Since *The Guardian* left Manchester in 1964 (Mayes 1999), all of the so-called national dailies are published in London. The quality broadsheets are often seen as papers of record, thus most universities in England, while subscribing to all of these London papers, eschew the papers of other parts of the UK, thereby perpetuating the metropolitan bias and myopia. This bias and parochialism, evident in broadcast media also, has been described pejoratively as 'metrovincialism'. According to *The Guardian* itself, this can take the form of not devoting enough coverage to 'devolution, to delineating and looking at the implications of the north/south divide, to looking at the regeneration of regional cities: Birmingham, Manchester, Newcastle' (Mayes 1999: 7).

National identities are catered for by output such as *This England* and *The Scots Magazine*. The former dates from 1968 and is aimed at 'all who love our green and pleasant land'. Judging by the content and the advertisements the audience is middle class and in older age groups, both in England as well as in the old Commonwealth and the USA. Typical content includes 'Garden flowers of England', 'There'll always be an England', 'Patriotic postcards to protect the pound', 'The Silver Cross of St George' and 'A Parliament for England?' The latter was established in 1739, and is again aimed as much at an expatriate audience as a domestic one. The content again reflects a similar, though slightly less nationalistic, view as that in *This England*. For example, 'What is Scottish?', 'Darien revisited', 'The great fire of Glasgow' and 'Speaking Scots'.

11.4.5 Religion

The new UK of 1801 was heterogeneous in terms of religion. Anglican was the established religion in England, Wales and Ireland, with Presbyterianism in Scotland and Northern Ireland, and Roman Catholicism in Ireland and the Highlands. And 'Methodism would do much to help rearticulate popular Welsh identities in the nineteenth century' (Brockliss and Eastwood 1997: 1).

The UK remains a Protestant Christian state, in legal terms at least. Yet, although there has been a decline of formal religion during the last few decades (see Chapter 9), 64 per cent of Scots described themselves as Presbyterian or Protestant (and 15 per cent as Roman Catholic) in 1987 (Marr 1995). Jones (1998) has also stressed that England remains a predominantly Christian country, despite the decline of church-going and adherence to formal organisations. Indeed, although membership of Christian churches has declined, the Christian community was much the same in 1995 as it was in 1975 at around thirty-eight million people (Davie 1994). In spite of this Christian hegemony there are significant numbers of observers of other faiths and evidence of this observance has an impact on

the urban landscape (Figure 11.5). For example, there are over 500 Buddhist groups or centres; there are estimated to be between 400,000 to 550,000 adherents of the Hindu faith and the first Hindu temple was opened in the early 1950s; there could be as many as two million Muslims and the first mosque was opened in 1890; the Sikh community is believed to number between 400,000 and 500,000 and the first Sikh temple was built in 1908 (see Chapter 5). The oldest non-Christian community, however, is the Jews who first settled in England at the time of the Norman conquest. The present community dates from 1656 and numbers around 285,000 (Pearce 1998). But formal religious adherence has declined dramatically since the Second World War, and the UK could not 'reasonably claim to be a Protestant country' (Davies 1999: 915). As we saw in Chapter 9, consumerism is the new faith.

Religious identity (with its heady overtones of ethnicity) is of course at the heart of the sectarian problems in Northern Ireland (Boyle 1991), though gender and kin ties are also important (Cecil 1993). However, this bigotry spills over into other parts of the country with a legacy of Irish immigration. It can be found, to some extent, in London, Manchester, Liverpool and Edinburgh (Cusick 1997), but it is

Figure 11.5 Non-Christian religious observance leaves its mark on the urban landscape: a Nottingham mosque

Source: David T. Graham.

best exemplified in the Glasgow area. The Glasgow Rangers Football Club, for example, employed no Roman Catholic players (or indeed, anyone married to a Catholic) until the mid-1990s; Paul Gascoigne was admonished for making pro-Loyalist gestures when he played for the club; the deputy chairman Donald Findlay was forced to resign for singing anti-Fenian (Nationalist) songs (Bell *et al.* 1999). According to some, bigotry is as deep-rooted throughout Scottish society as it is in Northern Ireland (Seenan 1999). Devine (1999b), however, disputes this, arguing that overt bigotry has declined recently.

11.4.6 Sport

Sport is a powerful source of culture and identity. For example, sport, at least as played in the English public schools, was seen as a way of reinforcing Englishness during Imperial expansion (Holt 1989). Indeed, there has been a 'wholesale sportification of our culture' (Perryman 1999: 20) in recent years as commercialism, consumerism and big business continue to dominate many sports (see Chapter 9). Sport is spearheading 'new' Labour's 'cultural policy around Young England', as part of its 'English cultural revolution' (Redhead 1999: 203). Multiculturalism and fuzziness of identities can also be seen in sports (Cohen 1995). In one of the world's great sporting events, the Olympic Games, the four countries compete as Great Britain (not the UK), but at the world's greatest sporting tournament, the World Cup, each country fields a team. This reveals 'the schizoid nature of the British state' (Carrington 1999: 75) and 'must be seen as a symptom of the wider failure to complete the construction of a British nation' (Davies 1999: 990). When it comes to playing in some of the teams the situation becomes even more complex, with dubious claims to lineage in order to play for a national side in football, rugby and other sports (Butler 2000).

This fuzziness extends to cricket, that most quintessentially English of games (Birley 1993) and the ultimate Imperial sport (Hutchinson 1996). In April 1990, in an interview in an American newspaper, the Conservative politician Norman Tebbit proposed

his infamous cricket test. When England played India or Pakistan, which side did the Indians and Pakistanis living in England support? Although Tebbit claimed his motives were in support of integration he was being exclusionary and trying to define England in terms of culture rather than a territory or even a 'race'. Why, asked the President of the Confederation of Indian Organisations, did Tebbit not criticise English settlers in Australia who supported Australia against England in test matches, or Scots, Irish or Welsh living in England who supported their own national football or rugby sides against England (Marqusee 1994)?

The confusion of Englishness and Britishness extends into sport and, indeed, is perhaps best exemplified here. For example, the recovering of the Cross of St George as the symbol of identity for the English football team is relatively recent (Hague 1999; Paxman 1998). Up until the mid-1990s, the Union Flag was the one most seen at events where the English side played. Plenty were in evidence during England's matches in Euro 2000, as was the singing of the British national anthem, even though 'Jerusalem' was the official anthem for the competition (Smith 2000b). This is perhaps not surprising since 88 per cent of English in a recent poll identified with the Union Jack, compared with only 38 per cent with the St George Cross (David 1999). The Union Jack itself is not entirely inclusive since it is made up of elements of the English, Scottish and (former) Irish flags; but has no place for the Welsh and, of course, most of Ireland left the Union some time ago. It serves as a reminder, as David (1999: 3) points out, that the UK is 'not so much a nation, and certainly not an ethnic nation, as a political union of separate nations'.

Sport has complex interrelationships with national, regional, local, class and gender dimensions of culture and identity. Although cricket is played in the other UK nations it has failed to capture the imagination and become as deeply ingrained in the national psyche as in England. In Ireland, where it was once widely played, it was rejected as too symbolic of England (Birley 1995) and became a victim of the world's first sporting boycott (Marqusee 1994). The Gaelic Athletic Association, begun in Ireland in 1884 and active throughout the UK, opted to develop traditional, rougher Celtic sports – Gaelic football, shinty, hurling and the like (Jarvie 1999). These are played as far afield as Argentina, Canada, New Zealand, the USA and Zimbabwe. In Northern Ireland, a fondness for Gaelic games is taken as support for Republicanism (Sugden and Bairner 1993). The most popular Gaelic sport in Scotland is shinty – marginally less suicidal than hurling. The much more familiar Highland Games, played in even more countries than Gaelic games, are governed by the Scottish Games Association. So peripheral are Gaelic and Highland sports as viewed through a metropolitan lens that they receive very little media attention. Even groundbreaking work by the sports geographer John Bale (1982, 1989) fails to devote much space to them, briefly mentioning Gaelic football and hurling in the latter work.

As well as spatial dimensions to sport that impact on culture and identity there are very clear class dimensions. The division of Rugby football into the Rugby Union and the Rugby League, for example, was a clear manifestation of such regional and class differences (Bale 1982) and the different codes of these games 'acted as the focus for fierce group identities' (Davies 1999: 799). Women's sport is rarely treated seriously in the popular media and sportswomen attract much lower fees than their male counterparts. This is in spite of the fact that female national teams often do much better internationally than the male teams – compare the records of the male and female England football teams, for example, then compare the respective media coverage.

The sports of cricket, athletics and rugby – the pursuits of the English public schools and therefore viewed as central to the promotion and reinforcing of Englishness – were introduced and supported in most of the colonised countries during Imperial expansion; football was too working class (Holt 1989). And as Perryman (1999: 28) argues, 'football is one of the most powerful symbols of our multi-cultural society'. So much is sport emblematic of culture and identity that the respected political scientist, David Marquand (2000: 131), was moved to state, 'I must start watching football very seriously if I want to study national cultures in Britain.'

11.4.7 Dress

Although it is common today for most people to dress in a globalised transatlantic uniformity, dress does and did play an important part in culture and identity in some parts of the UK. Even in the earlier decades of this century it was possible to determine a person's origin (whether in terms of status or area) from their style of dress. Dress was seen as such a strong symbol of culture and identity that, after the 1745 Jacobite Rebellion, the Hanoverian regime banned the wearing of tartan and the kilt in Scotland under Act of Parliament (Devine 1999a; Lenman 1980). Highland dress and the Gaelic language were only kept alive thanks to the efforts of the Highland Society in London and the British army. Later efforts of Walter Scott and others to 'tartanise' Scotland meant that a 'national tradition' was effectively invented as a 'gooey pastiche' – a 'Celtification of Scotland' (Marr 1995: 26–7). This 'Highlandism' has been one of the longest running and most successful myths of the modern period (Harvie 1989; Mackenzie 1978), yet it has endured as an international phenomenon through the diaspora – there are more kilts in Canada, for example, than Scotland.

For the UK's non-white ethnic minorities dress is seen as a way of retaining a sense of identity and reinforcing cultural norms. Conforming to religious mores also influences dress, as in the wearing of the veil or covering the head among Muslim women, or wearing the turban among Sikh men. Increasingly, more recent cultural groups, particularly from Africa, are identifiable by traditional dress. Hardill and Raghuram (1998) have highlighted the retailing of Asian fashion clothes – ready-to-wear *salwaar kameez*. These are hybrid clothes, 'modern stuff' (p. 258), which have been influenced by British designers. The substantial Asian middle class within the diaspora is the target market. Many members of the middle class now use Asian designer wear for leisure and social occasions, drawing on Asian cultures and traditions.

Dress is also particularly important among youth cultures. Some Asian youths, in particular, adopt elements of ethnic dress as a way of expressing collective identity (Gillespie 1998). Certain groups aside, such as the crusty/new age subculture (Croft 1997), spending large sums of money on identifiable, logofied attire is a mark of youth (see Chapter 9) – even though to the untrained eye the resulting uniformity, unimaginativeness and blandness of appearance seems symbolic of postmodern irony. Despite this general uniformity of dress, nowhere is identity better manifested than in gender dress codes. Sarong-wearing football icons aside, accoutrements of dress, including hair, cosmetics and so on, remain very conformist, even in these enlightened times.

11.4.8 Music

Despite globalisation, and all that has done to homogenise cultures and identities, music is still a powerful fabricator and symbol of culture and identity. As well as banning Highland dress and language the Hanoverians banned the bagpipes as being a source of military, as well as cultural identity. The militaristic music associated with Orangemen and the marching season in Northern Ireland is also a powerful cultural form and source of identity (Cecil 1993), especially the powerful and unforgettable sound of the Lambeg drums. The Celtic cultures still celebrate music, as well as literature and poetry, at such gatherings as the National Mod in Scotland and the Royal National Eisteddfod in Wales. There has also been a recent increase in Celtic music with international festivals organised among Celt communities in the UK, as well as in Spain, Canada, Denmark, France and Ireland (Jarvie 1999). As Cohen (1994: 10) reminds us though, these Celtic fringe events are 'completely ignored in the English popular media'. Within England, music plays a large part in determining culture and identity and is best exemplified by the 'Proms' (Albert Hall Promenade) concerts, with traditional renditions of the patriotic 'Land of

Hope and Glory', 'Rule Britannia' and 'Jerusalem'. Perryman (1999: 28) claims that popular music is very much 'part of what it means to be English'. And Blur is a 'band often accused of epitomising Englishness' (Peel 1999: 36). Indeed, one member of that band was a member of the group 'Fat Les', which was responsible for the spoof song 'Vindaloo', adopted as the unofficial song of the England supporters in the 1998 World Cup. This was not only a parody of 'In-ger-land' but recognition that curry is England's national dish (Smith 2000b).

Each UK country has its 'national' anthem, used particularly at sporting events, the exception being England, which still uses the British national anthem, though this has been challenged (Perryman 1999; Smith 2000b), with many opting for 'Jerusalem' or more light-heartedly 'Vindaloo', 'Three Lions' or 'Ingerland'. If the English Football Association does respond to increasing demands to find its own anthem for England then 'Land of Hope and Glory' and 'Rule Britannia' have been ruled out because of associations with racist elements (Campbell 2000c). 'God Save the Queen' is particularly disliked in Scotland, since one verse intones that rebellious Scots should be crushed – lyrics antipathetic to unity.

Black and Asian communities have their own distinctive musical traditions, and it is as common now in the streets of some English towns to hear the sound of the sitar or steel band as it is the music accompanying Morris dancers. Indeed, there has been a growth in popularity of 'ethnic' music in recent years, particularly as a fusion with popular music. This can take as its root some form of indigenous music as in the Irish folk-influenced Corrs or Celtic rock. More exotic influences have come from the Caribbean, in the form of calypso and reggae (Bakari 1989), or the various elements of Bhangra (Huq 1996) and other Asian-inspired music (Hutnyk 1997; Sharma 1996). Hybridised musical forms, such as Bungle, a mix of Bhangra and Jungle, or Gujarati Rock, a fusion of western and Indian instruments (Childs 1997), tend to be viewed as distinctively British (Dwyer 1999).

11.4.9 Food and drink

Despite 'McDonaldisation', food and drink are still potent cultural symbols in ethnic, regional and class terms. And there have long been strong 'associations between diet and identity, whether individual, class, cultural or national [and food] is on a par with language in terms of cultural definition'(Bishop 1991: 31–2). Many foodstuffs still retain national (Scotch egg, Ulster fry and Welsh rarebit), regional (Lancashire hotpot, Yorkshire pudding, Cumberland sausage, Cornish pasty), or local (Pontefract cakes, Dundee cake) identifiers, despite having a wider appeal. Drinks can also have geographical identities, for example, Scotch whisky, London gin, Irish stout, and so on. Food and drink can also symbolise class or gender identities – wine versus beer or pints versus shorts, and so on (Figure 11.6). Indeed, the relationships between food and cultural identity are very complex (Narayan 1995). Tikka

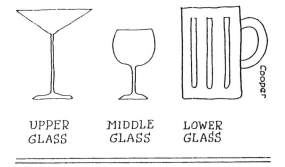

Figure 11.6 'Glass' distinctions

Source: *Private Eye*, 16 June 2000, p. 14. Reproduced by permission of *Private Eye*.

Figure 11.7 Multiethnic food outlet: one of the most obvious manifestations of the 'new' multi-culturalism

Source: David T. Graham.

Masala is said to be the most popular dish in England, overtaking fish and chips as the nation's favourite dish. This is not that surprising given that the dish originates in that country and not the Indian subcontinent, as is popularly supposed. The British Asian food industry is a bigger contributor to the English economy than the steel, coal and shipbuilding industries combined (Marr 1999a). 'Exoticised' catering is a physical manifestation of the changing nature of 'British' food (James 1995). The vast array of restaurants and food shops purveying 'exotic' wares is perhaps the most obvious tangible evidence of the UK's 'new' multiculturalism (Figure 11.7). Yet just as white 'Brits' are turning increasingly to 'ethnic' food, Ram *et al.* (2000) found that the younger generation of non-white ethnic minorities was much more likely to be found in McDonald's or Pizzaland.

Culture and religion combine in some dietary requirements – halal and kosher foods, for example. 'Indian' restaurants, or curry houses, are ubiquitous and sell food that would be unfamiliar to many people in the subcontinent. But the most widespread of the new ethnic food is to be found in the 'Chinese' restaurants. Many of these, especially in Northern Ireland, Scotland and Wales, are actually owned and run by Vietnamese refugees, many of whom are ethnic Chinese. However, the cuisine would be equally 'foreign' in Beijing as so-called 'Indian' food would be in Mumbai. Most 'Chinese' dishes have come from the USA, as indeed has most 'Mexican' and 'Italian' food. The longer an exotic food has been in a country the more it conforms to the tastes and expectations of that consumer base.

11.5 GENDERED CULTURE AND IDENTITY

As noted above, culture and identity depend on many variables. One of the most important of these is gender. Many of the symbols of identity outlined above are gendered, especially sport, music and dress. But gender differences also occur in religion, media and education. Even accents have a gender dimension (Smith 2000a). However, gender identity,

because of its universality, is 'inevitably more attenuated and taken for granted than other kinds of collective identity' (Smith 1991).

Male identity is something that has been challenged in recent years (Reeves 2000b; Segal 1999), and, as noted in Chapter 10, there are some serious health and well-being outcomes as a result of problems associated with male identity. Much of this centres around feelings of inadequacy and lack of a role. This is also reflected in poorer educational performance among males (Bright 1998). Reactions to this include violence, deviance, delinquency and **laddism**. Although most of these traits apply to marginalised and excluded males, laddism 'is often espoused by middle-class men who are conscious of their decision to be outrageously, provocatively, and supposedly ironically, sexist and racist' (Crolley 1999: 64). Icons of laddism, such as David Baddiel, Johnny Vaughan and Sacha Baron-Cohen (Ali G), who affect working-class credentials, come from relatively privileged backgrounds. 'New Lads' are devoted to 'footy, pop and fashion' and make no excuse about being 'English' (Carrington 1999). 'New Lads' were a reaction to the 'New Man' of the 1980s who was supposed to be a manifestation of gender equality. Laddism is a reaction to political correctness and unease about identity in a post-feminist era (Veash 1999). Devotees of laddism are the 'backlash boys' (Arlidge 1999d). As Andy King, managing director of Top Man put it: 'Lads have an identity crisis. Lad culture may not seem to have much to recommend it, but at least it's a role a boy can understand and in which he will enjoy the security of peer group approval' (quoted in Arlidge 1999e: 3).

Laddism and other manifestations of the male identity crisis manifest themselves in terms of media – television programmes such as *Men Behaving Badly*, *Fantasy Football*, *They Think It's All Over*, *TFI Friday*, and 'lads mags', like *Loaded*, *FHM* and so on, which have been termed somewhat pejoratively 'women's magazines for men' (Figure 11.8). As one critic put it

Men increasingly seek to insulate themselves against their changing gender roles in post-industrial

Figure 11.8 Lads mags: 'women's mags for men'?

Source: *Private Eye*, 30 June 1995, p. 24. Reproduced by permission of *Private Eye*.

societies, which are no longer founded on the need for hard manual labour and the masculinities that went along with this, by their consumption of such magazines.

(Carrington 1999: 79–80)

The success of these magazines is partly due to a 'backlash to feminism' (Jackson 2000). 'Young males, especially, fail to grow up because they no longer need to. The old disciplines of marriage, fatherhood and work have gone' (Hitchens 1999: 321). As Liz Crolley (1999: 63) notes on the feminisation of football, 'Perhaps it is the men who have become feminised.' The rise of the 'ladette' female version of this moronic, behaving-badly image suggests that these shifts in gender culture and identity are indicative of the complexity of cultural politics that defy easy explanation (Jackson 2000).

According to James (1998: 160) 'relations between the sexes have never been more acrimonious and rancorous' and that much of this relates to changing work patterns for both sexes (see Chapter 8). As we saw in Chapters 7 and 8, women are increasingly adopting independent lifestyles. Women now socialise on their own terms. When they go for a night out they are 'dressing up for fun' and other women as much as to gain the attention of men. This is particularly the case with the female subculture which dominates the city centre pub–club 'circuit' in

so many English towns (Croft 1997; see Chapter 9). While 'women are busy socialising, flat hunting, and generally living life to the full', men have become, according to Barbara Ellen (2000: 31), 'a bunch of spoiled, lazy, mummy's boys, who put CDs and beer before mortgages and groceries'.

Another aspect of gendered culture and identity is the thriving gay and lesbian culture in many of the larger towns. This began to emerge in the 1970s and 1980s after the decriminalisation of homosexual activities (see Chapter 6). The widespread adoption of the term 'gay', which denotes positive self-identification as opposed to 'homosexual' or more pejorative terms, implied a growing acknowledgement of gay identity as an alternative lifestyle choice, rather than just a manifestation of sexuality. Among younger age groups especially, the distinctions between gay and straight culture are more blurred than ever (Garrett 1997).

11.6 SUMMARY

It is clear from the foregoing that culture and identity in the UK are shaped and influenced by a variety of factors. Ethnic and national origin and upbringing, education, class, gender and age all play a part in moulding identity and informing culture. Identity and culture are often expressed via a range of symbols and emblems. Some of these have been explored above. In cultural terms, the UK has been enriched by the variety of indigenous national and regional cultures as well as by those of successive waves of immigrants, most notably and recently those from Africa, Asia, the Caribbean and elsewhere. These most recent additions to the cultural realm, though, have not been unproblematic in terms of identity – most notably in the form of racism (see Chapter 6).

The UK is a multinational and multicultural state. Because it is not a nation-state and was a product of political and economic expediency rather than some cultural primordialism, there have always been cultural friction and identity tensions. Identity tensions have been most apparent between the national and the state. Tensions have resulted in bloodshed, as in Ireland during the long-standing (and ongoing) 'liberation struggle', or in Scotland during the Jacobite Rebellion. The tensions and friction have centred on cultural issues, like language and religion, as well as political matters. But nowhere have these tensions impacted more than between Englishness and Britishness. Increasing devolution is seen in some quarters as a threat to identity, in others as a boost to it. To complicate matters of culture and identity in the UK there is the issue of increasing European integration and all that means for multiculturalism and regionalism (Robins 1999). What is clear is that even in a period of globalisation and seemingly ever-increasing homogenisation of cultures and identities, there remains within the UK an astonishing array of cultures and identities.

R E V I S I O N   Q U E S T I O N S

- How would you define your own identity? What factors influence your identity?
- To what extent do culture and identity matter in a globalised world?
- Describe and account for the ways in which the UK is a multicultural state.

KEY TEXTS

Bryson, B. (1997) *Notes from a Small Island*, London: Black Swan Books.

Crick, B. (ed.) (1991) *National Identities: The Constitution of the United Kingdom*, Oxford: Blackwell.

Davies, N. (1999) *The Isles: A History*, Basingstoke: Macmillan.

Marr, A. (2000) *The Day Britain Died*, London: Profile.

Nairn, T. (1997) *Faces of Nationalism: Janus Revisited*, London: Verso.

Paxman, J. (1998) *The English: A Portrait of a People*, London: Michael Joseph.

POLICY RESPONSES

12.1 INTRODUCTION

Since the election of the 'new' Labour administration in 1997 a new economic and social strategy has begun to shape the geography of opportunity and economic well-being in the UK. The Labour administration has adopted a radically different economic and social agenda to that of previous Labour administrations, no longer is the Keynesian welfare state the Labour mantra, with full employment or nationalisation, the key elements today are international competitiveness, notions of full employability and a productivist reordering of social policy (Peck 1999). In some respects the economic policy (both monetary and fiscal policy) of the administration of John Major has not only been maintained but extended by Labour. This is illustrated by the fact that in May 1997, shortly after the election, control of interest rates was passed to the Bank of England. The term 'fiscal conservatism' certainly applies as much to the Labour government today as it did to the Tories. Chancellor of the Exchequer, Gordon Brown, stresses fiscal prudence, signalling that this Labour administration, unlike previous Labour ones, will not spend, spend, spend. This prudence is designed to appeal to 'Middle England', those middle-class voters who deserted the Tory party in 1997 to sweep Labour to victory. Labour is very conscious that their future electoral success depends upon these voters as much as on traditional Labour voters.

In social policy, however, the agenda is different to that of the Tories; it is one of social inclusion (see Chapters 4 and 15), with the aim of reuniting the country and implementing one-nation policies. As was discussed in Chapter 4, EU debates on social exclusion have profoundly affected UK social policy post-1997. One of the key elements of achieving social inclusion is the New Deal in which Labour sees work as the solution; it is attempting to 'rebuild the welfare state around the work ethic' (Peck 1999: 345; see pp. 22–3, this volume). A second distinctive feature of Labour is in its 'joined up government/thinking', of looking more holistically at the likely impacts of policies such as new house building and car dependency. In the remaining part of this chapter we highlight a number of policy themes, including the countryside and urban areas; transport; regional policy for the English regions; welfare and labour market policy as well as EU economic development and regional policies. As will be revealed below different strategies are emerging for each country of the union in these policy areas following devolution (see Chapter 13)

12.2 RURAL AND URBAN ISSUES

The current framework for urban and rural policy in the UK developed in the early post-war period and was enshrined in the Town and Country Planning Act of 1947 and the Agriculture Act of 1947. These two acts, along with several others covering the provision of public services (the welfare state), have constituted the main foundations of domestic policy from the Attlee government to the present day (see Chapters 2 and 6). At the time, the government had a clear view of what the countryside and urban areas were for. It saw agriculture as the primary function of rural areas and therefore as their key economic sector; and it viewed agriculture's primary role as a supplier of food. It assumed that farmers, as stewards and shapers of the 'countryside', could be relied upon to protect the quality of the rural environment. It also assumed that the shift of population from rural to urban areas would continue and that the greatest threat to the rural environment would be around the fringes of urban areas. Through planning controls such as the **green belt** legislation an attempt was made to 'contain' urban sprawl, especially on the fringes of towns and cities.

The government believed that rising output from agriculture and other primary sectors in rural areas, combined with universal models of public service provision, would generate sufficient wealth to tackle rural poverty (see PIU 1999). In the late 1940s, the food security issues of the Second World War and memories of the agricultural depression of the 1920s and 1930s were still uppermost in the minds of policy-makers – food rationing was not phased out until the 1950s, for example. The government intervened heavily in agriculture to encourage the expansion of domestic food production, introducing price support, production subsidies and special treatment to farmers within the tax and land use system. Today agriculture receives some £5,000 million per annum in the UK, of which over £3,000 million is for England through EU and domestic programmes (PIU 1999). In relation to the environment, restrictions were placed on any development in what were viewed as the most valuable sites, National Parks, Areas of Outstanding Natural Beauty, green belt land and less protection for others. Throughout the post-war period the Ministry of Agriculture, Fisheries and Food (MAFF) has retained responsibility for agricultural policy in England while agricultural policy for Wales and Scotland and Northern Ireland, are devolved.

The values and beliefs on which post-war policy was based have changed. Social and economic trends have combined to produce a fundamental change in views of what the countryside is for and in values and priorities for public policy. Public concern about food has shifted from issues of quantity to issues of quality, notably safety (such as

salmonella in eggs and BSE and beef). There has been a marked increase in levels of interest and concern for the environment (Figure 12.1 and Box 12.1; see Chapter 6), so that many people now value rural areas more as sources of environmental goods than as places for food production. Moreover, more and more people want to live in the countryside (see Chapter 16) than in urban areas, and/or to enjoy the leisure and recreational opportunities it offers (PIU 1999).

Rural economies are facing a number of problems, and while some communities continue to prosper, others are experiencing problems of economic adjustment. The problems are social, technological and economic, as well as external, such as fiscal problems – exchange rate fluctuations that devalue Common Agricultural Policy (CAP) payments – and the government is unable to control structural and cyclical problems. Average net farm income is £16,266 from all sources (PIU 1999). Nearly two-thirds of farms now have some off-farm income, the result of the diversification of activities as a survival strategy (PIU 1999; see also Ilbery *et al.* 1998). But it is not just rural areas that appear to be in a state of crisis, the economic base of urban areas has changed

Figure 12.1 Tens of thousands of protesters joined the pro-countryside march in central London on 1 March 1998. Bearing banners such as 'No to the destruction of the Countryside' and 'Leave country sports alone', the marchers made their way to Hyde Park in one of the largest civil protests seen in London

Source: Michael Crabtree, courtesy Popperfoto.

**Box 12.1
Urban and
rural pressure
groups**

Council for the Protection of Rural England (CPRE) is a national charity, formed in 1926, which helps people to protect/enhance the countryside. This campaigning organisation opposes developments that threaten the countryside, supports beneficial change and campaigns on rural policy issues.

http://www.greenchannel.com/cpre/

Countryside Alliance exists to champion the countryside, country sports and the rural way of life. It seeks to preserve the freedoms of country people, to lead campaigns for country sports, cooperate with other bodies to promote and protect the rural way of life, promote conservation, develop education programmes and undertake research.

http://www.countryside-alliance.org/

Greenpeace is an international organisation with an office in Amsterdam where Greenpeace campaigns are coordinated and from where the fleet of campaign vessels is managed. In the UK, Greenpeace is divided into two parts. The first deals with campaigning work, direct actions, political lobbying; the second is the scientific arm that funds scientific research. It has 169,000 supporters.

http://www.greenpeace.org.uk/

Reclaim the Streets was originally formed in 1991 around the dawn of the anti-road movement. The organisation is more a network and was established when a small group of individuals got together to undertake direct action against the motor car. Their work has been small scale but effective, such as DIY cycle lanes and street parties, including the M41 street party in 1996.

http://www.urban75.com/Action/reclaim2.html

radically over the last two decades (see Chapters 15 and 16). And it is against this background of perceptions of urban and especially rural crises that the new Scottish and Welsh Assemblies have been particularly proactive.

It is not only rural areas which are facing problems. Northern towns and cities are grappling with problems which stem from the erosion of their economic base as a consequence of deindustrialisation (see Chapters 3 and 8). The loss of millions of jobs has been one of the critical issues in people, mainly the young and more able, deciding to migrate, especially to London and the South East. This population drift has resulted in the erosion of

the population base, especially in the larger metropolitan areas (see Chapters 7 and 14). In addition to the North–South drift there has been a more general urban exodus as part of the process of counterurbanisation, the strong preference of people especially in England for rural living (see Chapter 16). Those who are more socially mobile (more affluent and prosperous), with the economic resources to make choices over where to live, are more spatially mobile; one manifestation is the choice of a rural place of residence. We should also note that some urban areas are being upgraded by **gentrification** (see Chapter 16). On the whole the urban exodus and the erosion of the employment base have contributed to urban problems, but so too have the comprehensive development plans of the 1950s and 1960s. These slum clearance programmes in many cities created concrete jungles of high rise, poorly planned estates with few community facilities. As a result we see neighbourhoods in every British town or city characterised by a poor physical environment, poor schools, high dependency on welfare, high unemployment rates, lack of social and cultural facilities, lack of choice and quality of housing (see Chapter 16).

Turning firstly to England, an Urban Task Force under the Chairmanship of Lord Rogers of Riverside was established in 1998. The Urban Task Force's remit was to try and reverse the spiral of decline in the quality of life of English cities to regenerate urban areas, based on the principles of good design, social inclusion and environmental responsibility (see Chapters 15 and 16). While it is true that the population of cities is declining, especially the inner cities, most people in England still live in urban areas. The Task Force published their report *Urban Renaissance* in 1999 (DETR 1999a), and they will also help shape the Urban White Paper that DETR is drafting (autumn 2000).

The Urban Task Force feels that the planning system should be employed more strategically to achieve regeneration objectives by making it easier to develop on urban recycled land and harder to develop in out-of-town greenfield locations – especially as the government is committed to accommodate 60 per cent of the additional housing needed on recycled/**brownfield** land (see Chapter 16). Coupled with this vision is the call for strong municipal strategic leadership (see pp. 179–80). The Urban Task Force uses the term 'urban renaissance' (DETR 1999a) as its vision, but it fails to deal with social and economic issues at the heart of urban decline. An improvement in the built environment on its own is not enough to bring about an urban renaissance.

Since 1997 a plethora of initiatives for tightly defined, largely urban, areas have been implemented (Box 12.2). In a recent report (PIU 2000) the government acknowledged that many of these initiatives are generally run separately and are not

linked together. Coordination at the centre (Westminster) needs to be improved, and it is planned to strengthen the regional offices in England. Some of these new schemes are community based and illustrate Labour's 'joined up' thinking, such as the New Deal for Communities which involves the production of community based plans covering jobs, crime, health and housing.

Although rural England is primarily agricultural in terms of land use, farming is no longer the foundation of the rural economy today, nor is it the linchpin of rural society (MacFarlane 1998). But rural policy is at a crossroads. Over the last two decades, priorities for rural areas have been transformed. In agriculture the imperative to expand production has been replaced by imperatives to curb over-production, move towards world market prices and integrate environmental protection into farming supports as well as some encouragement of organic farming (Lowe and Ward 1998). At the same time, many rural areas have experienced profound changes as a result of rural in-migration. The weakening of agriculture and the growth of environmental and leisure demands have encouraged interest in the notion of a more diversified countryside in farming, conservation and rural development circles (Lowe and Ward 1998). This view is very much reflected in Labour's vision for the English countryside (outlined in the prospectus for the Countryside Agency created in April 1999) which encompasses five main themes:

- a *living countryside* – vital and viable with services for residents and visitors;
- a *working countryside* – with jobs for local people;
- *one nation, town and country* – recognising greater interdependence;
- *enhancing the environment* – protecting and enhancing the beauty and diversity of the countryside;
- a *countryside for all* – improving access for everyone (see Countryside Agency 2000).

The Countryside Agency for England resulted from a merger of the Countryside Commission and

**Box 12.2
Area-based
initiatives**

1   Schemes involving local and regional partnerships
Pioneer Community Legal Partnership (Lord Chancellor's Department)
Crime Reduction Unit Programme: Burglary Initiative (Home Office)
Crime Reduction Unit Programme: Targeted Policing (Home Office)
Employment Zones (DfEE)
Health Action Zones (Dept of Health)
New Deal for Communities (DETR)
New Start (DfEE)
Single Regeneration Budget (DETR/RDAs)
Sure Start Trailblazers (Sure Start Unit)

2   Schemes involving regional partnerships in particular regions
Coalfields (DETR)
European Regional Development Fund (DTI, DETR, GORs)

*3   Other government funded regeneration initiatives, local partnerships
in particular areas*
Better Government for Older People Pilots (Cabinet Office)
Drugs Treatment and Testing Order Pilots (Home Office)
Early Excellence Centres (DfEE)
Excellence in Cities (DfEE)
Healthy Schools Initiative (Dept of Health)
New Deal for Disabled People – Personal Advisor Pilots (DfEE, DSS)
Early Education Places (DfEE)

*4   National government initiatives being piloted in particular areas (without
local and regional partnerships)*
Local Authority Best Value Pilots (DETR)
Crime Reduction Programme (Home Office)
New Deal for Long Term Unemployed (12–18 months) aged 25+ (DfEE)
New Deal for Long Term Unemployed (over 2 years) aged 25+ (DfEE)
New Deal 50+ Pathfinders (DfEE)
Personal Medical Services Pilots (Dept of Health)
Youth Justice Pilots (Home Office)

5   Other local initiatives
New Commitment to Regeneration Pathfinders (Local Government
Association)
Drugs Action Teams (Cabinet Office)
Early Years Development and Childcare Partnerships (DfEE)
Lifelong Learning Partnerships (DfEE)
Local Transport Plans (DETR)
Renewal Areas (DETR)
Territorial Employment Pacts (Commission of the European Union)

Source: PIU (2000: annex F)

the Rural Development Commission (RDC) in 1999. Some RDC powers were transferred to the English regions. MAFF is the only UK ministry with a specific rural focus. The loss of its food regulatory functions to the Foods Standards Agency has been accompanied with speculation about some restructuring at the heart of Westminster, perhaps with a Ministry for the Countryside. At present, MAFF retains responsibility for agricultural policy in England, and an English rural White Paper is being drafted (late 2000).

Labour is drawing attention to the interconnectedness of urban and rural Britain, the wider environmental impacts of rural in-migration, as well as issues of social cohesion and competitiveness (Countryside Agency 2000). The Regional Development Agencies (see Section 12.4) now have responsibility for English rural economic policy. The role of rural areas in regional economies is being planned at a time when there is considerable debate in England on relations between 'town' and 'country'. The feeling that the countryside was not understood is well illustrated by the 'Countryside March' of 1 March 1998 (Figure 12.1), organised by the pro-hunting Countryside Alliance to coincide with the Committee Stage in Parliament of a Private Member's Bill to outlaw hunting wild mammals with hounds. This march became a vehicle for a large number of people to register their anger with the problems of rural communities. Some of these problems stem from the economic crises affecting a number of sectors in farming, and the feeling of neglect and lack of understanding of the needs of farming persists.

In both Wales and Scotland, research into rural issues, and the framing of policies to deal with them, has been a key feature of the new devolved governmental bodies. This is because of the importance of agriculture to their economies. For example, 89 per cent of Scotland's landmass is in rural areas with one-third of the population. In Scotland, measures introduced include a land reform action plan (Danson and Lloyd 2000). The land reform plan includes legislation to abolish feudal ownership, tenancy and crofting schemes, as well as countryside and heritage issues. There are also integrated planning schemes for rural land use. In common with England's emerging rural strategy, the strategy for rural Wales also involves a holistic approach with sustainable economic, community, social, cultural and environmental development aims. The strategy is designed to address both local and regional needs. Special arrangements exist in Scotland and Wales relating to agricultural policy and a concordat exists with MAFF. Regular meetings are timetabled regarding the formulation of policy with regard to the EU CAP.

12.3 TRANSPORT AND SUSTAINABILITY DEBATES

The aspiration towards reducing travel and thereby attacking the 'mobility explosion', particularly car travel, is at the heart of a number of policies including the White Paper *A New Deal for Transport*. It was published in July 1998 and applies to the whole UK, but implementation of the principles of integrating transport is devolved as different parts of the UK have differing transport needs (DETR 1998). The 1998 White Paper represents the first major policy statement on transport for over twenty years, and arguably since Buchanan in 1963 (Buchanan 1963). In the intervening years, for many people in the UK, using a car has become no longer a choice but a necessity.

Patterns of commuting have become increasingly complex in recent decades. Although there is still a marked pattern of commuting from peripheral areas to urban cores, along traditional lines, this is now paralleled by complex patterns of suburb to suburb movement, and even reverse commuting from urban homes to non-urban workplaces (Breheny 1999). The spatial ties between residences and workplaces have been reduced by car-based mobility. This is illustrated by the length of journeys to work. In England and Wales they rose by 15 per cent from 1981–91, while the proportion of longer trips (over 30 km) increased substantially (by 45 per cent) – again for England and Wales (Bannister and Gallent 1998). This has come about for a number of reasons: because of changing home and work locations; commuting is more fragmented and irregular; and a decline in the number of consistently defined travel to work areas, the result of falling levels of commuting self-containment, the net result of suburbanisation and counterurbanisation (Gillespie 1999; see Chapter 16). Not all car journeys are for the journey-to-work; many are related to caring tasks for the home, such as shopping, the school run and for leisure. While the majority of UK households have shared in the mobility revolution we should not forget that three in ten homes do not have a car – a total of thirteen million people (DETR 1998).

The DETR wishes transport to contribute to quality of life in the UK, not detract from it (pollution, mortality, injury, congestion and so on). The way forward is seen to be an integrated transport policy:

• integration within and between different types of transport;
• integration with the environment;
• integration with land use planning;
• achieving social inclusion (DETR 1998: 13).

The White Paper placed particular emphasis on:

• developing local transport plans which are integrated transport strategies, with local targets for improving air quality, road safety, public transport and road traffic reduction; more certainty of funding and greater use of traffic management;
• new powers of road user charging and levies on parking to tackle traffic jams and traffic growth;
• better interchanges;
• new airports policy;
• new independent Commission for Integrated Transport, which will advise on integration at the national level (DETR 1998: 14–15).

Although most of the focus is on urban areas, there is recognition that conventional public transport cannot always meet the diverse accessibility needs of populations in remote rural areas (DETR 1998). A new Rural Transport Partnership is pro-posed to get extra resources into rural transport. In terms of tackling the social exclusion agenda, 'inadequate and costly transport' has been identified as one of three main aspects of disadvantage in rural areas, alongside low incomes and lack of affordable local services.

In Scotland and Wales, the distinctive problems of dispersed communities and terrain are also being tackled by integrated transport plans. The Scottish Integrated Transport White Paper provides a framework for local transport strategies, and full strategies are being prepared in 2000. Scotland's transport challenges are the result of its geography, population and settlement pattern, peripherality and lower car ownership rates. Transport issues in Wales were tackled in *Transporting Wales into the Future* (published in 1999), and, like Scotland and England, integrated transport strategies are being prepared by local authorities. In 1999, extra funds for transport in Wales were announced, embracing extra bus subsidies, improved concessionary fares for pensioners, as well as a larger transport grant for integrated transport packages.

12.4 REGIONAL GOVERNANCE AND THE ENGLISH REGIONS

Regionalist movements in England have been small and relatively insignificant, but the issue of regional government has come to life as a response to Scottish and Welsh devolution (see Chapter 13). Despite the absence of a European-style directly elected regional government in England, the English regions are characterised by an extensive level of regional **governance** and regional structures, though the boundaries may not always coincide. Central government organisations in the regions have traditionally operated through a variety of different regional structures and boundaries, though a number of decentralised government offices were combined into the ten Government Office Regions (GORs) in 1994 (Mawson and Spencer 1997; see Chapter 14). It is anticipated that the GORs will be gaining a higher profile in monitoring how government policies are affecting local areas (PIU 2000).

Moreover, the English regions have been governed by a variety of regional institutions such as water authorities (now privatised), health authorities, a variety of quangos dealing with economic development and training, such as the Training and Enterprise Councils (TECs), and so on. In the 1990s, a number of the English regions also developed regional institutions, such as the Yorkshire and Humberside Assembly and the Campaign for a Northern Assembly, which campaigned for devolution for the English regions (Lynch 1999).

In April 1999, a new form of regional governance, Regional Development Agencies (RDAs), commenced operations in the English regions, with the task of promoting economic development (Foley 1998; Lynch 1999; Box 12.3). They represent the first strand in the government's English devolution proposals. The agencies' economic development role is modelled on the Welsh and Scottish Development Agencies (see Chapter 13). The RDAs represent a response to pressure from the English regions for similar bodies to allow the regions to compete with Wales and Scotland for inward investment, as well as being part of a broader strategy for regional economic development.

The RDAs have taken on several functions previously carried out by central government and government-sponsored bodies, including the Rural Development Commission's functions of stimulating job creation and the provision of essential services in rural areas. The respective role of regional institutions, and especially between the GORs and the RDAs, has come under scrutiny from a number of sources (Mawson and Spencer 1997), not least from the government itself (PIU 2000). Each region now has a Regional Chamber (see Chapter 13). The role of the Chambers is evolving, but they are seen as a regional partnership body. They focus on transport, economic development and land use planning and European funding. Such developments have had profound implications for regions and regionalism. The Chambers are also looking at other regional strategies such as health (see Chapters 13 and 14).

In contrast to the devolved types of regional government established in Wales and Scotland, the RDAs are fairly limited in their scope and capacity for autonomous activity.

The RDAs have five key functions:

- economic development and regeneration;
- the promotion of business efficiency, investment and competitiveness;
- the promotion of employment;
- to enhance development of employment skills in the region;
- to contribute to sustainable development.

In the key areas of powers, appointments, accountability and finance, the RDAs lack autonomy. The powers of the RDAs are either delegated from central government or transferred from other bodies such as the Rural Development Commission, English Partnerships and some GOR staff transferred to the RDAs, such as those responsible for the Single Regeneration Budget (SRB) (see Chapters 5, 13 and 14, and Box 12.4). Significantly, apart from the regional role of English Partnerships and the SRB, the RDAs exist alongside the GORs, which continue to exist as decentralised units. In April 1999, responsibility for the administration of the SRB passed from the GORs to the RDAs. Over the period 2000–3, nearly £2,000 million of resources will go to support over 500 existing and new local regeneration schemes. The RDAs are also responsible for drafting a regional training and skills strategy and an innovation strategy.

All English regions have urban and rural areas, and as a result of this integration it is intended that, 'rural interests [will] share centre stage with those of urban areas'(East Midlands Regional Development Agency 1999: 6). Despite claims that it will be a central focus of the new RDAs, there are fears that sustainable development is likely to remain marginal to the perceived needs of regional development. Sustainable development requires that human activities take place within the ecological limits of the planet and this requires the integration of environmental, economic and social decision-making. RDAs are required to take the

**Box 12.3
The RDA
regions of
England**

East Midlands
The counties of Derbyshire, Leicestershire, Lincolnshire, Northamptonshire and Nottinghamshire. The non-metropolitan districts of Derby, Leicester, Nottingham and Rutland.

Eastern Region
The counties of Bedfordshire, Cambridgeshire, Essex, Hertfordshire, Norfolk and Suffolk. The non-metropolitan districts of Luton, Peterborough, Southend-on-Sea and Thurrock.

London
Greater London.

North East
The counties of Durham and Northumberland. The metropolitan districts of Gateshead, Newcastle upon Tyne, North Tyneside, South Tyneside and Sunderland. The non-metropolitan districts of Darlington, Hartlepool, Middlesbrough, Redcar and Cleveland and Stockton on Tees.

North West
The counties of Cheshire, Cumbria and Lancashire. The metropolitan districts of Bolton, Bury, Knowsley, Liverpool, Manchester, Oldham, Rochdale, St Helens, Salford, Sefton, Stockport, Tameside, Trafford, Wirral and Wigan. The non-metropolitan districts of Blackburn with Darwen, Blackpool, Halton and Warrington.

South East
The counties of Buckinghamshire, East Sussex, Hampshire, Isle of Wight, Kent, Oxfordshire, Surrey and West Sussex. The non-metropolitan districts of Bracknell Forest, Brighton and Hove, the Medway Towns, Milton Keynes, Portsmouth, Reading, Slough, Southampton, West Berkshire, Windsor and Maidenhead.

South West
The counties of Cornwall, Devon, Dorset, Gloucestershire, Somerset and Wiltshire. The non-metropolitan districts of Bath and North East Somerset, Bournemouth, Bristol, North Somerset, Plymouth, Poole, South Gloucestershire, Swindon, Torbay and the Isles of Scilly.

West Midlands
The counties of Shropshire, Staffordshire, Warwickshire and Worcestershire. The metropolitan districts of Birmingham, Coventry, Dudley, Sandwell, Solihull, Walsall and Wolverhampton. The non-metropolitan districts of Herefordshire, Stoke-on-Trent, and Telford and Wrekin.

Yorkshire and Humberside
The county of North Yorkshire. The metropolitan districts of Barnsley, Bradford, Calderdale, Doncaster, Kirklees, Leeds, Rotherham, Sheffield and Wakefield. The non-metropolitan districts of the East Riding of Yorkshire, Kingston upon Hull, North East Lincolnshire, North Lincolnshire and York.

Source: http://www.hmso.gov.uk/acts1998/8045 – f.htm

Box 12.4 Single Regeneration Budget

The SRB, which began in 1994, brought together a number of programmes from several government departments with the aim of simplifying and streamlining the assistance available for regeneration. SRB provides resources to support regeneration initiatives in England carried out by local regeneration partnerships. Its priority is to enhance the quality of life of local people in areas of need by reducing the gap between deprived and other areas, and between different groups. SRB schemes include those that improve employment prospects, education and skills; or address social exclusion and improving opportunities, promote sustainable development; housing, growth in local economies and businesses and reduce crime and drug abuse.

Source: http://www.detr.gov.uk/regeneration/srb/index.htm

environment and sustainable development into account where appropriate (Gibbs 1998).

In addition, Labour has changed the governance of London (the Greater London Authority, GLA) following a referendum in 1998 (see Chapter 13). The GLA is a strategic authority, with a narrower remit than a unitary local authority in England and Wales. The key tasks of the GLA centre on economic and social development and the improvement of the environment of Greater London. The GLA has responsibility in eight main areas: transport, planning, economic development and regeneration, the environment, police, fire and emergency planning, culture and tourism and public health. The mayor is responsible for developing and keeping under review eight strategies covering transport, economic development, spatial development, three environmental strategies and cultural issues. The mayor has thus got a very wide portfolio, but the key policy area for most Londoners will probably be transport as over one million people enter central London daily between 7 a.m. and 10 a.m., and around 83 per cent use public transport (compared with 13 per cent for Great Britain as a whole). Yet many groups want the motor vehicles off the road so that people can 'reclaim the streets' (Figure 12.2; see Chapter 6). The legislation allows for four agencies to be set up (the London Development Agency, Transport for London, the London Fire and Emergency Planning Authority and the Metropolitan Police

Authority). These boards are a mixture of Assembly members and appointees. The mayor has responsibility for shaping policies for over seven million Londoners.

12.5 THE NEW DEAL FOR WELFARE TO WORK: FROM FULL EMPLOYMENT TO FULL EMPLOYABILITY

A key element of the government's strategy for tackling social exclusion is the New Deal. The New Deal has the labour market as its focal point, of reconnecting people to the labour market, and mainstream opportunities. It represents a Labour Manifesto commitment that allows young people, in particular, a choice of training and job opportunities. Most resources in the New Deal are targeted at the young: 18- to 24-year-olds who have been unemployed for six months or more; other client groups, but with far less resources, include the long-term unemployed (those unemployed for two years or more); lone parents and the disabled (Peck 1999). For the young unemployed there are four options: subsidised jobs with employers for the largest element of the client group; full-time education and training for a smaller group with basic skills needs; and voluntary sector work or places on an environmental task force with the

Figure 12.2 Reclaim the streets: Reclaim the Streets members protest near the Westway trunk road in London in 1996

Source: *The Guardian*, 22 April 2000. Courtesy Andrew Testa.

remainder. These measures are designed to act as the 'first step' on the employment ladder (Peck 1999).

But the success of the New Deal in getting large numbers reconnected to the labour market is likely to be very sensitive both to macroeconomic conditions and to local labour market factors (Peck 1999). The bulk of resources are directed at the young unemployed, the most employable client group, when compared with single parents, for example. Experience in the USA has shown that programmes directed at single parents are very costly to deliver. The New Deal in the UK was launched in the context of a relatively buoyant labour market. While one recent study suggests that it has helped reduce long-term youth unemployment (Employment Service Research Division 1999), the scheme has received much criticism (Peck 1999; Turok and Webster 1998).

Criticism centres on the length and quality of training on offer, its effectiveness in weak local labour markets as well as the number of New Dealers who remain connected to the labour market after they leave the scheme. The success of the New Deal depends on the capacity of local labour markets to absorb New Dealers, and generally speaking youth unemployment is acute in the northern and western regions of the UK, where job growth remains the weakest. Some of the highest concentrations of persons eligible for the New Deal are found in chronically weak labour markets in the north and west, such as Middlesbrough, Rotherham, Liverpool and Bradford (Peck 1999). Moreover, an analysis of the spatial distribution of the New Deal client group reveals a strong urban bias, with one-third of the client group located in just five large cities: London, Birmingham, Glasgow, Liverpool and Manchester (Peck 1999). Most young New Dealers are men (73 per cent), a reflection of the collapse of traditionally male held manual employment in depressed local economies, especially in northern cities, coupled with the fact that the relative buoyancy of the low grade service employment, even in these areas, makes it easier for young women to enter employment than men (Peck 1999: 347; see Chapters 3 and 8).

The New Deal, while it represents a major financial and political commitment, perhaps more fundamentally reflects a change in Labour's thinking about the underlying causes of, and appropriate remedies for, unemployment. The aim of the programme is not to create jobs (as it was for old Labour) but to recreate a work ethic: to raise employability especially amongst young people (Peck 1999: 345). The New Deal is part of the government's Third Way policy (see Chapter 6), which includes the minimum wage, new National Insurance contributions, increases in child benefit, a guaranteed minimum pension, as well as a national childcare strategy (it is widely recognised that the UK has the worst childcare provision in the EU). Together these policies represent a concerted strategy for addressing poverty and social exclusion in the UK (see Chapters 15 and 16).

12.6 THE UK REGIONS AND EU ECONOMIC DEVELOPMENT POLICIES

In 1973, the UK became a member of the then EEC (see Chapter 2). Membership means that the UK government cannot act alone in a whole array of policies. This applies to aid packages to failing industry, such as the proposed aid package for the ill-fated Rover car plant, as well as to aid packages for struggling UK farmers. Moreover, any major change to the framework for agricultural reform has to be agreed among the partner member states. As a result the policy framework for agriculture has been 'Europeanised'. European policy for rural areas is the CAP and Structural Funds. CAP is a combination of financial support and market regulation. CAP direct payments are unevenly distributed around the country, with most funds going to eastern England.

CAP has been incrementally reformed since the mid-1980s, initially in response to chronic food surpluses and the consequent budget costs of storing and disposing of them. Milk quotas were introduced in 1984, followed by the agreement of agricultural budgetary guidelines in 1988. Subsequently, in response to growing international pressures for the liberalisation of agricultural trade and reduction in subsidies coupled to farm production, a wider package of change – the so-called MacSharry reforms – was agreed in 1992. These reforms reduced some support prices, introduced direct income compensation for those cuts, and developed further the concept of a European rural policy in the form of accompanying social and agri-environmental measures. The 1992 reforms have been criticised because their environmental objectives were not explicit (Winter 2000).

There are also world pressures on CAP. For example, the 1995 Uruguay Round Agricultural Agreement brought agriculture under the ambit of the General Agreement on Tariffs and Trade (GATT). The trend is for a progressive reduction of tariffs and market support. The commitment to enlarging the EU and the next round of the world trade negotiations mean that CAP is economically and politically unsustainable in its current form. In July 1997, the European Commission published proposals in its Agenda 2000 document to move CAP away from an emphasis on production support and towards a more 'Integrated Rural Development Policy (Lowe and Ward 1998). Since 1988, the expansion in EU Structural Funds has delivered significant new resources to and procedures for the development of Integrated Rural Development Programmes, especially in 'fragile regions' (Objective 5b and LEADER). These programmes have imported innovative systems of partnership, co-financing, strategic planning and community participation into rural development policy.

Agenda 2000 sets out the Commission's proposals for CAP and Structural Funds. The document seeks to continue the reforms initiated in 1992 (MacSharry reforms) by reducing price support and increasing direct compensation payments to farmers. Measures for rural development are proposed to continue in a similar way as at present to those regions currently designated as Objective 1 (that is, those regions lagging behind) under the Structural Funds. In addition, the areas designated for Objective 5b programmes (in more peripheral or agriculturally dependent regions) have changed. But those areas that have lost their 5b status will receive transitional funding for the period 2000–6.

12.7 SUMMARY

In a number of respects 'new' Labour has to date made little impact on everyday life for many people in the UK, except those in receipt of welfare payments. Labour ministers increasingly deploy a discourse of welfare dependency in their justification for welfare to work, thereby casting 'welfare lifestyles' and the 'benefit culture', rather than a lack of jobs or poverty *per se*, as the underlying basic problem (Peck 1999). The new mantra is that 'work is the best form of welfare'; and talk of full employability rather than full employment. It is very difficult to see any difference between them and the Tories with monetary and fiscal policy, and privatisation continues – for example with the Housing Benefits Agency and with the proposed privatisation of Air Traffic control. We therefore have deflationary macroeconomic management combined with an advocacy of flexible labour markets, backed up by an active welfare to work strategy concerned not only with effecting transitions from benefit to welfare but also with

transforming the UK's welfare settlement. This represents a significant shift in Labour's approach to the policy of poverty and unemployment.

But in other respects there are changes as regional policy and regional planning are back on the agenda. No longer is capital favoured at the expense of regional, urban or rural planning. There is also an easier relationship with our European partners, and the European Social Chapter has been signed, which is shaping labour and social policy in the UK, benefiting part-timers and extending paternity rights, for example. The Human Rights charter will apply to the whole of the UK in October 2000 (it was incorporated into Scottish law before England), and we are already seeing its impact, for example over bullying in schools and with asylum seekers.

The Labour administration has been criticised for its lack of understanding of the countryside. Indeed, there appears to be an 'urban–rural collision' of division and intolerance, especially in England, yet the reality today is one of complex and multiple links between town and country, as exemplified by a number of pressure groups (Box 12.1). Rural communities are part of a wider society, and this perhaps explains why the RDAs are being charged with regional economies, and emphasising urban and rural interdependencies. It is interesting to note that in both Scotland and Wales within the first year of the new assemblies the problems of the rural economy were addressed head on.

R E V I S I O N   Q U E S T I O N S

- In what ways has devolution affected policy formation in the UK?
- Why should there be separate policies for urban and rural areas?
- In what ways is transport policy central to the human geography of the UK?

KEY TEXTS

Amin, A. (ed.) (1994) *Post-Fordism: A Reader*, Oxford: Blackwell.

Countryside Agency (2000) *State of the Countryside 2000*, Wetherby: Countryside Agency.

DETR (1998) *A New Deal for Transport: Better for Everyone*,
http://www.detr.gov.uk/itwp/paper/index.htm

—— (1999) *Urban Renaissance: Sharing the Vision*,
http://www.detr.gov.uk/regeneration/urbanren/index.htm

Peck, J. (1999) 'New Labourers? Making a New Deal for the workless class', *Environment and Planning C* 17(3): 345–72.

PART 3

3

THE UK

A society and state divided?

CONSTITUTIONAL AND POLITICAL CHANGE

13.1 INTRODUCTION

As a multinational state with a convoluted constitutional history, constitutional issues are a recurrent topic in political and wider debates in the UK. Nationalism within the UK has resulted in violence, most notably in Northern Ireland. However, organisations in Scotland (Scottish Liberation Army) and Wales (Free Wales Army [Meibion Glyndwr]) have engaged in acts of violence, if not full-scale terrorism. To some extent, the aspirations of nationalists and others disillusioned by the centralist UK state have been assuaged by recent devolution packages for Scotland, Wales and Northern Ireland. This, however, has raised the spectre of English nationalism (see Chapter 11) and the prospect of devolution for English regions, as part of the programme of devolved government, instituted by the Labour government, was an assembly for Greater London, headed by an elected mayor. Similar policies have been promised to other big cities.

As well as devolution of power from the centre of Westminster, the Blair regime has also put into place mechanisms for reform of the House of Lords. The election of 'new' Labour, by a large majority in 1997, meant that these constitutional and other reforms could be put in place. In many ways these were part of the legacy left by Blair's predecessor, John Smith, who, as leader of 'old' Labour was firmly committed to modernising the constitutional and political arrangements in the UK. Blair reiterated the Labour Party's commitment to reform in a recent speech:

> . . . devolution is a necessary part of keeping Britain together; more regional decentralisation in England makes sense; City mayors with real power have their place; hereditary peers in the House of Lords don't; and a constructive, engaged attitude towards Europe reflects the best of British values . . .
>
> (Blair 2000)

This chapter explores some of these constitutional issues and looks at some of the more recent developments in national, local and European dimensions of UK politics.

13.2 DEVOLUTION

The UK is very much an artificial state. The dissolution began in 1921–2 when the twenty-six counties of Ireland withdrew from the UK, and Northern Ireland got its own devolved parliament. Most of the calls for devolution have taken place in what some writers (e.g. Mohan 1999; Pattie 2000) erroneously call the 'Celtic fringe'. In none of the three smaller countries can anything but a minority of the population be described as Celtic (see Chapter 11). These countries' claims for devolution are based on history and identity (see Chapter 11). In 1999, after referenda, Scotland, Wales and Northern Ireland (the Stormont parliament was suspended in 1972) were granted some form of devolution, as was Greater London. Calls for greater devolution of power or independence have come from the other parts, including England (Gorman 1999).

During the successive Conservative governments from 1979, the UK became the most centralised of European states (Bradbury 1997; Sharpe 1997). The devolutionary reforms of the Labour administration elected in 1997 were an attempt to decentralise power within the country. The ideas behind devolution are not new, and ever since the Union there have been calls for various forms of devolution for different parts of the UK. Home Rule, particularly for Ireland, was a recurrent theme throughout nineteenth century politics. In Scotland, these feelings were partly assuaged in 1894 by the setting up of the Scottish Grand Committee to discuss Scottish affairs. After the First World War there were proposals for a federal UK, with separate Irish, Scottish and Welsh parliaments. The Prime Minister at the time, Asquith, called for 'Home Rule All Round' during the second reading of the Government of Ireland Bill, soon to be followed by a Home Rule measure for Scotland. England rejected this federalism, because English nationalism was the least publicised but most potent of all in Britain (Ferguson 1978). But the federal idea was taken up again in the 1920s when a separate parliament was set up in Northern Ireland, and this proved to be a source of inspiration for the Scottish and Welsh home rule movements during the interwar years.

After the Second World War, nationalist sentiments died out until the 1960s. Pressure for reform in Scotland and Wales, coupled with the success of the Scottish National Party (SNP) in the 1974 general elections (February and October), saw the Labour government issuing a White Paper in 1975, which proposed a legislative assembly with executive powers for Scotland and an executive assembly for Wales (Ferguson 1978). The Labour Party became increasingly swayed by devolutionary arguments to stave off the growing support for the nationalist parties (SNP and Plaid Cymru) in their traditional electoral heartlands in Scotland and Wales. Labour governments have almost always depended on Scotland for their majorities, and the Labour Party has the most to lose if devolution leads to secession (David 1999; *The Economist* 1999a). The Liberal Party, and later Liberal Democrat Party, has always been supportive of devolution and other constitutional reforms. It was the Conservative and Unionist Party – the party of a strong, centralised, unified United Kingdom – that was most antipathetical towards anything that would threaten the stability of the Union. Although, interestingly, it was the Conservative and Unionist Party that was most supportive of the Stormont regime in Northern Ireland, notorious for the undemocratic principles and discriminatory policies that led to the 'modern troubles', which in turn demonstrated the weakness and injustice of a multinational state so heavily unbalanced and dominated by one overpowering hegemon.

The example of Ireland and other small countries within the EU has also helped the cause of devolution (Roberts 2000; *The Economist* 1998). There has even been a revival of the notion of a kind of federalism as a panacea for the 'break up of Britain' crisis (Barnes 1997; Keating 1982). However, federalism based on the four home countries would be unworkable given England's hegemonic demographic, economic and political power (Walker 1999). Only if there was true regional devolution in England would any federal arrangement work. Scottish Nationalists would go further. They point to Scandinavia as a model for the British Isles. That is, 'a confederation of sovereign nations mutually

bound together through common historical and cultural ties' (Kay 1986: 147).

The Labour Party was elected on a platform of devolution and a commitment to reverse the over-centralisation inherited by the Thatcher–Major governments. Devolved governments are now in place in Scotland, Wales, Northern Ireland and London. Prior to this there had been a policy to devolve more powers to Scotland and, to a lesser extent, Wales in the areas of housing, nature conservancy, training and enterprise, higher education, arts and environmental protection (Hogwood 1996). We now look at devolution in the four nations.

13.2.1 Ireland–Northern Ireland

For several centuries, the island of Ireland existed as a colony of the English and British empires. Throughout this time, politics in Ireland was characterised by successive uprisings and rebellions and the attempt of the sovereign power to dominate or subdue the 'troublesome' Irish by means of transportation and the imposition of armed force. Oliver Cromwell's infamous use of armed force, in the 1650s, had the express intention, via extreme violence, of pacifying, and converting, the indigenous Roman Catholic population of Ireland. Despite centuries of draconian and violent attempts to suppress the Catholic Irish such measures eventually proved unsuccessful.

One significant policy, which began early in the seventeenth century and which aimed to dilute the power and influence of the Catholic Irish, was the policy known as the 'Plantation of Ireland'. This policy involved 'planting' a significant Protestant population in Ireland, which would, it was hoped, be faithful to the British Monarchy. This population came predominantly from Scotland and northern England and was established mainly in the north of Ireland. After several centuries the plantation brought about a significant Protestant majority in the northern Province of Ulster. More importantly, however, the plantation of Protestants created to this day a population which strongly identifies itself with the UK and the British Crown. On the other hand, many members of the minority Catholic population, who remained in the north of Ireland, were vehemently opposed to this Protestant power base (Shirlow and McGovern 1997).

The loss of Catholic land during the plantation and the growth in the power base of Protestants was further intensified by the partition of Ireland in 1921. Prior to this, the growth of the Irish Republican Party, Sinn Fein, and the ability of its armed wing, the Irish Republican Army (IRA), to wage war against Britain led to the removal of British authority in the twenty-six counties in Ireland. In 1921, via the Anglo-Irish Treaty, these twenty-six counties (the Irish Free State) gained dominion status and joined the British Commonwealth. However, in 1949 the Free State seceded from the Commonwealth and became a fully fledged republic. Partition, in 1921, effectively meant that Northern Ireland was the first part of the United Kingdom, since 1801, to receive a devolved government with executive powers.

Northern Ireland was very much the result of gerrymandering of the borders to exclude three of the predominantly Roman Catholic counties of the traditional nine-county Province of Ulster. It, therefore, had an in-built Protestant majority in its

devolved parliament at Stormont. The militarised police force, which was always armed, wore green uniforms and operated out of 'barracks' rather than police stations, was almost exclusively drawn from the Protestant community. The Protestant majority was able to practise discrimination against the Roman Catholic minority with impunity, especially in the fields of housing and employment. The central government in Westminster turned a blind eye to these abuses of human rights. The Protestant community lived in perpetual fear of a threat from the Roman Catholic community within, and of those in the Republic of Ireland, whose constitution from 1937 to 1998 laid claim on the six counties (see Chapter 16). This siege mentality is best seen during the marching season and in the Apprentice Boys' march, which celebrates the siege of Derry of 1689, although it must be remembered that the biggest Orange parade is in Donegal in the Republic of Ireland (Valentine 2000).

By the 1960s, and influenced by what was happening in the USA and elsewhere, the Civil Rights Association began a series of peaceful protests against the regime. Civil law and order broke down and the threat of civil war forced Westminster to

Figure 13.1
A barricade of burnt-out vehicles in Northern Ireland after a 1997 riot. Note the list of Nationalist demands and the painted kerb stones – in white, green and gold in the original

Source: Peter Shirlow.

**Figure 13.2
Nationalist sentiment
on the Northern
Ireland situation**

Source: Peter Shirlow.

**Box 13.1
Nationalist
and Unionist
fears**

Nationalist fears	*Unionist fears*
Nationalists are wary of any slippage towards the bad old days of Unionist domination.	Unionists are vigilant with regard to any slippery slope towards the spectre of a united Ireland.
Nationalists tend to identify more with an all-Ireland entity and view that as their natural nation.	Unionists identify with the British state and relate to it as their nation.
Nationalists claim that they have suffered in this way since the introduction of the state of Northern Ireland in 1921.	Increasingly, and especially over recent years, Unionists have become similarly aggrieved. Many now believe that the British state is appeasing Catholics to their disadvantage.
Nationalist arguments for preventing Orange marches echo the 'not an inch' slogan which has been the clarion call of Unionists for years.	The Orange Order claim that Nationalists are denying their traditional right to demonstrate their culture.

send British troops into Northern Ireland, and later, in 1972, to suspend the undemocratic Stormont government and introduce direct rule from London. Thus, until recently, Northern Ireland was unique within the UK, having had a long tradition of fully devolved power from 1921 to 1972. Efforts to restore self-government were thwarted and armed gangs assumed control – the Provisional IRA, the Continuity IRA, the Real IRA and Irish National Liberation Army on one side, and the Ulster Defence Association, Ulster Freedom Fighters, the Loyalist Volunteer Force and Ulster Volunteer Force on the other. The result was one of the most sustained and bloody episodes of civil unrest in the history of a modern democracy. Bombs and bullets became a way of life – and death (Figure 13.1). Murder and mayhem persisted despite the best efforts of politicians from all parts of the British Isles and further afield. As Ignatieff (1994: 165) reminds us, 'there is more death by political violence in Great Britain [of course, he means the UK] than in any liberal democracy in the world. Since 1969 there have been 3,000 killings and more than 50,000 people have been seriously injured.' For both communities the future was in Britain's hands (Figure 13.2). The ethnic cleavage in Northern Ireland is based on mutual fear and distrust; some of these fears are outlined in Box 13.1.

The Anglo-Irish Agreement (1985), the Downing Street Declaration (1993), and the Belfast Agreement (1998), known as the Good Friday Agreement, are examples of how governance has altered in Northern Ireland in recent years (see Box 13.2). The capacity of the Irish and British states to work together in order to remove conflict has undoubtedly created a political climate which provides for some form of political progression. In combination with paramilitary cease-fires, alterations in modes of governance also underpin a wider realisation that conflict resolution is multidimensional and does not rest solely with the British state. The transnational nature of peace building, which includes the UK, Ireland, South Africa and the USA, paramilitary groups and the European Commission, is identified as 'the totality of relationships' (Shirlow and McGovern 1997).

The drafting of the Belfast (Good Friday) Agreement in 1998, and its subsequent support in referenda north and south of the Irish border, indicates a further shift towards an institutionalised political agreement (see Boxes 13.3 and 13.4 for the 'yes' and 'no' arguments). The referenda held in May 1998 won support in favour of creating a devolved government in Northern Ireland. In Northern Ireland, 71 per cent of those who voted did so in favour of the Agreement. In the Republic of Ireland, 94 per cent backed the Agreement, which includes dropping the claim – Articles 2 and 3 of the Irish Constitution – to Northern Ireland. In overall terms 85 per cent of those who voted on the island supported:

- a devolved government in Northern Ireland in which power would be shared by all political parties;
- decommissioning of paramilitary weapons and the demilitarisation of Northern Ireland;
- significant reform of the Royal Ulster Constabulary, especially in relation to encouraging Catholic membership;
- equality of provision and the endorsement of British and Irish cultural rights;
- the principle of consent and the ruling that Northern Ireland's constitutional status within the UK cannot be altered without majority support.

In terms of political identity, the aim of the Belfast Agreement was to achieve a workable political structure that recognised the contested nature of Irish and British sovereignty. The idea was to endorse the principle of rights and consent by the pro-British and pro-Irish sections of the electorate. This will mean closer involvement between the two states over policy decisions in Northern Ireland, a theoretical form of joint-authority, cross-border trade including cultural exchange, and an effective veto over future policies in Northern Ireland. The key actors in this are the elected representatives of Unionism/Loyalism and Irish Republicanism/Nationalism.

The Good Friday Agreement is a highly ambitious but much needed project. It is, at best, seen by the

**Box 13.2
Peace
process
timetable**

1994
August: Provisional IRA announces cease-fire and a desire to negotiate
 a new political future.
October: Loyalist paramilitaries (UVF, UFF, RHC) announce a cessation
 of all military activities.

1995
December: British and Irish government Framework Document issued,
 calling for a devolved Northern Ireland Assembly and cross-border
 political and economic bodies.

1996
January: Multi-party talks begin and include Sinn Fein.
February: Senator Mitchell establishes 'Mitchell Commission' which
 calls for decommissioning of paramilitary weapons in parallel to all-
 party talks. The British government calls elections to a 'Peace Forum'.
May: IRA bomb Canary Wharf and kill two. Sinn Fein are excluded from
 multi-party talks.
June: 200 injured by large IRA bomb in Manchester.

1997
May: New Labour government take office. Within two weeks of election
 Blair confirms that officials will hold exploratory talks with Sinn Fein.
July: New and second IRA cease-fire announced.
August: Independent Commission on Decommissioning, led by General
 John de Chastelain, established.
August: Northern Ireland Secretary of State Mo Mowlam states that IRA
 cease-fire sufficient for Sinn Fein to be permitted to all-party talks.
September: Ulster Unionist Party enters talks.

1998
March: Senator George Mitchell sets 9 April deadline for peace deal.
April: Belfast (Good Friday) Agreement struck.
May: 71 per cent vote in favour of Belfast (Good Friday) Agreement. In
 the Republic of Ireland, 94 per cent back agreement, which includes
 dropping constitutional claim to Northern Ireland.
June: Assembly elected.
July: Assembly meets and elects David Trimble as First Minister, SDLP's
 Seamus Mallon becomes his deputy.
August: Real IRA car bomb kills 29 in Omagh.
October: SDLP leader, John Hume, and David Trimble awarded Nobel
 Peace Prize.

1999
July: Ulster Unionists refuse to join devolved government before IRA
 starts to decommission. Ulster Unionists boycott Stormont, Mallon
 quits as deputy.

October: Peter Mandelson replaces Mo Mowlam.

November: Anti-agreement Unionists mount pressure on Trimble. IRA states that it will appoint representative to de Chastelain body to consult on arms handover. Mitchell concludes review. Unionist Party Council votes 58 per cent for the leadership going into a developed executive with Sinn Fein, but only on condition that the IRA decommissions. Executive formed, with Unionists, Nationalists and Republicans sharing power.

December: Irish and British governments sign treaty establishing North–South ministerial council and cross-border implementation bodies.

2000

February: Mandelson suspends Province's nine-week-old power-sharing Executive and announces he is returning Northern Ireland to Direct Rule. De Chastelain commission says it has received 'no information' from IRA on when decommissioning will begin. IRA pulls out of disarmament talks with de Chastelain.

March: Trimble says he is prepared to re-enter government with Sinn Fein – if there is a firm guarantee on decommissioning. Trimble narrowly defeats Revd Martin Smyth to retain party leadership.

May: British and Irish governments put forward proposals to restore devolved government in Northern Ireland by 22 May. IRA responds with a statement saying it is prepared to 'completely and verifiably put IRA arms beyond use'.

June: Devolved government restored.

two governments as a process on the road to consensus. In many ways Northern Ireland is now dictated to by the tyranny of pluralism and the process for devolution, a political exercise based upon brushing certain problems under the carpet. As such there is a marked failure to explore the realities of social and cultural discrimination (see Chapter 11).

13.2.2 Scotland

In some ways Scotland has always had an element of devolved government. The Act of Union made it clear that the distinctive Scottish educational and legal systems would be retained, as would the Scottish Kirk and the Scots language. In 1885, the Scottish Office – based in London – was founded

to help govern the country. From 1907 there was a Scottish Grand Committee within the House of Commons to debate matters purely affecting Scotland. More recently, since 1926, Scotland has had a representative in the Cabinet in the form of the Secretary of State for Scotland. Since 1939, the Scottish Office has been a separate organisation, based in Edinburgh, dealing with much of the routine administration of the country (McCarthy and Newlands 1999). Scotland, then, has had an 'institutional identity' for a long time (Nairn 1997).

Ever since 1707 there have been calls for Scotland to withdraw from the Union. This movement gained momentum in the nineteenth century. The Scottish Home Rule Association was founded in 1886 as part of the movement for 'Home Rule All Round'. In 1927, this body demanded that Scotland have the

**Box 13.3
The 'Good
Friday'
Agreement:
arguments
for 'yes'**

Unionist/Loyalist viewpoint	Nationalist/Republican viewpoint
The Belfast Agreement endorses the principle of consent and keeps Northern Ireland within the UK.	The Agreement endorses a role for the Republic of Ireland in Northern Ireland's affairs and dilutes the significance of British sovereignty.
The Agreement endorses the need for decommissioning and in particular will lead to the removal of Republican terror groups.	The Agreement endorses the need for reform of the Royal Ulster Constabulary and in so doing promotes impartial policing.
The return of devolved government promotes Northern Irishness.	A devolved government is a step closer to a United Ireland.
The Agreement means that the Republic of Ireland must alter Articles 2 and 3 of its constitution.	The rights and aspirations for a united Ireland are endorsed within British sovereignty.

**Box 13.4
The 'Good
Friday'
Agreement:
arguments
for 'no'**

Unionist/Loyalist viewpoint	Nationalist/Republican viewpoint
A devolved government with power-sharing rejects a democracy based on majoritarianism.	A return to devolved government is an acceptance of partition.
The Royal Ulster Constabulary does not need to be reformed.	Reforms of the Royal Ulster Constabulary will be tokenistic.
Power sharing undermines the Unionist veto.	The Unionist veto over partition remains.
The Republic of Ireland should have no more than a consultative role in Northern Irish affairs.	The alteration of Articles 2 and 3 of the constitution undermines Irish sovereignty.
The early release of prisoners undermines victims' rights.	Unionists still control the police and other armed forces.

same status as the Irish Free State. In 1934, the Scottish National Party was formed. The first SNP MP was elected in 1945 (Harvie 1998).

During the 1960s there was a revival in Scottish nationalism. The expansion of the Scottish oil industry, coupled with disillusionment with London control, saw resurgence in nationalism again in the 1970s. Thus the SNP won seven Westminster seats in the general election of February 1974, with 21 per cent of the Scottish vote. After this, the Labour Party committed itself to setting up a Scottish assembly with unspecified powers. This did not convince voters in Scotland, however, since the SNP won eleven seats in the October 1974 election with 30 per cent of the vote. This resulted in the referendum 'fiasco of 1978–79, when Whitehall imposed on Scotland a devolution scheme tailored to its own requirements' (Harvie 1982: 19). After 1979 the SNP shifted to the left politically. This gained them greater support in the industrial heartlands, particularly in Clydeside and other Labour strongholds (Brand 1990).

In 1980, the Campaign for a Scottish Assembly (CSA) was established as a cross-party forum for a Scottish legislature, but failed to gain any significant support (Brand and Mitchell 1997). The issue came to a head after the 1987 general election, when the Tories were reduced to ten seats and only 24 per cent of the vote (Bradbury 1997). The introduction of the poll tax in Scotland a year before the rest of the UK further alienated people and was grist to the Nationalists' mill. The community charge, as Brand and Mitchell (1997: 44) point out, 'provided a focus for the claims that the Tories had no mandate in Scotland'. Many other Thatcherite policies boosted the cause of devolution (Marr 1995). These related to the so-called 'democratic deficit' 'where the majority of voters supported parties . . . in favour of constitutional change' (Brown 1997: 671). The Conservative government rejected this notion.

In 1988, the CSA-inspired *Claim of Right for Scotland* was produced. This called for an elected convention to debate Scotland's future (Devine 1999a). This proved to be too expensive so, in 1989, a Scottish Constitutional Convention (SCC) was set up composed of people from all aspects of Scottish life

to discuss devolution. The key demand was for a Scottish Parliament with law-making powers. The SCC, and the work it did, was rarely alluded to in the 'national', that is London, media. The same can be said of the work of the new legislature (Linklater 2000). Although London ignores what is going on in the minority countries, the work of the SCC and the new legislative bodies is being keenly followed in some of the English regions, which seek to emulate Scotland's path towards devolution (see pp. 179–80).

The SCC report formed the basis of the devolution policy presented in the Labour Party manifesto for the May 1997 general election. The Liberal Democrat Party also supported the report. After the election, the Labour government arranged for a referendum on its proposals, which were set out in a White Paper of July 1997, *Scotland's Parliament*. Thus, on 11 September 1997, a referendum voted 74.3 per cent for 'yes' 'for the Scottish Parliament to resume business' and 63.5 per cent voted for tax varying powers for the Parliament (Nairn 2000: 192). Because Scottish politics advocates civic rather than ethnic nationalism all registered voters were able to vote in the referendum (Miller 1998; *The Economist* 1999c).

For the Scots, devolved powers are more than just regionalism. On 12 May 1999, a simple declaration stated that the Scottish Parliament, 'adjourned on 24 March 1707, *is hereby reconvened*' (Nairn 2000: 254). That is, business as usual despite the three-century gap. The Parliament has tax-raising powers and a considerable range of powers over Scottish affairs, but no control over defence, foreign affairs, employment, transport and other economic policies. The voting system allowed some seats to be allocated according to the proportion of the votes cast, so that while Labour had fifty-six seats after the election of May 1999 (Table 13.1), it did not gain overall control. This resulted in a coalition of Labour and Liberal Democrats. As a price for Liberal Democrat support the Labour Party in Scotland has had to reverse its decision on tuition fees in higher education. This has brought the Edinburgh body into direct conflict with that in London, an inevitable consequence of this process (Hassan 1999; Linklater 2000).

Table 13.1 Summary of Scottish Parliamentary election results, 1999

Party	Constituency MSPs elected	Share of constituency votes (%)	Regional list MSPs elected	Share of regional list votes (%)	Total MSPs elected
Labour	53	38.8	3	33.6	56
SNP	7	28.7	28	27.3	35
Liberal Democrat	12	14.2	5	12.4	17
Conservative	0	15.6	18	15.4	18
Green	0	0.0	1	3.6	1
Independent	1	0.8	0	1.2	1
Scottish Socialist Party	0	1.0	1	2.0	1
Total	73	99.1	56	95.5	129

Source: The Scottish Parliament Information Centre 1999, *Scottish Parliament Election Results 6 May 1999*, Research Paper 99/1, Tables 1 and 2
Note: Percentage shares do not sum to 100 due to votes for other parties and candidates.

It is predicted by a number of observers (e.g. Hague 1999; Marr 2000; McSmith 1999b) that devolution in Scotland will lead to calls for complete secession from the Union. Recent opinion polls suggest that a majority of people in Scotland (not all Scottish) wanted Scotland to leave the Union (Heffer 1999). As David Maclean, Scots Tory MP representing an English constituency observed, 'Let us make no bones about it – Scotland will be independent sooner rather than later. One cannot be 95 per cent pregnant: one cannot be 95 per cent independent' (quoted in McSmith 1999b: 8). As Heffer (1999: 30) puts it, with no exaggeration, 'To exist, the Union as a political entity needs the ancient nation and Kingdom of Scotland. If Scotland goes, then there is no such thing as a United Kingdom, even if the Welsh and the Six Counties are clinging on to their differing forms of association with the Crown.' Davies (1999: 1043) uses the example of the Tsarist Empire and the Soviet Union as a reminder of what happens to artificial states when they break up under the weight of nationalist–separatist sentiments. The hegemon is 'left in a state of total bewilderment, not knowing what to believe about its past or its present or its future'. As we saw in Chapter 11, this is very much the condition of England now. But before we turn our attention to England we will look at the situation in the remaining nation, Wales.

13.2.3 Wales

The Welsh Assembly has even fewer powers than the Scottish Parliament. However, it does represent a recognition of the differences of culture and identity outlined in Chapter 11 and gives Welsh people a more direct say in Welsh affairs. But the referendum for an assembly was approved by the smallest of margins from a turnout of just over half of the electorate. This suggests that even where there is a strong sense of national identity it does not always readily translate into a political identity. One reason for the relatively low interest might be that, unlike Scotland and Ireland, what is now Wales was absorbed into England before it was a coherent unity. As Heffer (1999: 20) put it somewhat contentiously: 'Welsh nationalism is a preposterous concept because there is no Welsh nation.'

This rather narrow view will be disputed by many people in Wales, where there has long been a strong and vociferous nationalist, rather than separatist, sentiment. In 1925, Plaid Cymru was formed, following a nationalist renaissance, to promote Welsh nationalist aspirations (Harvie 1998), and since 1929 the party has contested elections (Balsom 1990). It has had some success in Westminster elections, but the party has had little support in the Labour heartlands of industrial South Wales, mainly because of its strong association with the Welsh language, which few in the south can speak (Jones

1997). Because of the lack of widespread support for a potent nationalist force, Welsh devolution cannot be explained in the same terms as that in Scotland. Indeed, unlike Plaid Cymru, traditional support for the SNP is not strongly correlated with social and cultural characteristics (Taylor 1973).

Although Wales never had the degree of self-regulation that Scotland had long enjoyed, during the last thirty-five years it has undergone a 'process of regionalisation . . . leading to the . . . Assembly' (Benneworth 2000: 23). Since 1965, Wales has had an integrated territorial office with a Secretary of State in the Cabinet (Bogdanor 1999). The Welsh Office grew in strength from a budget of £250 million in 1965 to one of £7,000 million in 1995. This represented 70 per cent of public expenditure in Wales (Jones 1997). From 1976, the Welsh Development Agency tried to stimulate the economy of Wales and was also seen as part of the devolution plans of the Labour government under Callaghan. Thus there was a strong administrative apparatus in place in Wales when the Assembly was created. By 1997, Wales had a Minister of State with extensive powers to run Welsh affairs. The Welsh elements of the position of Secretary of State were vested in the new Assembly, while the London elements remained with the Secretary of State's reduced office (Benneworth 2000).

Like Scotland, Wales had a chance for devolution in 1979. The referendum in that year was defeated by a large margin. Even in the heartlands of Welsh Wales there was no great support for devolution (Balsom 1990). But in 1997, the Welsh voters endorsed the government's proposals for a directly elected Welsh Assembly in a referendum. The Government of Wales Act, passed in 1998, allowed for power to be devolved to Cardiff, although laws passed by the UK Parliament still apply to Wales. The Secretary of State for Wales and the MPs from Welsh constituencies still retain seats in the House of Commons. The National Assembly of Wales (Cynulliad Cenedlaethol Cymru) comprises sixty members, thus:

- Labour 28 seats – 27 directly elected and 1 additional member;

- Plaid Cymru 17 seats – 9 directly elected and 8 additional members;
- Conservative 9 seats – 1 directly elected and 8 additional members;
- Liberal Democrats 6 seats – 3 directly elected and 3 additional members.

As in Scotland, the Conservatives have a reasonable presence only because of the additional member system. Although the Labour Party does not have a majority, unlike the Scottish case, it has failed to create a Liberal Democrat coalition and the Cabinet, which consists solely of Labour AMs (Assembly Members), continues as a minority government

13.2.4 England?

One overriding question of devolution is the West Lothian question, so called because it was first posed by the MP for West Lothian, Tam Dalyell, during the 1978–9 devolution round. He asked why Scottish MPs could vote on purely English matters in Westminster while English MPs would have no say in purely Scottish matters raised in an Edinburgh parliament.

One of the underlying corollaries of the West Lothian question is devolution for England. Devolution for Scotland and Wales 'is a constitutional revolution but in the middle there's an English hole' (Walker 1999: 15). The progression of devolution in the other countries has led to calls for an English Parliament or English Grand Committee, notably from the political right, although it is interesting to note that there has recently been a 'restoration of the Parliamentary Standing Committee of (English) Regional Affairs, in which the needs of the regions are conflated to the needs of England' (Benneworth 2000: 35). The government, though, has made it clear that this will not form the basis of an English Parliament. This is part of the 'English backlash' to devolution in other parts of the country (Hague 1999). Much of this has come from right-wing commentators and politicians. According to one such politician (Gorman 1999: 21), 'an English Parliament might be just the tonic required to restore our countrymen's faith in their ability to govern themselves . . . We may even be able – like the Scots and the Welsh – to celebrate our national saint's day, our national traditions, and even our language, without being accused of xenophobia!' But, according to Andrew Marr (1999c), there is a need for the left in England to address this very important issue.

If such a body does come into being it will be the first English Parliament since 1546. Those who favour such a body are mainly of the opinion that it be sited away from London, York being the most favoured location, though Ripon, Wakefield, Lancaster, Chester or Crewe have also been mentioned (Engel 2000). A more central site would make sense insofar as this would negate the complaints that London and the South East have too much political and economic clout in England, as much as in the UK. Whether there will be an English Parliament in the foreseeable future, there are calls in some part of that country and from the EU for greater regional devolution in England. The issue of a separate governing body for England and devolution for the English regions is part of the vexing and long-standing – at least for people outside England – *English Question* (Chen and Wright 2000).

13.3 REGIONALISM

Regionalism is generally regarded as having a tripartite rationale: reform of local government, regional development and planning, and regional decentralisation (Smith 1965). Regionalism was first advocated as a solution to local government problems. The novelist and futurist H.G. Wells proposed urban-based regions as early as 1902. In 1915, the geographer-polymath and visionary Patrick Geddes argued for regions to be based around the seven great cities (Gilbert 1960). But it was the geographer C.B. Fawcett who, in 1916, first argued for an abolition of the system based on historical counties and the establishment of 'natural regions'. He later developed his ideas into *Provinces of England* (Fawcett 1960). He was influenced by the 'Home Rule All Round' political and constitutional debate that was raging at the time throughout the UK. Fawcett wanted to combine much-needed local government reorganisation with devolution from central government, with twelve provincial parliaments. His far-sighted plans were ignored by central government, and it is tempting to imagine what the social, economic, cultural and political histories of the UK would have been during the twentieth century had they been implemented.

However, the regionalist lobby remained active throughout the interwar years, especially through the work of H.J. Mackinder, G.D.H. Cole, the Geographical Association and a number of special inquiries and Royal Commissions. The exigencies of the Second World War saw considerable decentralisation of government functions to the regions. This was supported, according to Gilbert (1948: 180), by Regional Commissioners and 'the miniature Whitehalls which sprang up in a dozen provincial capitals'. After the war, the government disbanded this system, though it was to be used as the basis for the 'standard' regions (see Chapter 14). Regionalist adherents became more involved in the growing town and country planning movement. The Greater London Council (GLC) is probably the best example of this new approach to regionalism. More *ad hoc* agencies, such as the North East Development Council and the Joint Committee for the Economy of the South West, were also set up. These latter were largely quangos based on economic expediency and firmly rooted in the regional planning philosophy – there were no plans for wholesale devolution of power from Westminster and Whitehall.

In the 1990s, a number of English regions began to develop regional institutions, such as the North West Regional Association, North East Constitutional Convention, Yorkshire and Humber Assembly, West Midlands Forum, East Midlands Regional Planning Forum, Northern Development Company and the Campaign for a North of England Assembly, which campaigned for devolution for the English regions (Mawson 1997). These are based on the successful SCC, which was a precursor to the Scottish Parliament. These bodies are made up of people with a wide range of interests and backgrounds and it is interesting that both the

Constitutional Conventions in Scotland and the North East were chaired by senior clerics. Campaigns now exist in four English regions to establish elected and representative regional government. The Campaign for the English Regions, an organisation based in the North East, consists of the North East Constitutional Convention, Campaign for Yorkshire, and the Campaign for a West Midlands Assembly. The North West Constitutional Convention is an associate member.

Regional identity and demand for devolution tend to increase with distance from London and the Home Counties (Gilbert 1939; Hogwood 1982). The arguments for English regional government relate to the relative over-representation of Welsh and Scottish MPs in Westminster, the relative over-funding of these countries and the lack of democratic control at the regional level through the over-use of quangos, as well as the perceived need to develop a formal regional structure along European lines so as to compete more effectively for EU regional funding (Sharpe 1997). Indeed, it was the European dimension to regionalism that led to the setting up of the Government Office Regions (GORs) in 1994 (Mawson and Spencer 1997). There is also an increasing feeling of being marginalised – wedged between the hegemonic South East and a Scotland 'increasingly confident of its own identity' (Esler 1999: 13).

As we discuss in Chapter 14, the new regional structure in the UK was introduced in 1994 (see Figure 14.1). In April 1999, a new form of regional governance, Regional Development Agencies (RDAs), commenced operations in the English regions, based on the GORs, with the task of promoting economic development (Lynch 1999). The GORs and RDAs work closely, particularly in the formulation and execution of central government's Regional Economic Strategies (Benneworth 2000). The RDAs represent the first strand in the government's English devolution proposals (see Chapter 12). Central government's policy is to allow elected regional government where there is a demand for it (Dungey 1997). Thus some regions may have devolved elected governments before others, although there is likely to be a domino or bandwagon effect once one or two take the plunge. Both John Prescott, the Secretary of State for the DETR, and the Prime Minister, Tony Blair, are enthusiastic supporters of regional devolution (White 2000). According to Tony Blair, 'when you go to the North-west or the North-east, or you go down to, say, Cornwall, people are talking about greater democratic accountability' (quoted in Hetherington 1999a: 4).

Just as the GLC was an early example of regionalism in England so the creation of the Greater London Authority (GLA) with its own assembly and elected mayor is the first example of modern English regionalism. It is envisaged that other cities will get their own mayors and greater local democratic control (Bagehot 1999; BBC1 2000c).

One major difference between regionalism and nationalism has been highlighted by Brett (1976). Scotland and Wales are nations, whereas regions, such as the North East, are not. The boundaries of the nations are clear, those between the regions less so (see Chapter 14). There is in England a lack of the strong sense of regional identity found in the other nations of the UK and much of Europe (Sharpe 1997). Thus, although regional identity may be rather nebulous, 'geographers have insisted that it be employed as a determinant of regional boundaries' (Smith 1965: 10). It is this identity issue that is at the heart of English regionalism (Harvie 1991). Regional identity is weak, fuzzy or non-existent in the English regions and regional government has tended to become an issue as a response to the allocation of regional aid (Keating 1982) or national devolution to Scotland and Wales during the 1970s and 1990s, 'rather than as a popular expression of support for autonomous regional government' (Lynch 1999: 73). The whole process of regionalisation is very gradual. As David (1999: 6) reminds us, 'it has taken France and Spain between ten and twenty years to introduce regional assemblies, and in neither country has the system yet settled down'.

We now look at regionalism in England through the examples of London, the North East and Yorkshire and Humber. These are some of the areas most advanced in terms of regional government, either actual or proposed. Other parts of the country also have a strong identity. The strongest claim is that of Cornwall, where Mebyon Kernow (Sons of Cornwall), formed in 1951, has been seeking self-government. The party has been contesting elections since the 1960s (Anthony 1999).

13.3.1 London

As we saw above, the GLC was an early example of practical regionalism in England. It replaced the London County Council, which was set up in 1889 (Cole 1947). However, Mrs Thatcher had a deep-rooted dislike of local government and the GLC in particular – especially since this body was not only socialist in structure but also radical in nature. Legislation was put into place and the GLC was abolished in 1986. Mrs Thatcher also abolished the regional economic planning councils and the six metropolitan councils (Sharpe 1997). Thus, between 1986 and July 2000, when the GLA assumed power and Ken Livingstone (the last leader of the GLC) was sworn in as mayor, there was no effective, democratically elected body to oversee this major global city. Each borough had its own elected body, as they still do, but overall planning rested with a government minister, numerous quangos and the London Regional Board based in the DoE (Hogwood 1996). Further, the newly elected mayor should not be confused with the Lord Mayor of London – a purely ceremonial role.

London clearly dominates the UK (see Chapter 14), but this only happened after the First World War (Harvie 1982). Now the Metropolis so over-shadows the social, economic and political life of the UK that the imbalance is detrimental to the country:

> The inflation of London's economic, political, and cultural role has not just sapped the rest of Britain, it is blighting life in London itself. Its once-famous institutions – transport, police, housing – are in disrepair; the expenses of London life are becoming crippling. It is probably West Europe's sleaziest, most over-priced tourist trap.
>
> (Harvie 1982: 20)

Though written in the early 1980s, these words are arguably truer now. This is one reason why London was chosen as England's first experiment in regional devolution – the Greater London Authority (GLA). This body consists of the new mayor and his office, as well as the Greater London Assembly.

Following a referendum in 1998, in which 72 per cent voted 'yes' but only on a 33.6 per cent turnout (Lynch 1999), the first direct election for a mayor for London was held, along with the GLA, in May 2000. The Assembly is designed to act as a check on the mayor by scrutinising his activities. The mayor is required to consult the Assembly regarding strategies and the Assembly can reject the mayor's proposed budget with a two-thirds vote (see Chapter 12). When elected mayor, in May 2000, Ken Livingstone quipped, 'As I was saying before I was so rudely interrupted 14 years ago'; a reference to his leadership of the abolished GLC (*The Guardian* 2000b: 3).

13.3.2 North East

As we saw in Chapters 11 and 14, there is a strong regional identity in the North East (Townsend and Taylor 1974, 1975) and the question of regional government has been raised for many years in the area (Colls 1992). Thus, of all the English regions, it has one of the most advanced plans for regional devolution. Although initially antagonistic towards a Scottish Parliament because of the economic threat from a potentially stronger neighbour, 'now many Geordies regard the shift of power to Scotland as something of an inspiration' (*The Economist* 1999d: 28). The hope is that an elected regional assembly can replace the existing unelected one, plus the various other regional bodies and quangos, thereby making the governance of the area more accountable democratically. The moves towards an elected assembly are well advanced in the North East. The Campaign for a Northern Assembly has been active since 1992. The North East Regional Assembly, which is made up of representatives of the local authorities, MPs and MEPs, as well as regional stakeholders, is a non-elected body that functions as the Regional Chamber to oversee the RDA. In

October 1998, the North East Constitutional Convention (NECC) was launched based on the Scottish Constitutional Convention. The goal of NECC is to reach widespread agreement amongst the stakeholders and people of the region on proposals for directly elected regional government for the North East region. This should be done in a manner that persuades central government to hold a referendum that is likely to be successful. The NECC has already brought together disparate interests to draw up proposals for democratically elected regional government (Campaign for the English Regions 2000).

The North East is particularly vexed by neighbouring Scotland not only having devolved powers and a relatively larger apportionment of Treasury funds, but also having its own well-entrenched machinery for dealing with the EU and inward investors. Some observers point to the strong regional structure of Spain as a model for England generally (Roberts 2000) and the North East in particular (Tomaney 2000). A poll, in December 1999, showed 56 per cent support for regional government, with nine out of ten respondents believing that the government does not pay enough attention to the problems of the North East (cited in Tomaney 2000). Thus it is likely that the North East will have the first of England's provincial regional assemblies. However, as Taylor (1991) points out, while there may be support for devolution in the northern areas there is little support for separation from the rest of England.

13.3.3 Yorkshire and Humber

As we saw in Chapter 11, there is a relatively strong sense of national identity in Yorkshire. So, like the North East, Yorkshire and Humber is at the forefront of English regionalism. The region's RDA has styled itself as Yorkshire Forward since 1999. An unelected Regional Assembly has been in operation since 1996 and has played an active part in shaping policies affecting local authorities and the region, both nationally and in Europe. It has developed strong working relationships with key partners in the region and has offices in Wakefield and Brussels (Yorkshire Forward 2000). A campaign for an elected Regional Assembly has now been launched in Yorkshire and Humber. The Campaign for Yorkshire launched its Claim of Right in March 1999, with £75,000 from the Rowntree Reform Trust and the Archbishop of York as it president (Wainwright 1999). It is holding a number of public meetings across the region in the run up to establishing a Constitutional Convention in October 2000 (Campaign for the English Regions 2000). There is seen to be a need to move towards a democratically elected Regional Assembly since, according to the Campaign chairman, 'Yorkshire has as clear a sense of identity as Scotland or Wales' (quoted in Wainwright 1999: 11).

13.4 OTHER CONSTITUTIONAL REFORM

A number of other aspects of the British Constitution have been under scrutiny for a long time and have wider implications for the human geography of the UK.

13.4.1 The Monarchy

Throughout the history of the UK and its constituent countries one underlying theme runs deep – the position of the Monarch in the Constitution. Most of the internal conflicts in these countries have surrounded the claims of various dynasties, each claiming the 'divine right' to rule over subjects. Much of the arguments revolved around blood ties and/or religion. Occasionally, there would be resistance to the whole idea of Monarchy at all – for example the Republic of Oliver Cromwell in the mid-seventeenth century, and various other republican revolts, the most successful of these being that of the Irish Republic as it came to be known. The withdrawal of Queen Victoria from public life after the death of Prince Albert in 1861 led to an increase in republican sympathies – more than fifty republican clubs operated (Nairn 1988). Throughout this century there has been recurring interest in reform of the Monarchy as the apex of the British Constitution. The behaviour of Victoria's son Edward VII, when Prince of Wales, attracted even more criticism than that of the current Prince of Wales. The abdication crisis of 1936 forced Edward VIII to abdicate and relinquish his reign to his younger brother, George VI, husband of the Queen Mother. This dealt another severe blow to the prestige and standing of the institution.

Now the popularity of the Monarchy is at an all time low (Channel 4 2000). Indeed, there is growing discontent with the Monarchy, especially among Her Majesty's younger subjects, and all it stands for, in terms of privilege, deference and lack of democracy. Yet three out of four Britons like the idea of the Crown as Head of State (Reeves and Zinn 1999). Despite adverse publicity surrounding the lives of most of the Royal Family a curious amount of affection is still retained for certain members of the Windsor family among large sections of society, particularly in England.

There have been recent calls for radical reform, short of a republic. This has led to some institutional change. For example, there has been an increasing willingness on the part of members of the Royal Family to be less dependent on the taxpayer and indeed themselves pay taxes. Recently, the Civil List – the money paid to the Queen's household by the government – has been frozen. Prince Charles has let it be known that he would willingly support a referendum on the future of the Monarchy even though such a democratic notion is unthinkable to party leaders (Reeves and Zinn 1999).

There are spatial implications in this, insofar as support for the Monarchy, as with support for the Union, tends to be highest in the Home Counties in what Taylor (1991: 150) calls 'Upper England' and Nairn (1988: 194) calls the 'heartland area'. Support for and identification with the Monarchy tends to weaken outwards from this core area. The Queen's image does not appear on Scottish bank notes, for example (Grant 2000). It is no coincidence that the most vociferous opponents of Monarchy are from Scotland – Willie Hamilton (1975) and Tom Nairn (1988). Republican sympathies are stronger in the former coalfields of Wales or Hamilton's native Fife than in the drawing rooms of Berkshire or the clubs

and pubs of Surrey. In Wales, republicanism reared its head at the Investiture for Charles – the 'English Prince' – in 1969 (Nairn 1988). Though it is arguable that Her Majesty's most loyal subjects are to be found in Northern Ireland among the Unionist community, at the same time the most fervent republicans are among the Nationalist community – after all, it is the Irish *Republican* Army. The whole issue of the Crown and its place in UK society all links with politics, culture and identity.

13.4.2 The House of Lords

Another anachronistic element of the British Constitution is the House of Lords. It too is symbolic of deference and negation of democratic principles, and is a 'primeval . . . relic of a caste system' (Brook 1997: 6). Most democracies have a **bicameral** system. In the UK, the second chamber is the House of Lords made up entirely of unelected members. These are either hereditary, relying on dubious blood ties and historical legacy, are government appointees, or sit in the Lords by virtue of being senior clerics or judges. The most serious challenge to the Lords came in 1911 when the Prime Minister, Welshman Lloyd George, threatened to abolish it. However, the Lords survived, though its powers were severely curtailed. In 1958, an Act allowed for the creation of life peers. The House of Lords Act of 1999 removed the right of most hereditary peers to sit and vote in the House, but an amendment allowed ninety-two hereditary peers to remain until the House was fully reformed. In January 2000, a Royal Commission report was published and offered three options for composition involving varying numbers of appointed and elected members. There have also been suggestions that the elected mayors of cities be appointed as members of a reformed upper chamber. Whatever the final outcome, any reform should take account of the geographical source of members of the second chamber. As with the Monarchy, there has been a traditional bias towards the 'heartland area' in the House of Lords. There has been over-representation from the rural areas and the Home Counties – thus the fiasco over the government's hunting with dogs legislation.

13.4.3 Voting reform

The UK has had a long love affair with the first-past-the-post (FPTP) voting system. It is argued that it makes for clear strong government since coalitions, common in many European countries, are rare in the UK – the National governments in the 1930s and the Coalition governments during both world wars being notable exceptions. But coalition governments are largely successful, as the relative prosperity of many European countries demonstrates. The FPTP system is also said to be straightforward and easy for voters to understand. But that insults the intelligence of the electorate, increasingly uninterested in the current state of the political scene. For example, fewer than one in fifty people belongs to any political party – lower than membership of some charities (Phillips 2000). This is further reflected in the fact that turnout rates have been poor in recent elections – general, local and European. There are various reasons for this – for example, ideological decline, weakness of class allegiance, cynicism about politics and politicians, little difference between parties and decline in media coverage of politics and important issues (Cox 2000). Turnout varies geographically. In the 1997 general election it was highest in the South West (75.1 per cent) and lowest in Greater London (67.8 per cent). In the 1999 European elections, the turnout was less than 25 per cent. In recent council by-elections in Merseyside and South Yorkshire it has been as low as 6 per cent (Ahmed 2000a). Turnout in Northern Ireland is always higher than in the rest of the UK, for national, local and European elections (Esler 1999). Turnout also varies by age and class, with younger people and the socially excluded tending not to vote or even register to vote. The young tend to be cynical and dismissive of conventional party politics, though many develop single-issue attachments. But voter turnout tends to increase when electors feel that their candidates have a chance of success and some systems allow for this (Bell 1999; Younge 2000b).

It has been argued that the FPTP system, favoured in the UK, helps fuel apathy with voters feeling that their votes do not always count. In the 1951

general election, for example, the Conservatives won 321 seats with 48 per cent of the vote, while Labour won just 295 seats with 48.8 per cent of the vote. Furthermore, it took just 42,731 votes to win a Conservative seat but 121,759 to win a Liberal seat (Sked and Cook 1990). In 1983, the Liberal–Social Democrat Alliance attracted 25.4 per cent of the vote and won only twenty-three seats, while Labour got 209 seats with just 27.6 per cent of the vote (see Table 13.2 for these and other injustices). A more recent example of this was in the 1997 general election in Scotland, when Labour got fifty-six seats with 45.6 per cent of the vote. Yet, with only 13 per cent of the votes, the Liberal Democrats won ten seats. In contrast, the SNP won only six seats with 22.1 per cent of the vote, while the Tories won no seats even though they got 17.5 per cent of the vote (Brown 1997). This unfairness is evident in local elections also. In the May 1999 local elections in Glasgow, for example, Labour won 54 per cent of the vote, but took seventy-four of the seventy-nine seats, while the SNP, with 32 per cent of the vote, won only two seats (*The Economist* 1999a).

Table 13.1 demonstrates the 'disproportionality' of the FPTP system. Labour won 73 per cent of the 73 constituency seats with only 39 per cent of the constituency votes. The additional member system corrects for this by allocating additional seats from the regional vote, so that Labour's final allocation of seats works out at a more equitable 43 per cent (Scottish Parliament Information Centre 1999). A lot might be said in favour of the FPTP system, but fair it patently is not.

There are a variety of systems given the generic title of **proportional representation** (PR). Most of these are not true proportional systems but a variation on the majority system; for example, second ballot, alternative vote. The party list system is the one most favoured in the UK because it gives the political parties a great deal of control. Here, a political party draws up a list of candidates that cannot exceed the maximum number of total seats for the area. Seats are won according to the number of votes cast. If party A achieves enough votes for three seats then the first three candidates on the list get the seats. The arguments against this are that the

Table 13.2 Election results, seats won and (percentage share of vote) UK, 1945–97

Year	Conservative	Labour	Liberal[a]	PC[b]	SNP[c]	Others[d]	Prime Minister[e]
1945	213 (39.8)	**393** (47.8)	12 (9.0)	—	—	22 (2.8)	C. Attlee
1950	298 (43.5)	**315** (46.1)	9 (9.1)	—	—	3 (1.3)	C. Attlee
1951	**321** (48.0)	295 (48.8)	6 (2.5)	—	—	3 (0.7)	W. Churchill
1955	**344** (49.7)	277 (46.4)	6 (2.7)	—	—	3 (1.2)	A. Eden
1959	**365** (49.4)	258 (43.8)	6 (5.9)	—	—	1 (0.9)	H. Macmillan
1964	304 (43.4)	**317** (44.1)	9 (11.2)	—	—	0 (1.3)	H. Wilson
1966	253 (41.9)	**363** (48.0)	12 (8.5)	0 (0.2)	0 (0.5)	2 (0.9)	H. Wilson
1970	**330** (46.4)	287 (43.0)	6 (7.5)	0 (0.6)	1 (1.1)	6 (1.5)	E. Heath
1974 Feb.	297 (37.9)	**301** (37.1)	14 (19.3)	2 (0.6)	7 (2.0)	14 (3.1)	H. Wilson
1974 Oct.	277 (35.9)	**319** (39.2)	13 (18.3)	3 (0.6)	11 (2.9)	12 (3.1)	H. Wilson
1979	**339** (43.9)	268 (36.9)	11 (13.8)	2 (0.4)	2 (1.6)	13 (2.8)	M. Thatcher
1983	**397** (42.4)	209 (27.6)	23 (25.4)	2 (0.4)	2 (1.1)	17 (3.1)	M. Thatcher
1987	**375** (42.2)	229 (30.8)	22 (22.6)	3 (0.3)	3 (1.4)	18 (2.7)	M. Thatcher
1992	**336** (41.9)	271 (34.4)	20 (17.8)	4 (0.5)	3 (1.8)	17 (3.5)	J. Major
1997	165 (30.7)	**419** (43.3)	46 (16.8)	4 (0.5)	6 (1.8)	19 (3.5)	T. Blair

Source: Sked and Cook 1990; King *et al.* 1993; Norris and Evans 1999b

Notes: [a] Liberal Party 1945–79; Liberal/Social Democrat Alliance 1983–7; Liberal Democrat Party 1992–7. [b] Plaid Cymru. [c] Scottish National Party. [d] Independents and Northern Ireland MPs. [e] Prime Minster at time of election. Changes took place during the lifetime of some governments; for example, Callaghan replaced Wilson in 1976 and Major replaced Thatcher in 1990.

geographical link between elector and elected is weakened, electors are voting for parties rather than individuals and, although independent candidates can stand, it is difficult for them to do so. The single transferable vote is widely used in the UK outside politics. Those students who have bothered to vote for student union officers will be familiar with the system. The single transferable vote is the system least favoured by party politicians since it is less concerned with 'the fortunes of parties than with giving greater freedom to the individual voter' (Lakeman 1970). This system is based on multi-member constituencies and the voter ranks candidates in order of preference. The votes of the candidates who do not meet a prescribed quota are reallocated to the most popular until clear winners emerge.

The arguments frequently put forward against PR in the UK include:

- the link between constituency and representative is lost;
- political instability is likely;
- it results in permanent coalitions driven by small parties;
- it is confusing to the elector;
- party lists are undemocratic.

PR has long been used in the UK in the Northern Ireland European Parliament elections. In June 1999, European elections in Great Britain used a system based on party lists for the first time. The geographical base for these are the GORs plus Scotland and Wales, so that there has been a drastic reduction in the number of constituencies but the same number of MEPs. The London mayor was elected using a second preference system. Voters voted for a first and second choice. In the event, Ken Livingstone did not win 50 per cent of the votes outright so the votes of the lowest polling candidates were redistributed between Livingstone and Steve Norris the Conservative. After the second preference votes were counted Livingstone was declared

winner, having won twelve of the fourteen constituencies. Londoners also got two votes to elect the twenty-five seat Assembly. One candidate was chosen for each of fourteen London constituencies. The remaining eleven were elected from a London-wide list of independent candidates and party acolytes. These seats were distributed in proportion to the votes cast. The elections for the Scottish Parliament used a similar mix of FPTP and party lists based on the Euro-constituencies to top this up. The regional vote allocated seven additional members after seven rounds of calculation. It is clear from Table 13.1 that the regional list system allowed a Tory presence in the Parliament. In Wales, the mixed system was repeated. Forty members to the Welsh Assembly were elected by FPTP in the same constituencies used for the UK Parliament. The remaining twenty additional members were elected on the basis of lists drawn up for the five 1994 Euro-constituencies. Each of these elected four AMs. The 108 members of the Northern Ireland Assembly were elected, in June 1998, by the single transferable vote system from the existing eighteen Westminster constituencies.

In terms of Westminster politics there is one camp, led by Foreign Secretary Robin Cook, which favours PR, and another, led by the Home Secretary Jack Straw, for which PR is anathema (*The Times* 2000). The Royal Commission on Voting Systems, led by Roy Jenkins, has recommended a number of alternatives to FPTP. The favoured system is Alternative Vote (AV) top-up with traditional constituency votes, based on ranked preferences, and second vote from city-wide or county lists. This retains the present constituencies but allows voters to express second and third preferences on their ballot paper. No individual could be elected until they had secured the backing of at least half of the electorate. The top-up should reflect the standing of weaker candidates and parties (Wintour 1998). This is simply a refined version of the traditional FPTP system and designed to cause minimum damage to the main parties.

13.5 NATIONAL GOVERNMENT AND POLITICS: 'NEW' LABOUR, NEW POLITICAL GEOGRAPHY?

After the Second World War it became clear that the Labour Party was now a major political force nationally. For most of the second half of the twentieth century, general elections tended to be two-horse races between Labour and Tory. The fortunes of both parties waxed and waned over the years. Of the fifteen general elections since 1945 seven have been won by Labour and eight by the Conservatives (Table 13.2). To a considerable extent, the Conservatives and Conservative ideals have dominated post-war UK politics. Indeed, the Tories were very much a 'one nation' party, throughout most of the period, but after the 'Thatcher experiment' came to be seen as a purely Middle England, or even Little England, party. In 1955, for example, the Conservatives gained over half the vote in Scotland. The irony is that the Conservative and Unionist Party, the party of the Union, had become so unpopular that it lost all parliamentary seats outside England in 1997. Elements within the party, and many supporters, experiencing the 'English backlash' (see Chapter 11) have sought to reinvent the Tories as the party of English nationalism.

For historical reasons, the Labour Party has always done well in both Scotland and Wales. There have been a disproportionate number of Labour Prime Ministers from these countries and witness the large number of Cabinet members in the Blair administration from Scotland, including the Prime Minister, Chancellor of the Exchequer and Foreign Secretary. There has also traditionally been a greater interest in politics in these countries (Marr 2000), compared with England where 'there is widespread lack of interest' (Pattie 2000). As Heffer (1999: 11) reminds us, 'The English (unlike their Scottish cousins) are a simple and politically unsophisticated people.' He goes on, 'It had long been respectable for the middle classes to vote Labour in Scotland, something it did not become respectable for the middle classes in England to do until Mr Blair was invented' (p. 28). 'People', Lindsay Paterson (2000)

points out, 'who would have been Conservative in England between the mid-1970s and the mid-1990s vote Labour in Scotland.' Indeed, the Scots 'have always been more sympathetic to socialist ideals than the English' (Hitchens 1999: 313). It was the long spell of Thatcherist governments, under the eponymous grocer's daughter and John Major, elected by English constituencies on policies anathema to most Scots and Welsh (as well as many English from outside 'Middle England'), at a time when Conservative MPs from Scotland and Wales were an anomaly, that convinced many in these countries that the current centrist system had to be reformed. By ignoring the deep-rooted political and social differences between the Union's two main partners (Brown *et al.* 1998; Surridge *et al.* 1999) Mrs Thatcher unleashed forces that would help lead to her undoing, as well as undermining the Union that she so vociferously supported.

After the long Thatcher–Major years (1979–97), the electorate, especially in Middle England wanted a change – though critics might argue that the Blair administration has been 'more of the same' (Cohen 1999). Thus in 1997, the Labour Party, after eighteen years in the wilderness, was voted in with a large majority, in an election that Pattie *et al.* (1997: 253) claim, 'had geography written all over it'. Labour returned 419 MPs (including the politically neutral Speaker). This was the highest number of Labour MPs ever, and the majority of 179 was the largest for any administration since the National coalition government of 1935, and the largest in Labour Party history (Table 13.2), though, with just 43.3 per cent of the UK vote, the victory was in seats, not widespread support (Norris 1997). The Conservatives were the biggest losers, returning only 165 MPs – the lowest since 1906. The Tories did particularly badly in Scotland where 'the scale of the electoral meltdown that overtook them was quite unprecedented' (Election UK 1997). They returned no MPs and their share of the vote declined to just 17.5 per cent, behind the SNP at 22.1 per cent. Three Conservative Cabinet Ministers were unseated – two by Labour and one by the SNP. In Wales, the Tories were trounced, squeezed between Labour in the south and north-east, Plaid Cymru to the north-west and the Liberal Democrats to the east. In every English region, there was an increase in the share of the vote for the Labour Party and the Liberal Democrats, while there was a decrease in that for the Tories. The biggest shift to Labour, as well as the biggest shift away from the Tories, was in London. The lowest shift to Labour was in Wales – already a Labour heartland. The lowest shifts away from the Conservatives were in Wales and Scotland where support for the party was already down to the hard-core solid Conservative anyway. That is, there was not the great mass of uncommitted voters that there was in Middle England. The Tories are now seen as a purely English party. But even in England they have tended to become marginalised to mostly the suburban and rural south (Pattie *et al.* 1997). For example, the whole of Lincolnshire, with the exception of Lincoln, remained Conservative, as did Dorset and West Sussex. The Conservatives have been largely rejected by the Middle England that sustained the Thatcher–Major administrations. London is now almost entirely Labour with a few Liberal Democrat seats, and many urban areas in the South East have switched to Labour from Conservative. This rural–urban, Conservative–Labour divide was repeated across England. Of the defeat of the Conservative Party:

> It is not merely that they have been wiped out of Scotland, Wales and many entire counties of England, but that in the South East, their natural stamping ground, they have become almost purely a party of the rural shires, with a few recalcitrant suburbs to their name. They will have a long twilight struggle ahead to be regarded as the national party of England, let alone the party of the Union.
>
> (Election UK 1997)

That said, the Labour Party has suffered due to Conservative gains in subsequent local elections, European elections and a by-election in Ayr (Hencke *et al.* 2000).

The Liberal Democrats have sound bases in parts of Wales and Scotland and the South West of England, where they got 31.3 per cent of the vote

in 1997 (McAllister 1997). Their fortunes vary geographically, with no MPs from any East Midlands constituency in either the 1992 or 1997 elections. However, with forty-six MPs this was their best result since 1929, despite a drop of 1 per cent in their vote (Norris 1997). In Northern Ireland, the Great Britain parties are excluded by the politics of sectarianism. Sinn Fein – the political wing of the IRA – is strong in Nationalist areas, West Belfast and the border areas. The various Unionist parties fare best in the Loyalist areas of the east. The largely Roman Catholic, but non-sectarian Social Democratic and Labour Party (SDLP) is strongest where ethnic tensions are least evident.

There have been significant shifts in the way voters choose candidates in recent years (Denver 1998; Evans *et al.* 1999). There has been a marked decline in the traditional class alignment in voting behaviour and a weakening of party attachment (Tunstall *et al.* 2000; Webb and Farrell 1999). This might have made the electorate more volatile but new voting patterns have emerged in which geography is becoming more important (Norris and Evans 1999a; Pattie *et al.* 1997). For example, there has been an increase in the importance of regional voting (Johnston and Pattie 1998; McAllister 1997). There is even evidence of more local patterns of voting behaviour emerging (Pattie and Johnston 2000). This clearly helped the Labour Party in 1997. There has also been evidence of some successful tactical voting, where voters temporarily switch their allegiance to another party simply to oust the least desirable party. This type of voting often favours the Liberal Democrats (Pattie and Johnston 2000) and does most damage to the Tories (Curtice and Park 1999). The Conservatives have also been hit by a weakening of female conservatism in recent years, particularly among younger women (Norris 1999). However, there is evidence that recent dissatisfaction with the Labour government is strongest among women who are most directly affected by the failure to deliver on health and education promises (Ahmed 2000b). The size of non-white ethnic minority voting populations concentrated in specific areas has not been lost on the political parties who have recently courted the 'ethnic vote'.

In electoral terms, the Labour Party has done best in gaining most of the 'ethnic vote' (Saggar and Heath 1999), though Asians are more sympathetic towards the Conservatives than are Black groups (Saggar 1997). All this made the 1997 election a 'critical election' and 'represents a critical realignment in the established pattern of British party politics' (Norris and Evans 1999b: xx).

13.6 LOCAL GOVERNMENT AND POLITICS

Traditionally there have been two levels of local government in the UK. This has usually been at county and more local level. County councils, or their equivalents, are responsible for certain levels of service provision – education, roads and the like. The more local bodies, such as town or borough councils, have been responsible for more local service delivery. To confuse the situation, local government takes different forms in each of the constituent countries, though Wales and England have been the closest in terms of structures. By and large, the system in England and Wales has remained much the same since the Local Government Acts of 1888, 1894 and 1899 (Painter 2000). There have been a number of local government reorganisations in the past hundred years or so, some of which have been far-reaching, particularly those of 1974/5. These major reforms saw the retention of a two-tier system, including the subsequently abolished metropolitan counties. In Scotland, the upper tier comprised regional councils and island authorities, while the lower tier comprised district councils. The Conservative administration at the time rejected the advice of the Maud Commission that single-tier local units, based on urban 'cores' and surrounding functional regions, was the best way to reform local government. The national government, as Mohan (1999: 169) notes, was criticised for letting party politics cloud its judgement and over-enthusiastic in retaining the historic counties, rather than creating 'appropriate regional units for the late twentieth century'.

Local democracy was dealt a blow by the long spell of Conservative power from 1979 to 1997, when an ever-increasing number of quangos took over many of the functions of local government (Marr 1996). Mrs Thatcher was particularly antipathetical to local government and did everything she could to emasculate local authorities in her attempts to draw as much power as possible to the centre. She also wanted to reform local government financing arrangements. Central government took greater control of local government spending through rate-capping, which prevented local authorities from raising rates above a certain level (Mohan 1999). The old system of local rates based on property values was seen to be unfair since it ignored household income. An elderly widow in a detached, five-bedroomed house might pay several times more rates than a family of three or four wage earners in a local authority terrace. The fact that the former case would much more likely be a Conservative voter than the latter case was not lost on Mrs Thatcher, or indeed the voters. So the community charge – a flat-rate charge for each adult – was introduced in Scotland in 1988, and a year later in England. It was met with widespread civil disobedience and non-payment in Scotland – most of which went unnoticed in the London media. Only when its effects were fully understood in England – leading to full-scale rioting in London – did the government look for an alternative in the form of the council tax, a rates type system based on property values.

The logic behind the most recent reforms was in 'relating the structure of local government more closely to communities with which people identify' (Department of Environment, quoted in Werther 1997: 24). The 'artificial' counties of Avon, Cleveland and Humberside were seen as exemplars of this mismatch of local identity and local government (Mohan 1999). The argument was that a two-tier system blurs political identity, as well as causing waste due to duplication, and that a unitary structure would foster a sense of community identity and be more effective in dealing with local issues (Paddison 1997). The general consensus was to move towards unitary authorities (UAs) – stripping out the upper-tier altogether. This is what happened

Table 13.3 Local government change, counties, districts and unitary authorities, England, 1994–8

Year	Non-metropolitan areas			Metropolitan areas	
	Counties	Districts	Unitary authorities	London boroughs	Metropolitan boroughs
1994	39	296	0	33	36
1995	38	294	1	33	36
1996	35	274	14	33	36
1997	36	260	27	33	36
1998	34	238	46	33	36

Source: ONS 1999a, *Regional Trends 34*, p. 216

in Scotland and Wales, while in England new UAs were created but the two-tier structure was retained in many cases.

The Local Government (Scotland) Act, passed in 1994, abolished the ten regional councils and fifty-six district councils and created twenty-nine unitary councils. The existing unitary island councils were retained so that the new structure consists of thirty-two unitary local authorities. The new councils were elected in April 1995 and became active a year later (Paddison 1997). In April 1996, the eight counties and thirty-seven districts of Wales were replaced by twenty-two UAs. The procedure has been more convoluted in England and, as Mohan (1999: 175) points out, 'The contrast with Scotland and Wales . . . could not be clearer.' The reorganisation took place in four phases from April 1995 to April 1998 and the process is summarised in Table 13.3. The new local authority pattern can be seen in the maps in Chapter 7. UAs have replaced the two-tier system of County Councils and Local Authority District Councils in parts of some shire counties and, in some cases, across the whole county. The two-tier system has been retained in some, mostly rural, areas (Chisholm 1995). There are thirty-five County and 241 Local Authority District Councils (Painter 2000). Legally all UAs in England are counties. Many have been created without any geographical change. Some have had boundary changes at district and ward level, especially Herefordshire UA and Peterborough UA (for a fuller account see ONS 1999a). The five former metropolitan counties consist of

thirty-six metropolitan boroughs such as Newcastle, Manchester, Liverpool, Sheffield and Leeds. It would seem that local government reform is a work in progress in England, since the three main political parties are committed to the concept of unitary authorities. Moreover, if full-scale, demo-cratically elected regional assemblies go ahead a single tier of government below this would make much more sense than the county–district two-tier set up (Chisholm 1995) with all that entails for waste, duplication, red tape and civic confusion. Northern Ireland was unaffected by this latest round of local government reform. The main unit is the district council area of which there are twenty-six. In light of devolution for Scotland and Wales, the abolition of the upper tier of local government makes sense. Restructuring of local government, relying as it does on the redrawing of boundaries, can lead to some major shifts in the fortunes of political parties at local elections (Graham 1995).

Party politics at local level leads to even more apathy than at national level. For example, in the May 2000 round of local elections in one district only 14 per cent of voters turned out (Brown 2000). All of the national parties are represented at local level – save in Northern Ireland. However, at local level the party machine is weaker and there is more chance for candidates from smaller parties and independent candidates to get elected. Thus local government tends to be characterised by more councils dominated by coalitions of minority parties, or councils where no party has overall control

(Graham 1996). Nevertheless, most councils tend to be dominated by one of the main parties. And, as in the national political scene, the Conservatives have fared worst in local elections in the post-Thatcher era. In 1979, when Mrs Thatcher first won power, the Tories had 12,143 councillors, and control of such major cities as Greater London, Birmingham and Edinburgh, as well as metropolitan counties like Greater Manchester, West Midlands and Mersey-side. After the May 1996 local elections there were only 4,400 Tory councillors, compared with 11,000 Labour and 5,100 Liberal Democrat. Labour controlled 212 local authorities in Great Britain and the Liberal Democrats held fifty-five, the Conservatives were left with just thirteen (Norris 1997). By the May 2000 local elections there was a considerable predicted victory for the Tories who had gained control of councils at the expense of the Labour Party (Hencke *et al.* 2000). In fact the Labour Party lost 546 council seats and the Conservatives won 542, representing a swing back to a more natural balance in each party's natural heartland (Brown 2000), though the Tories still did not control any of the large metropolitan districts or unitary councils in England (*The Economist* 1999e).

Like by-elections, local elections are often seen as an opportunity to register discontent with the party of the prevailing government and both major parties have become victims of this type of protest vote (Graham 1995, 1996). In Northern Ireland, as with national politics, the whole process is dominated by religion and local parties tend to align along the religious/ethnic cleavage rather than class or other allegiances seen in the other countries (Graham 1994a).

13.7 EUROPE

As we saw in Chapter 11, the UK's position in the EU has become the focus of a strong and often acrimonious debate. In general, most opposition stems from English 'nationalists', with most people in the smaller nations supportive of the EU (Bell 1999; Brown *et al.* 1998). Opponents of greater ties with the EU are fearful of a loss of sovereignty and

the dominance of the EU hegemon, Germany. The irony of this is not lost on the Scots, Welsh and many Northern Irish, since this has already been their fate when these nations joined – not altogether of their own free will – that earlier and very successful common market, the United Kingdom. Sovereignty was sacrificed, and political and cultural dominance by a much larger state-nation followed – many writers (e.g. Goodwin 1999; Ignatieff 1994) still erroneously refer to the UK as a nation-state, something it has never been. Rather, the countries of the UK are what McCrone (1992) calls 'stateless nations'. There were none of the mechanisms in place to protect the rights of minorities that exist under the various EU treaties when the various Acts of Union were drawn up.

No British government has been fully committed to the kind of political integration that is the goal of the key players in the EU – particularly Germany and France. Successive governments have sought to renegotiate the original deal. The Labour administration of Harold Wilson sought a new package from Brussels, mainly the British budgetary contributions, access to EEC markets for Commonwealth products and reform of the Common Agricultural Policy – always a bone of contention, since it was geared towards the relatively underdeveloped agriculture of the continental countries. Wilson sought to legitimise the renegotiated package with a referendum in 1975 – the first in British Constitutional history. What followed was 'the unusual spectacle of a Prime Minister recommending voters to support continued membership, while it was official Party policy to oppose it' (Fay and Meehan 2000: 214). The result was 67.2 per cent voting 'yes' to remain in the EEC. Ironically, given the present situation, support was strongest in England and Wales and weakest in Scotland and Northern Ireland. That position has changed, with most Eurosceptics in England and fewer in the other countries (Bell 1999). Although the British public voted in favour of staying in the EEC, withdrawal remained official Labour policy until the election of Neil Kinnock in 1983. Indeed, it is interesting that the once Europhile Conservatives, the party that took the UK into the Common Market, is now the most Eurosceptic, while the

Euroscepticism of the Labour Party has, officially at least, switched to a Europhile stance (Evans 1999).

Even with a renegotiated settlement the UK continued to pay a disproportionate share of the EEC budget and this became one of the main platforms of Mrs Thatcher's policy towards Europe until 1984. Thatcher was also bitterly opposed to further integration, especially further Economic and Monetary Union (EMU) and more Brussels-directed social legislation. Her successor, John Major, managed to negotiate an opt out of the Social Chapter of the Maastricht Treaty. But, by the 1997 Treaty of Amsterdam, the newly elected Labour administration opted in to the Social Chapter. The failure of the Exchange Rate Mechanism (ERM), which the UK entered in 1990, further split the Conservative Party and heightened the debate on the UK's role in Europe. The main issue is EMU – abandoning sterling for the euro – and the inevitable closer political integration of the partner states. The Eurosceptics are not entirely convinced by the rhetoric of both major parties that UK sovereignty is non-negotiable and there will always be the option of a British veto on anything that imperils that sovereignty.

As we have seen in Chapter 11, much of that notion of sovereignty is subsumed within the backward-looking vision of Englishness. This is perhaps best seen in the 'Keep the Pound' campaign (Figure 13.3). Here imperial or martial images are evoked as Britain is 'under attack' from the Continent in the form of the euro. Yet much of UK law is now subject to overview by the European Court of Justice, and the EU ban on UK beef during the BSE fiasco signalled how ineffectual the London government was in the face of German and French intransigence. Most of the Eurosceptic support, as we have seen, centres on England. One political expression of this is the UK Independence Party (UKIP), which has little if any support in the other nations. Adherents to the UKIP and the Eurosceptic line are particularly fearful of 'subsidiarity' (the doctrine that decisions should be taken at the lowest possible tier of government), or the surrender of sovereignty. The Scots and the Welsh, and to a lesser extent the Northern Irish, are less anxious about this because of historical circumstance and the knowledge that peripheral and smaller units have tended to do well economically within Europe. In the 1997 general election, the Eurosceptic Referendum Party got 3 per cent of the vote, but only after spending £20 million, or £24.68 per vote (Norris 1997). In recent years, there have been serious calls for the UK to withdraw from the EU and join the

Figure 13.3 'Keep the Pound' campaign postcard, 'Don't let Europe Rule Britannia': Britannia on the 'white cliffs of Dover', with her 'back to the wall' ready to take on the European enemy

Source: *This England*, Summer 1999, p. 66. Reproduced, by permission, from *This England* magazine.

North American Free Trade Agreement (NAFTA) – a looser alignment of the USA, Canada and Mexico (*The Economist* 2000d). Within the greater UK context, the EU is likely to have far-reaching consequences in terms of political – for example, destruction of the Conservative Party (Fay and Meehan 2000) – and constitutional change – for example, the break up of Britain (Nairn 1977) – in the coming years.

13.8 SUMMARY

There have been considerable constitutional and political changes in the UK in the last few decades. The Labour government, elected by a landslide majority in May 1997, has put in place a number of revolutionary reforms that it will be politically difficult – though constitutionally possible – for a subsequent government, with a different set of policies, to reverse. The Scottish Parliament and the Welsh Assembly are likely to become a permanent part of the constitutional machinery of the UK. The only change that can come is if there is a popular democratic demand to dissolve these bodies or, the more likely scenario, by an extension of their powers into legislatures completely independent of Westminster, within a Wales and Scotland withdrawn from the British Union. The situation in Northern Ireland, ever in flux, is less predictable. The new Northern Ireland Assembly, within the framework of the links between the UK and Ireland, should, in theory, be successful. But nothing in Northern Ireland is clear cut. In the longer term, given the demographic situation, and current political aspirations of the Nationalist community, it is likely that Ireland will once more become a thirty-two county nation. Whatever happens to the UK there would likely be close ties between the nations even in the form of some loose 'Association of British States' as mooted by the SNP. The Greater London Authority is an extension of the former Greater London Council, except this time Ken Livingstone is the elected mayor rather than simply the leader of the majority party. The government is also committed to regional devolution, and will countenance

and encourage directly elected regional assemblies where there is demonstrable desire for them. A number of regions, especially the North East and Yorkshire and Humber, have borrowed from the success of Scotland and set in motion processes towards this end.

As well as putting in place these promised reforms, the Blair government has tackled the vexing subjects of the House of Lords and voting reform. Both have long been in need of reform and the reform process has been instigated. There is a need for a second chamber in the British Constitution, given the lack of a written constitution and a bill of rights. The House of Commons is no longer an effective check on the executive, especially where the government has a majority of 179. How to create an effective second chamber which is not an electoral clone of the first or full of government lackeys, but is democratic, independent and constructive, is a very difficult task. There are a number of options for voting reform aimed at removing the obvious lack of fairness of the first-past-the-post-system. The chosen system is likely to be a compromise, such that change can be effected but the status of the main parties remains largely unaltered. One aspect of the British Constitution left undisturbed by the Blairite revolution is the Monarchy. Cynics suggest and hope that it will destroy itself through bad behaviour and bad publicity. Realists of a democratic, if not republican, bent hope that a future government will tackle the issue head on. Monarchists and some constitutionalists argue a monarch is as effective a head of state as some failed politician or clapped-out 'celebrity'.

The most recent general election signalled an end to the failed Thatcher–Major project. In this critical election (Norris and Evans 1999a) Middle England rebelled against the excesses of that style of government, if somewhat belatedly, although some recent local, European and by-election results have shown that not everyone is happy with the Blair government, especially women. Region has replaced class as the main predictor of electoral success. Yet the decline in interest in party politics has been reflected in ever-decreasing voter turnout at elections at all levels. It is to be hoped that the many constitutional reforms, particularly the proposals for electoral reform, can increase turnout at elections and restimulate an interest in party politics at all levels, especially among the young.

R E V I S I O N   Q U E S T I O N S

- Is devolution a positive or negative process for the UK?
- Should there be a Parliament for England?
- Is regionalism the answer to South East/London hegemony in England?
- How important is local government at a time of increased globalisation?
- Why should there be reform of constitutional institutions and practices, such as the Monarchy, the House of Lords and the first-past-the-post voting system, when they have worked well for centuries?

KEY TEXTS

Bradbury, J. and Mawson, J. (eds) (1997) *British Regionalism and Devolution: The Challenges of State Reform and European Integration*, London: Jessica Kingsley.

Election UK (1997) *Election UK 1997*
http://www.election.co.uk

Evans, G. and Norris, P. (eds) (1999) *Critical Elections: British Parties and Voters in Long-Term Perspective*, London: Sage.

Mohan, J. (1999) *A United Kingdom? Economic, Social and Political Geographies*, London: Arnold.

Painter, J. (2000) 'Local government and governance', in V. Gardiner and H. Matthews (eds) *The Changing Geography of the United Kingdom*, London: Routledge, pp. 296–314.

Pattie, C., Johnston, R., Dorling, D., Rossiter, D., Tunstall, H. and MacAllister, I. (1997) 'New Labour, new geography? The electoral geography of the 1997 British general election', *Area* 29(3): 253–9.

GEOGRAPHICAL DIVISIONS

A tale of two regions

- **Introduction**
- **Regions in the UK**
- **A tale of two regions: the North East and South East**
- **Intra-regional variation**
- **Summary**
- **Revision questions**
- **Key texts**

14.1 INTRODUCTION

The region as a unit of analysis has gone in and out of fashion in geography. Regional geography was very much influenced by the work on the 'natural region' of the French geographers Vidal de la Blache and Elisée Reclus at the end of the nineteenth century. For them and others of the French School of geography, the main unit of analysis was the synthetic region. The idea of using synthesis in regional studies was taken up in Britain by such pioneers as Gilbert (1960) and Ogilvie (1928).

Even though the 'natural region' ceased to be a central tenet of geographical study, some geographers continued to use the region, albeit more administratively defined, as a focus of analysis. The contributors to a book exploring the North–South divide in Britain, for example, adopted the Standard Region as the main unit of analysis (Lewis and Townsend 1989). The contributors to another book of the same period (Damesick and Wood 1987) explicitly addressed regional problems and problem regions. There has recently been a re-emphasis of the region as a crucial component in the study of geography (Allen *et al.* 1998). The translation of Claval's (1998) *An Introduction to Regional Geography* into English is further evidence of the increasing interest in the geography of the region. This revival of interest is not confined to academics but is also evident among politicians, policy-makers and the media. Devolution of power to Scotland and Wales has focused attention in some of England's regions on regional devolution and regional identity (see Chapter 13).

One boost to the region has been the 'Europe of the Regions' philosophy favoured by the European Union (EU). As the European Commission (1995: 3) has stressed, 'The dynamic diversity of the European Union is embodied in its regions.' One of the reasons for the push towards regional devolution in the more centralised countries, including the UK, is that European funding is geared towards regions. Regions bid directly to Europe, rather than central governments disbursing monies to the regions. Thus, a country with a well-defined regional structure is in a stronger position when bidding for funds from Brussels. It is the administratively defined regional approach that this chapter takes. By looking at differences between two contrasting regions of England – the North East and the South East – we show just how stark the contrasts in the human geography of the UK are. London is included in the analysis since it is very much part of the South East geographically, though it appears to be very much apart from it in other ways.

14.2 REGIONS IN THE UK

A major problem with the study of regions is in terms of definition and boundary determination. Defining regions will depend on many factors, particularly the rationale for the region. The problem with drawing 'a boundary around somewhere like the "east midlands" is that it implies a delimitation that does not exist and fosters ideas of inclusion and exclusion' (Massey 1994: 152). The main unit of analysis used in this chapter is that defined by the government. While recognising the drawbacks of these administrative regions, they still offer the best source of data for comparative analysis.

Some regions are easier to define than others. It is easier to define the South West than the East Midlands, for example. The South West is clearly bounded by the sea on two of its sides and the East Midlands merges into the West Midlands, Yorkshire and Humberside and East Anglia. The East Midlands does not 'have a natural geographical fit' (Tresidder 1999). Because of this different geography and a different history regional identity in the South West is much stronger than in the East Midlands (see Chapter 11). The demand for regional assemblies and greater devolution also comes from the regions of England with the greatest sense of identity and distance from London, such as the South West, the North East and the North West (Abrams 1999; see Chapter 13). There is, as McCrone (1969: 239) says, 'no simple way to work out the regional division of a country and clearly much sterile argument could be devoted to the merits and demerits of alternative schemes'.

Some areas will be defined as regions for one purpose and others for a quite different purpose. The Standard Statistical Regions of England (SSR) differ somewhat from the Department of Trade and Industry regions, which differ from the Environment Agency regions, which differ from the National Health Service (NHS) Regional Office areas, which differ from the Tourist Board regions, and so on. Confusion is added as these change over time. For example, since 1946 the main base for regional statistics in the UK was the SSR of which there were eight in England, plus Scotland, Wales and Northern Ireland. Although London was technically in the South East region it was often treated separately for statistical purposes. Data were given for London as well as for the rest of the South East. This makes regional analysis and planning particularly problematic and there are moves towards coterminous definitions currently taking place. In 1994, new government Offices for the Regions were established, based on the functions of the then four departments of Trade and Industry, Environment, Transport, Employment. Since April 1997, the primary base for regional statistics is the Government Office Regions (GORs), which were set up in 1994. The GORs also form the basis of the electoral regions for the European Parliament. Figure 14.1 shows the pre-1997 Standard Regions and the new GORs.

Governments have long recognised regions so as to formulate and implement regional policy (see Chapters 12 and 13). The development of regional policy in the UK began during the 1930s depression and has continued in some form since. Most of regional policy has been a reflection of the North–South divide that exists in the UK. Historically only northern regions were targeted for aid. Only during the recession of the early 1990s, when unemployment rose so much in parts of the South, were areas such as the Isle of Wight, Southend and Thanet considered for regional aid (Hudson and Williams 1995).

14.3 A TALE OF TWO REGIONS: THE NORTH EAST AND SOUTH EAST

To demonstrate geographical divisions within the UK in more detail we will focus attention on the two contrasting English regions of the North East and the South East. Although London is treated separately here, as it often has been in the regional literature and statistics, it is very much part of the

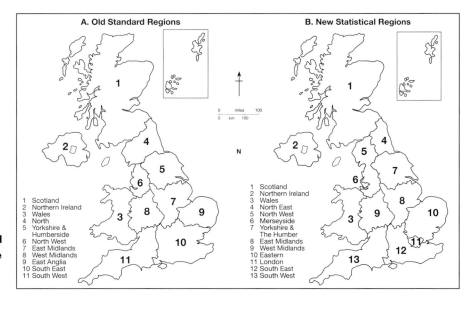

Figure 14.1 The pre-1979 regional structure and the new Government Office Regions

South East. London not only dominates the South East economy but that of the UK also. As Keeble (1980a: 123) puts it, 'Greater London is the functional heart of the South East.' As we will see, in terms of many labour and economic indicators, London performs in much the same way as the rest of the South East. However, in some other indicators, notably aspects of housing and health, the Metropolis is very distinct from the South East. As a large, global, capital city London has several problems either peculiar to itself or which are exacerbated by this status. On a whole host of indicators the South East and the North East demonstrate the geographical divisions within the country. More than any other contrasting regions they exemplify the North–South divide that is so characteristic of many social, political and economic variables under consideration here. This divide has been well documented (e.g. House 1978; Hudson and Williams 1986) and has a long history. Recent evidence shows how the North–South divide is a persistent part of the geography of the UK (Browne 1998).

> The North–South divide is firmly grounded in the economic organisation and the class structures of the UK, even if it is also formed by differences in housing markets, state intervention and in culture. It is in many ways an all-pervading influence when it comes to any analysis of the geographical dimensions of divided Britain.
>
> (Hudson and Williams 1995: 268)

Needless to say, defining the North East and the South East is problematic. The official boundaries of both have shifted over time, though the North East is easier to define as an area centred on Tyneside and Teesside and bounded by the Scottish border to the north, the Pennines to the west, the North Sea to the east and North York Moors to the south. There is a very strong sense of regional identity in the North East (Beynon *et al.* 1994; Colls 1992). Hetherington and Robinson (1988: 189), writing of Tyneside in particular, are of the opinion that both physically and culturally it is 'certainly distinct from the rest of northern England, while there is no doubt that many people in Tyneside regard London and southern

England as essentially another country'. The South East is more difficult to define, leading Allen *et al.* (1998) to ask 'where is the South East?' Meaningful borders are impossible to delimit and years of in-migration and relative affluence have meant that a south-east identity is weak – 'Essex man' notwithstanding (McDowell 1997). The same could be said of London. It merges into the surrounding area in such a way that administrative boundaries are the only way to define the city. Any London identity is largely the result of stereotype or very localised – the Cockney wideboy, the City gent, the Sloane Ranger – rather than any affinity with such a heterogeneous, disparate, amorphous mass.

The main units for analysis used here are the new Government Office Regions (GORs) set up in 1994 as a response to EU insistence that all regional aid be based on well defined regions. Thus the North East GOR is more clearly delimited, since it now consists of all of the counties of the former SSR of the North (Cleveland, Durham, Northumberland and Tyne and Wear) except Cumbria, which is now in the North West GOR. The South East GOR is smaller than the previous South East SSR. Bedfordshire, Essex and Hertfordshire are now in the Eastern GOR and London is treated separately as its own GOR. Thus, the South East GOR consists of Berkshire, Buckinghamshire, East Sussex, Hampshire, Isle of Wight, Kent, Oxfordshire, Surrey and West Sussex.

The North East has long been seen as a 'problem' region in terms of its relatively pervasive high unemployment, poor economic performance, bad health, low incomes and the like. The government continues to support the North East to a greater extent than London and the South East. In 1998, £38.1 million went to the region in the form of regional preferential assistance to industry, compared with only £2.7 million to London and £5.4 million to the South East. The North East was also allocated £86 million worth of EU structural funds for 1999, compared with £23 million for London and £4 million for the South East (ONS 1999a). But a contrasting view is that the South East, including London, is the *problem* region insofar as it draws people, resources and tax revenues away from other regions (Harvie 1982; Hudson and Williams 1995). The sheer scale of London 'has become a unique liability' and the huge costs of dealing with this 'are of national significance' (Wood 1987: 64). Indeed, the 'London Problem' and the 'Chaos of Commuterland' that is the South East, have long been recognised in the planning literature (Self 1967). Government economic policies have historically favoured the south-eastern part of the country at the expense of the rest (R. Martin 1995). As the following shows, there are wide discrepancies in the social and economic fabric of the North East and the South East, and in the majority of indices the North East and the South East are often at opposite extremes in the regional league. Unless otherwise stated the data in the following sections have been drawn from the 1999 edition of *Regional Trends* (ONS 1999a).

14.3.1 The people

The population of the South East was around 7,959,000, in 1997, and that of London 7,122,000. The North East had a much smaller population of around 2,594,000. The North East has been losing people for many decades. There were around 32,600 fewer people in 1996 than in 1971. The South East, on the other hand, experienced an increase in population of around 553,700 over the same period. However, London also saw a decline in population with some 136,500 fewer people in 1996 than 1971, most of these moving out of the city to the neighbouring South East or Eastern regions. Much of the South East's gain has been at London's expense, with the 'counter-magnets' of the new towns taking most of London's overspill (Keeble 1980b). In terms of overall population change, including births, deaths and net migration, the North East experienced an overall loss between 1971 and 1996. In the South East there were consistent gains during the same time period. London, on the other hand, suffered an overall loss of 9.6 per cent between 1971 and 1981, but a gain of 1.2 per cent between 1981 and 1991 and a further gain of 2.7 per cent between 1991 and 1996. Thus, London has made a demographic recovery, largely due to urban

regeneration, while the North East continues to decline.

Over the earlier periods there was natural increase (more births than deaths) in each of the regions but the rate was lowest in the North East. However, between 1991 and 1996 while there was a natural increase of 1.5 per 1,000 population in the South East and 5.6 in London, compared with 1.6 in the UK as a whole, the North East actually suffered natural decline of 0.1. That is, more people died in the area than were born. Much of this is due to the North East having an exceptionally high mortality rate. Between 1995 and 1996, 37,000 more people were born in London than died, in the South East the figure was 9,500, but in the North East 500 more people died than were born. The higher fertility in London is due to a relatively young adult population and to the presence of large numbers of ethnic minorities (see Chapter 7). In 1996, the total period fertility rate in the UK was 1.72, well below the replacement level of 2.10 children per woman. In the North East, it was 1.67, in the South East 1.69 and in London 1.75. However, some London boroughs, such as Hackney, Barking and Dagenham, Newham and Tower Hamlets had rates above replacement levels at 2.24, 2.11, 2.66 and 2.36, respectively.

The most notable component of population change, however, is migration. It is indicative of relative economic decline or prosperity. For many decades the north of the UK has been losing people to the south. The north loses 23,500 people every year to the south and the trend is accelerating. Since 1980, a third of a million people have moved from north to south. Most of the migrants are young adults and the majority of these are female. For many young women the south appears more attractive and liberating. As one young female migrant generalised, 'The further north you go, the greater the pressure to cook, clean, iron and generally look after your man' (quoted in Reeves 1999b: 14).

In terms of net migration, the North East lost 4,100 people between 1995 and 1996 while London gained 30,200 and the South East gained 38,600 during the same period. London has the most fluid population of all regions. As a **global city**, it has always attracted migrants from within and without the British Isles, and continues to do so. Economic, social and cultural factors act as magnets. However, while it is possible to measure and account for the economic factors, it appears that the 'social and cultural properties of the capital region are so difficult to measure that researchers have been deterred from considering them in their migration studies' (Fielding 1993: 210). Nevertheless, many people leave London as soon as economic or domestic circumstances allow. Most emigrants from the city do not move far. The South East and the Eastern regions absorb most – 87,000 and 57,000 out-migrants in 1997, respectively. One professional person leaving London for Suffolk cited pollution, urban decline, noise, crime and health threats, particularly as they affected her children, as negative factors (Gerrard 1999b). This is a reflection of how life-cycle changes can influence migration patterns as much as purely economic ones, as households progress through their 'housing career' (Hamnett 1999). The age group forming the largest exodus from the city in 1996 was the 30- to 59-year-olds. Many people, particularly professionals, leave London around the time of the family building stage. Many also leave on retirement.

Most inter-regional migration takes place in the south of England, between the South East, and Eastern regions and London. This process, whereby the young and less experienced workers are drawn into metropolitan areas, while more senior experienced workers move into the surrounding region, is consistent with the **'social escalator'** function of metropolitan areas identified by Fielding (1992). That is, people get off the 'escalator' later in their careers (Champion and Ford 1999) and, to some extent, slow down. Apart from some of the younger age groups, people are less migratory in northern regions, despite lower wages and poorer employment prospects. Some of this relates to a stronger sense of identity the further a region is from London, alluded to above and discussed in more detail in Chapters 11 and 13. Furthermore, unskilled and manual workers, of which there are more in the northern regions, are less willing and able to migrate than professional people. Also, migration has a financial and social cost. Many in the North East are

unwaged or on low wages and have little or no savings. This makes migration problematic. Social networks tend to be stronger in northern regions and people are reluctant to break these. The constraints of the public rented sector of the housing market also reinforce this immobility (Fielding 1993).

As well as internal migration, London has also been the main focus of international migration to the UK. In 1997, 101,000 overseas immigrants came into London and 68,000 people left, resulting in a net balance of 33,000 which was half the UK net international migration. As a global city, London attracts a wide range of immigrants – professionals from multinational companies posted to London for short or medium spells; diplomats and their families; the young who move for 'overseas experience' (classically Australians and New Zealanders, but also increasingly young members of the EU); most of the UK's refugees, asylum seekers and undocumented immigrants. The South East had the second highest number of international immigrants, after London, and the region had a net gain of 2,000 people from international migration. The North East attracts relatively few international migrants but it had a net gain of 1,000 in 1997.

In terms of structure of the population it is more difficult to observe patterns of age and gender at the regional level than at a more refined spatial resolution. However, it is evident that the population of London is younger than average. This is a reflection of high fertility and the attractiveness of the city to young adults. In 1996, the only age group in London which showed a net increase was the 15- to 29-year-olds. The age structures of the South East and North East are very closely matched and close to the UK pattern. London has a higher proportion of its population in the 16 to 44 age group than any other region. Less than half (49.4 per cent) of this group was female. Both the North East and South East have a below average percentage of their population in this age group. In each region, females outnumber males, but only after age 44. Only Inner London has an excess of females in the age groups 15 to 19 and 20 to 24. What can be said with some accuracy, however, is that the North East and the

South East have very ethnically homogeneous populations, but for different reasons. Only 1.7 per cent of the population in the North East, and only 3.1 per cent of that in the South East, was from a non-white ethnic group in 1998. However, more than a quarter of London's population was composed of non-white groups in 1998. Not only is London home to most of the UK's non-white ethnic minorities – almost 50 per cent – it is the area most favoured by non-white immigrants.

The smaller population of the North East is also reflected in political terms. Both the South East and London have considerably more Members of Parliament than the North East. Further, while having a large number of MPs from Scotland in the Cabinet served Scotland's purposes in terms of devolution, the North East has profited less from having five MPs (the so-called Geordie Mafia) in the Cabinet, including the Prime Minister, Tony Blair. While the North East returns four Members of the European Parliament the South East and London return eleven and ten, respectively. London and the South East have more corporate headquarters and house more senior civil servants and policy-makers than the North East. In regional terms, both the South East and London have more political clout than the North East. It has always been thus. At the height of the interwar depression a delegation of over 200 formed a 'hunger march' to London from the Tyneside town of Jarrow to protest at the conditions caused by an unemployment rate of almost 73 per cent (Mowat 1968). Nobody from London or the South East has ever marched to Tyneside on an empty belly.

14.3.2 Work

High unemployment in the North East was not just a feature of the interwar depression. The area has long been plagued by higher than average rates of unemployment since then. As Fothergill and Guy (1990: 47) have noted, 'the unemployment rate of the North East was twice that of the South East of England as far back as 1932'. Large-scale decline in the coal mining, steel making and shipbuilding industries (Beynon *et al.* 1991; Warren 1980), as well

as widespread factory closures (Robinson 1988a), has created a 'culture of unemployment' and unemployment in the region has almost become institutionalised (Pimlott 1985). Brazier (1999: 16) has observed that young men in the North East, 'far from following in their fathers' footsteps [into the shipyards, steelworks, coal mines], have only unemployment to look forward to'.

Not so long ago there were 120 working mines in County Durham. Now there are none. Between 1958 and 1973, 117,000 jobs were lost in mining, 25,000 in shipbuilding and 13,000 in metalworking. Further jobs were lost during the economic recession in the late 1970s and early 1980s. Manufacturing was particularly hard hit, with a net loss of 90,000 jobs between 1978 and 1982. This was catastrophic for certain communities. For example, Consett lost 4,500 jobs when the steelworks closed in 1981, and Shildon 2,600 when British Rail Engineering closed its works in 1982 (Buswell *et al.* 1987). Although much of this surplus labour was absorbed by inward investment in the form of foreign multinationals (which is essentially underwritten by government grants and generous tax relief), such as Fujitsu (subsequently closed) and Nissan, this work is extremely insecure, given the peripatetic nature of multinational enterprises. In 1998, for example, the German-based Siemens abandoned its state-of-the-art microchip factory in Tyneside, throwing 1,100 people on the dole. As one trade union official from Sunderland put it, 'Most companies here are satellites, whatever the industry. If it has to be their own country or England, it's England that goes. If it's a national company and it has to be the North East or London, it's the North East' (quoted in Ryle 1998: 3). This, much publicised, inward investment of multi-national companies might offer some local and temporary palliative to the North East's unemployment problem, but the deep-rooted structural problems must be addressed to effect long-term stability in the region. As a former shipyard worker from Jarrow explained to Ryle (1998: 3), 'It's a myth that the men laid off by the shipyards and the pits have found jobs in the new industries. Most of the older ones are on disability benefits.' Despite the loss of manufacturing jobs, over 20 per cent of employees in the North East were still employed in manufacturing at the end of 1998, compared with 13.7 per cent in the South East and only 7.9 per cent in London (NTC 1999b). Service industries have never been particularly expansive in the North East, although political expediency meant that Newcastle became the home of one of the main Inland Revenue centres. More recently, service providers, such as Barclays Bank, BT and Orange, have established themselves in the region – mostly in the form of low-wage call centres. It is a particular irony that London Electricity's call centre is located in the North East. On the other hand, the service economy, particularly the higher paid professional and financial sectors, is much healthier and better established in the South East and London than it is in the North East.

In 1997, the unemployment rate of 9.8 per cent in the North East was nearly double that of the South East, and the region suffered the highest redundancy rates in the UK. In 1998, there were thirteen redundancies per 1,000 employees in the North East, compared with nine in Great Britain as a whole, ten in the South East and seven in London (NTC 1999b). The North East also had above average levels of long-term unemployment, as did London. Long-term unemployment in the South East was lower than average (Table 14.1). Of the thirteen regions in the UK, the North East had the second highest percentage (after Merseyside) of households where nobody was in employment, whereas the South East had the lowest percentage. The figure for London, however, was slightly above the UK average. People were more likely to be self-employed in London and the South East than in the North East. In general, economic activity rates for

Table 14.1 Employment and social class indicators, UK, North East, South East and London, 1998–9

	UK (%)	North East (%)	South East (%)	London (%)
Unemployed more than five years 1999	4.9	6.0	3.7	5.6
Households with no one in employment 1998	17.3	24.1	11.6	18.7
Self-employed of those in employment 1998	7.1	4.6	8.4	8.5
Male economic activity rates 1998	83.9	79.4	88.7	82.6
Female economic activity rates 1998	71.5	65.3	75.7	69.3
Employees with a second job 1998	4.5	3.9	5.0	4.3
Days lost to labour disputes per 1,000 employees 1998	12.0	9.0	1.0	13.0
Trade union membership of all employees 1998	30.0	40.0	22.0	25.0
Working age persons in social classes I and II 1998	30.2	23.0	36.5	36.9
Working age persons in social classes IV and V 1998	19.6	23.2	16.7	14.6
Other classifications 1998	12.6	16.0	9.3	14.7
	£	£	£	£
Male average weekly full-time gross earnings 1998	426	377	454	566
Female average weekly full-time gross earnings 1998	309	274	324	403
Average full-time weekly hours worked 1998	44.0	43.2	44.6	43.9

Source: ONS 1999a, *Regional Trends 34*, Chapters 3 and 5
Note: Other classifications include members of the armed forces, those whose previous occupation was more than eight years ago, or those who never had a job.

both men and women were higher in London and the South East than in the North East. Not only were there more people in work in the South East and London than in the North East, there were more people with second jobs. One indication of the legacy of high unemployment, and a reflection of its industrial heritage, is the fact that the North East was the region with most days lost to labour disputes in 1997. Associated with this is the above average trade union membership in the North East – almost double that of the South East. The North East, as Pimlott (1985: 359–60) points out, is 'not only the region with the longest and most severe experience of unemployment, but also a region with an unusually homogeneous, strongly organised working class'.

Both male and female average earnings are highest in London and among the lowest in the North East. However, people in the South East work longer hours. Only in Northern Ireland do people work fewer hours than those in the North East (see Chapter 8). What is also of interest is the fact that many more men and women are involved in some type of flexible working arrangement, such as flexitime, annualised hours, short-term or job sharing, in the North East than in the South East or London. It would seem that this flexibility of working arrangements in the North East is a response to the fragile job market in the area and the low skills base. Morris (1995: 5) also cites the very large informal sector in the North East where some very well established firms offer poorly paid and uninsured jobs, via contractors, 'to those prepared to use such work to supplement their benefits claims'.

Closely linked to employment is social class. Because of the higher incidence of professional and related occupations in London and the South East there are more people in social classes I and II. The North East, on the other hand, has more people in unskilled and semi-skilled occupations and thus in classes IV and V. The residual category 'other' includes members of the armed forces, those whose previous occupation was more than eight years ago or those had never had a job, so it is not surprising that the rates in the North East and London

are above average given the nature of long-term unemployment in these areas (Table 14.1). The position of London as 'the world's most truly global city' (Hall 1998: 28) ensures that it attracts people of professional and managerial status. This spills over into the South East as one of London's main commuter belts.

Social class is also a reflection of education and training, and in a number of indicators the North East fares less well than both London and the South East. For example, in 1996 there were more pupils per teacher (19.3) in primary and secondary schools than in London (17.2) and the South East (17.4). In the North East, only 20.3 per cent of pupils achieved GCSE grades A–C in all core subjects, compared with 24 per cent in London and 30.8 per cent in the South East. This poorer level of education in the North East is reflected in the fact that as many as 23 per cent of people of working age had no qualifications, compared with only 15.8 per cent in London and 14.4 per cent in the South East. The North East also had the lowest percentage of 16-year-olds remaining in education in 1995/6. Only 69.7 per cent opted to continue in education, compared with the UK average of 79 per cent. In both London and the South East the rates were higher than average at 79.7 and 80.7 per cent, respectively (NTC 1999b).

14.3.3 Housing

Not only are economic conditions worse in the North East than the South East, so are social conditions. Although London has a much better employment profile than the North East some aspects of its social conditions are closer to those of the North East, and this tends to set it apart from the rest of the South East. One particularly strong indication of general social conditions is housing.

Generally speaking, the housing market is closely linked with the labour market. If the labour market is buoyant in an area then there will usually be greater demand for housing since more people will move into an area where work is available. More people in employment will also mean that household economies will be better able to afford housing through mortgages and rents. Despite the welfare state, all the actors in the housing arena look more favourably on those in work than those out of work, save for the fortunate minority with some capital. Both the private rented and mortgage lending agencies act as gatekeepers for entrants into the respective markets. They also control movement within these markets. Public authority housing has stagnated since the Thatcherite 1980s when the 1980 Housing Act established the 'right to buy' for sitting tenants in local authority housing. The best houses have been sold at 'bargain basement' prices and many buyers have subsequently sublet or resold at considerable profit. That stock is lost to the public sector. Further, little new stock has been constructed. In 1985, for example, 21,800 new permanent dwellings were built for local authority rental in the UK. In 1997, the corresponding figure was 1,400 (ONS 1999a). Access to what stock remains is also controlled, albeit by much more complex, and often esoteric, mechanisms than in the private sector.

Housing associations are also required to take account of social as well as economic criteria when allocating housing, but are generally more interested in the economic circumstances of tenants than most local authorities are. Although housing association provision is the smallest of the rented sector it is the fastest growing in some areas, especially as an increasing volume of local authority stock is transferred to housing associations. Also, with its relatively more liberal access criteria and affordable rents the housing association sector offers the best solution for many home-seekers, especially in London. Since housing stock is very inelastic in economic terms the price of housing, whether expressed in terms of rent or selling price, will closely match local economic conditions.

Table 14.2 shows a number of housing indicators. It is clear from this that, while the North East–South East contrast obtains in housing as much as it does in terms of employment, the Metropolis is very much not of the South East when it comes to housing. House building is one variable that we would expect to respond to housing demand, and while stock has increased overall between 1981 and 1996 it has not increased evenly across all regions at the same rate. In the North East, the rate of house building was the lowest of all regions and almost half that of the UK average. The South East, on the other hand, had a

Table 14.2 Housing indicators, UK, North East, South East and London, 1981–98

	UK (%)	North East (%)	South East (%)	London (%)
Increase in stock of dwellings 1981–98	15.6	9.9	20.1	14.2
Owner occupied 1998	67	63	75	56
Local authority rented 1998	19	26	8	20
Private rented 1998	10	7	11	17
Detached dwellings 1996–7	20	14	26	4
Terraced dwellings 1996–7	28	32	25	28
Increase in house price 1997–8	6.0	5.3	9.9	7.6
	£	£	£	£
Average weekly private sector rent 1998	84.0	64.0	99.0	133.0
Average weekly local authority rent 1998	40.9	35.3	48.4	55.3
Average house price 1998	84,557	57,409	107,035	127,814

Source: ONS 1999a, *Regional Trends 34*, Chapter 6
Note: Tenure and rents for Great Britain only. Price increase and house price for England and Wales only.

higher than average rate of increase. House building in London was lower than average over the period, but nevertheless impressive, given the physical restrictions and very strict planning considerations in that city.

The 'counter-magnets' or growth areas to work against the pull of the capital, much favoured by the planning community in the 1960s, have failed. These were promoted to ameliorate the 'shameful mess in the region's outer zone' and the 'acute housing problem within the conurbation' (Self 1967: 141). Since the Thatcher government abolished the Greater London Council in 1986, and virtually abandoned regional planning policy in favour of free market mayhem, these problems have become more acute (see Chapter 6). Not only is there a marked difference in the rate of house building there is also a difference in the speed of house building. In the North East, the completion rate was only 3.2 per cent in 1997, the second lowest after Merseyside, while the rate in the South East was the highest of all regions at 13.2 per cent. The completion rate in London was more modest at 6.4 per cent (NTC 1999b). The price of land is a major component in the housing equation. Land zoned for housing is twice as expensive in the South East than the North East, while that of London is more than three times the price.

Housing tenure can also be indicative of economic and social conditions in an area, but usually at spatial resolutions much finer than that of the region. Tenure can be as much a reflection of tradition and culture as wealth. Levels of owner occupation, for example have traditionally been lower in Scotland than in England and Wales. Indeed, levels of owner occupation in England and Wales and Northern Ireland are high by international standards. However, within England tenure patterns vary by region. Owner occupation is lower than the Great Britain average in the North East, and higher in the South East. Indeed, the North East has the lowest level of owner occupation after London. The North East's high level of local authority housing is beaten only by Scotland, although London also has an above average percentage of houses rented from local authorities. The

South East has the lowest regional level of houses in local authority ownership. Both London and the South East rely on the private rental sector more than average, while the North East has a relatively low percentage in this tenure. In both London and the North East, renting from a housing association is above average while in the South East it is below average. Many aspects of housing in the North East have traditionally been untypical, compared with the rest of England. Tenement houses and public provision were more common in the North East than anywhere else in England. This was the result of a history of poverty, as well as the influence of neighbouring Scotland, where flats were the typical housing style in urban areas and, after the First World War, public housing became more popular (Cameron and Crompton 1988).

Housing is provided in a variety of types and sizes. The detached house is usually seen as the most desirable type of dwelling. Predictably, the South East has an above average stock of detached housing and the North East below average. Because of the cost of detached housing in terms of land requirements, more than actual capital cost, this type of housing is rare in London where more cost effective flats are much more common. Indeed, flats make up around 32 per cent of London's housing stock, the highest in the UK after Scotland. Most of these London flats are purpose built and for public rental. However, a large proportion of the privately rented flats are in Victorian and Edwardian houses which have been subdivided because of rising property prices. After semi-detached housing, terraced houses are the most popular type in the UK. Although more demanding of land than flats, terraced housing is relatively inexpensive. The North East has a higher than average share of terraced houses, whereas London has the same percentage as the UK average and the South East has a below average share (Table 14.2).

The demand for housing will be reflected in the price or rental which a property can command. However, since housing is fixed geographically and the supply is very limited, housing prices are affected by a more complex range of factors than other common goods. Within the UK, there are two main systems at work where house purchasing is concerned. There is the more controlled, contractual, logical and less stressful system which operates in Scotland under Scottish law and there is a more haphazard and potentially disastrous system operating in the other parts of the country. Although there have been promises made by government to adopt the Scottish system, or at least some of the procedures, throughout the UK, these have not yet been put into place (Lane and Ryle 1999). There is still scope, then, for all the less attractive practices associated with the 'English' system, such as **'gazumping'** or withdrawing from a deal at the last minute. The former practice becomes more common during a housing boom and is now epidemic in parts of the South East and London. 'Property rage' is a term attached to the aggressiveness in the housing market there (Segall 1999) which has led to intimidation, corruption and violence. Indeed, the situation in London is such that key workers, such as nurses, care assistants, teachers and bus drivers, are unable to afford London housing costs. Many key workers are on nationally agreed wages that take little account of high housing costs in London. In mid-1999, the London Housing Federation reported that the average price that first-time buyers were paying across London was £97,300, and the average weekly private sector rent was £133. This is beyond the means of someone on a low wage and severely curtails the quality of life of those on a medium wage. London prices are 20 per cent higher than those in the South East and 200 per cent higher than in many parts of the north (McGhie 1999).

This is demonstrated in Table 14.2, which shows both rents and house prices higher in London and the South East than in the North East. The rate of price increase is also much higher in the South East and London. Assuming the average UK house price as an index of 100, the index in the North East was 68.9 at the start of 1999. The index for the South East was 131.9 and that for London 164.6. These last two figures so distorted the pattern that no other region had an index above 100 (NTC 1999b). In some parts of London demand is concentrated on 'hot spots', which can be as narrowly focused as individual streets. One such hot spot is Clapham in the popular south-west of the city. There, a four-bedroom house with a garden can cost as much as £400,000 (Segall 1999). While prices have risen by as much as 30 per cent a year in parts of London, in parts of the North East prices have fallen (Figure 14.2). In Newcastle, in mid-1999, flats were being sold for *fifty pence* each in one unpopular area (Browne 1999e) – although the average price was £57,409 (Table 14.2). At the same time, terraced housing in some parts of London was selling for £1 million. The housing problems in London are exacerbated by the city's capital, corporate and financial functions. Significant numbers of people, such as business executives, MPs, media people and others, who need to stop over with some frequency, maintain a second home (a *pied-à-terre*), using their London home for only part of the time. Such is the discrepancy between housing costs in the North East and London that for many professional and managerial people long-range commuting is the preferred option (Green *et al*. 1999). Hogarth and Daniel (1988) have called such commuters Britain's 'new industrial gypsies'.

It is hoped that that the 'urban renaissance' as proposed by Lord Roger's Urban Task Force could be of benefit to the North East and London. However, the problems in the two areas are fundamentally different. In London, the main problem is lack of land and too many people. In the North East, it is lack of money and too few people. The Urban Task Force has recommended the gentrification of inner city areas (see Chapter 16). This has been significant in retaining people, particularly young professionals in London, but also in other cities. Some areas of London, such as Islington and Notting Hill, became gentrified in the 1970s. Clapham, Wandsworth and Highbury became gentrified during the 1980s when young professionals realised how well-located these areas were in relation to the rest of the city. More recently, Clerkenwell, Hackney and Spitalfields have gone through the process. While gentrification might be good for the physical fabric of an area it is coupled with socioeconomic change which alters the social characteristics and forces house prices up. Houses in Islington which cost £8,000 in 1966 will now cost

**Figure 14.2
Unwanted housing
in Benwell,
Newcastle and very
desirable property
in Fulham, London**

Source: *The Sunday
Telegraph, Review*, 4
July 1999, pp. 18–19.
Courtesy Mark Pinder
and Alexander Brattell.

between £500,000 and £1 million (McGhie 1999). Small-scale gentrification has taken place in the North East such as in Jesmond in Newcastle (Robinson 1988b). But the urban areas continue to lose people, particularly the better-off. In parts of the region, such as the former steel making town of Consett, former pit villages like Horden or inner city areas like Benwell in Newcastle, it is impossible to sell houses. Though what happens once the impact of the Urban Task Force and Single Regeneration Budget funds begins to be felt remains to be seen (McGhie 1999).

Housing is closely linked to households. Given the contrasting economic fortunes and demographic trends of the two regions we would expect to see differences in household formation. The growth in the number of households in the North East between 1996 and 2016 is projected to be the lowest in England and Wales at 5.2 per cent. At 16.3 per cent, the South East is projected to have the largest growth, with London experiencing the second highest growth rate of 14.8 per cent (NTC 1999b). Household structure tends not to vary greatly at the regional level. Nevertheless, at 31.1 per cent of all households, London had the second highest incidence of one-person households after Scotland. In some central London boroughs as many as half the households consisted of people living alone in 1998. London, at 5.2 per cent, also had the highest level of households with two or more unrelated adults, compared with 2.1 per cent in the North East and 2.5 per cent in the South East.

14.3.4 Lifestyle

Lifestyle is linked to the issues discussed above. Social class, education, occupation and income contribute to lifestyle, as do age, gender and ethnicity. One fundamental determinant of lifestyle is income. The average gross weekly household income is highest in the South East and lowest in the North

East. People in the North East are most likely to live in households in the bottom fifth of the income distribution, and least likely to live in the top fifth, while the reverse is true for people in the South East. Some areas of west Inner London – Kensington, Chelsea and Mayfair – have the highest average incomes in the UK and west central Londoners are more than four times wealthier than the national average. However, London as a whole is only the fourth richest city in the country (Browne 1998). The source of income also varies by region. The South East had the highest percentage of households deriving income from investments in 1998 (Table 14.3). The North East had the highest percentage (19.6) of households deriving income from social security. The South East, on the other hand, had the lowest at 9.6 per cent. The North East also has the lowest percentage of households deriving income from self-employment.

In terms of expenditure, there are also regional differences. Generally speaking, total expenditure is linked to total income. Incomes are higher in the South East and London so we would expect that more will be spent in these areas. With a UK average as an index of 100, average weekly household expenditure in the North East was 84.3 in 1998, whereas it was 107.4 in London and 118.8

Table 14.3 Lifestyle indicators, UK, North East, South East and London, 1996–8

	UK (%)	North East (%)	South East (%)	London (%)
Average weekly household income from investments 1998	5	4	6	5
Individuals in top fifth income bracket 1996–7	20	14	30	26
Individuals in bottom fifth income bracket 1996–7	20	24	14	19
Households in receipt of any benefit 1997–8	70	77	63	68
Households with dishwasher 1997–8	22	13	31	21
Households with satellite dish 1997–8	27	30	26	28
Households with home computer 1997–8	28	21	37	31
Weekly household expenditure on food 1998	100	85.9	108.4	107.2
Weekly household expenditure on alcohol 1998	100	101.5	98.5	97.7
Weekly household expenditure on tobacco 1998	100	104.9	93.4	101.6

Source: ONS 1999a, *Regional Trends 34*, Chapter 8; NTC 1999b, pp. 88–9

Note: Households in receipt of any benefit, national figure for Great Britain, includes housing benefit, council tax benefit, retirement pension, family credit or income support, unemployment benefit/job seeker's allowance, incapacity or disablement benefits and child benefit or one-parent benefit.

in the South East. The bulk of this 'overspend' in London and the South East goes on housing. The housing expenditure index (UK still 100), in the North East was only 82.1, while that of the South East was 126.6 and London was 128.3 (NTC 1999b).

Although London is the country's biggest market-place, Tyne and Wear comes a respectable fifth, ahead of the West Midlands, West Yorkshire, Central Clydeside, Edinburgh and Bristol (CACI Information Solutions 1999). Thus, commodities and activities are as readily available in the North East as the South East, yet patterns of consumption and expenditure are different (Table 14.3). It seems that those in the North East prefer satellite television more than those in the South East and London, but are less lazy when it comes to washing dishes. In terms of home computers, and all that that means in terms of access to the so-called information super-highway, the South East fares best with the highest regional rate and London with the second highest. The North East, on the other hand, has one of the lowest home computer ownership rates (see Chapter 9). This relates to class, education, income and social, economic and cultural exclusion.

Expenditure on alcoholic drink is higher than average in the North East and more is spent on beer and cider, compared with higher expenditure on wine in London and the South East (Table 14.3). Nevertheless, and contrary to the popular image, people from the North East make more use of wine bars than those in London. With a Great Britain index of 100 for weekly visits to wine bars the index in Inner London was 144 in 1999 and in Outer London 109; that is, higher than the national average. However, in the main urban areas of the North East – Tyne and Wear and Cleveland – the indices were 243 and 224, respectively, or more than twice the national average (CACI Information Solutions 1999). Expenditure on tobacco is also higher in the North East. Altogether, 7.6 per cent of the average household weekly household budget was devoted to alcohol and tobacco in the North East in 1997,

Table 14.4 Health indicators, UK, North East, South East and London, 1993–8

	UK	North East	South East	London
Standardised mortality ratio[a] males 1997	100	112	90	97
Standardised mortality ratio[a] females 1997	100	109	95	94
Male mortality from coronary heart disease[b] 1997	34.7	40.7	29.7	28.7
Female mortality from coronary heart disease[b] 1997	20.8	27.2	16.7	16.4
Infant mortality rate[c] 1993–7	6.1	6.3	5.1	6.1
Limiting long-standing illness[d] 1997	22	25	19	21
Percentage of adults undertaking any sport 1997	64	57	67	62
Percentage male heavy smokers[e] 1997	39	42	38	35
Percentage female heavy smokers[e] 1997	30	36	24	24
Percentage male problem drinkers[f] 1997	27	33	27	25
Percentage female problem drinkers[g] 1997	14	13	14	13
Serious and fatal road accidents[h] 1997	69	46	63	90
Expenditure on industrial injuries benefit per head 1997	100	204.8	8.6	99.6
Expenditure on incapacity benefit per head 1997	100	165.3	58.2	48.9

Source: ONS 1999a, *Regional Trends 34*, Chapters 3, 7 and 10; NTC 1999b, *Regional Marketing Pocket Book*, pp. 11, 66
Notes: [a] Standardised for age, UK=100. [b] Rate per 1,000 patients, national rate for England and Wales, NHS Regional Office areas Northern and Yorkshire, North Thames and South Thames. [c] Per 1,000 live births average for 1993–6. [d] Percentage reporting a long-standing illness. [e] Smoking more than 20 cigarettes per day. [f] More than 22 units per week. [g] More than 15 units per week. [h] Per 100,000 people.

the highest in the country. On average, people in the South East and London spend least on these items. The people of the North East also appear to be the nation's biggest gamblers (NTC 1999b).

The North East had the lowest ratio of cars to people in 1997, while the South East had the highest. Only 308 cars were licensed per 1,000 people, compared with 334 in London and 463 in the South East (NTC 1999b). But that is reflected in the lower incidence of serious or fatal road accidents in the North East, compared with London and the South East (Table 14.4).

14.3.5 Health

Generally speaking, the people in the South East are healthier than those in the North East with London occupying a more intermediate, though better than average, position. This has long been the case (Holohan *et al.* 1988; Spence *et al.* 1954). By looking at the mortality of both sexes, standardised by age so that the effects of age structure are cancelled out, we can see how the South East and London have ratios below the UK average of 100 and the North East has a higher ratio (Table 14.4). This affects life expectancy, and people from the South East and London can expect to live longer and healthier lives than those in the North East. One of the biggest causes of disease and death is heart disease. The geography of health data is based on the NHS Regional Office areas which do not coincide with the GORs (see pp. 194–5). However, it is clear from Table 14.4 that the North East, which is contained within the Northern and Yorkshire area, has a poorer record for coronary heart disease than those areas which best approximate to the South East and London. The incidence of respiratory disease is also higher in the North East, particularly lung cancer, with rates in the North East only being exceeded by those in Scotland. Respiratory disease is less common in the South East, although the presence of large numbers of ethnic minorities, immigrants and homeless people in London means that some respiratory diseases, especially tuberculosis, are a cause for concern. Areas particularly affected are the Inner London boroughs of Brent,

Tower Hamlets, Hackney, Newham, Southwark and Haringey (Bardsley and Morgan 1996). Suicide is a complex phenomenon. However, it is closely linked with population density, unemployment, deprivation, loneliness and one-person households. It is no surprise, then, that rates are relatively high in both London and the urban areas of the North East, and comparatively low in the South East. In London, suicide is higher in central boroughs where it is more common among women. The incidence is lower in the peripheral London boroughs, and here more men than women commit suicide (Congdon 1996).

Infant mortality is recognised as a good indicator of socioeconomic conditions. Here, averaging the figure over several years, it is clear that the North East has the least favourable rate and one of the highest rates in the country (Table 14.4). The South East consistently has one of the most favourable rates. Linked to infant mortality is low birth weight. Babies born underweight are less likely to survive than full-weight babies, and the conditions which predispose to low birth weight are much the same as those which are linked to infant mortality – poverty, low social class, mothers working late into pregnancy, poor housing, poor maternal diet, maternal smoking and alcohol consumption (Coleman and Salt 1992). In 1996, the percentage of live births under 2.5 kilograms was 7.3 in the North East and 6.6 in the South East. London, however, had a relatively high rate at 7.8 per cent. This is due to the large ethnic minority population there. Mothers born in the Caribbean and Pakistan are particularly prone to high infant mortality rates (Endean and Harris 1998).

High mortality and poor health are related to a number of factors, such as social class, lifestyle, environment, unemployment, attitudes to health, housing conditions, housing tenure, education and income (Coleman and Salt 1992; Curtis 1995; see Chapter 10). Because these all vary from place to place so does health. As we have already seen (Table 14.1), the North East has a lower than average percentage of people in the higher social classes and a higher than average percentage in the lower social classes and 'other' classifications. The higher a person's social class, the better their health and life chances, other things being equal. Social class is closely linked with income and lifestyle issues. Again, the North East has lower than average weekly incomes. Unemployment is also higher than in the South East and London. Furthermore, it is not just unemployment *per se* which has an adverse impact on health, but also job insecurity and the threat of unemployment (Wilkinson 1996).

Weekly expenditure on food is proportionately less in the North East, while more is spent on alcohol and tobacco than is the case in the South East and London (Table 14.3). There are more heavy smokers and heavy drinkers in the North East (Table 14.4), and since smoking is the biggest single cause of premature death, as well as being a contributor to respiratory and heart diseases, this will be reflected in the health of that area. This has long been recognised as a major cause of excess mortality in the North East (Holohan *et al.* 1988). Smoking has fallen dramatically among the higher social classes and is retreating to the bottom of the income distribution. This helps explain not only why there are more smokers in the North East but also why the habit is declining more slowly in that region. According to Thomas (1998: 18), the 1990s smoker is 'female, poor and excluded'. The adverse health effects of excessive alcohol consumption are also well known. Another lifestyle factor closely associated with health is exercise. It appears that adults are more likely to take part in sport in London and the South East than in the North East (Table 14.4). Diet, too, is closely linked to health. Generally speaking, those in the North East eat a less healthy diet than those in the South East and London. A 1994 survey showed how people in the South East had more fruit and vegetables and high fibre cereal in their diet than those in the North East. Of the people surveyed in the North East 19 per cent used solid cooking fat, compared with only 5 per cent in the South East. However, added sugar in the diet was much the same in both regions, and added salt was higher in the South East (ONS 1997).

In 1997, fewer than one in five people in the South East reported having a long-standing illness, a lower proportion than any other region (Table 14.4). Only Wales had a higher incidence of long-standing

illness than the North East. This is clearly related to the health differentials already mentioned but is also influenced by past industrial practices. The areas with the highest rates of long-standing illness tend to be mostly former coal mining or heavy engineering areas. This is further reflected in the figures for industrial injuries benefits. While London had a figure close to the UK mean, that of the North East was double the mean and far higher than that of any other region – the second highest was Wales at 157.5. The index for the South East is very low, and much lower than the second lowest, the South West with 64.2. The figures for incapacity benefit paint a similar, but less drastic picture.

As well as high alcohol consumption, drug use in the North East is high. In 1996, 25 per cent of 16- to 29-year-olds admitted to having used an illegal drug in the previous year. However, London and the South East surpassed this, with about 30 per cent and 27 per cent taking them (ONS 1999a). The South East and London also have worse records for serious and fatal road traffic accidents. Road traffic in London is particularly associated with ill health in the form of atmospheric and noise pollution (Health of Londoners Project 1996). One problem for the epidemiology of infectious disease in London is the size of the young transient population which falls through the public health net. As a result, some sexually transmitted diseases have rates 60 per cent in excess of the national average (Bardsley and Morgan 1996). The large number of international migrants in the city also has implications for health, particularly among those entering illegally who are reluctant to seek medical help for fear of alerting the authorities. Most of the increasing incidence of multidrug resistant tuberculosis in London is among the homeless and international migrants (see Chapter 10). In terms of mental illness, the North East would appear to fare less well than London and the South East. Around 90 per 1,000 women patients in the North East were being treated for depression in 1996, compared with 74 per 1,000 in London and 65 per 1,000 in the South East. Rates for men were much lower at 40, 35 and 27 per 1,000, respectively (ONS, EOC 1998). High levels of long-term unemployment in the North East, with all that entails for stress, despair, rejection and loss of self-esteem, have been associated with both mental and physical health problems (Holohan *et al*. 1988).

14.4 INTRA-REGIONAL VARIATION

It must be borne in mind that within such macro-geographical units as regions there will be considerable internal variation. Indeed, as R. Martin (1995: 24) has noted, it is 'a well-known fact that *intra*-regional disparities are inevitably more pronounced than *inter*-regional differences, since the latter are weighted averages of the former'. Nowhere in the country are intra-regional variations more apparent than in London. Within the Metropolis are to be found the most affluent and most deprived areas in the country, often within the same **ward**. While we have seen that the South East is generally a fairly affluent region, areas of relative deprivation exist (Mohan 1995). Similarly, the North East is not uniformly deprived.

Within the North East there is considerable social and economic variation between the more prosperous rural districts, such as Castle Morpeth and Tynedale, in Northumberland, and the older declining, industrial, urban areas like Newcastle, Sunderland and Middlesbrough. For example, in terms of the index of local deprivation, the variation in the North East in 1998 ranged from 19 to 299. These are rankings based on an index of relative multiple deprivation produced by the Department of Environment, Transport and the Regions (DETR) and based on twelve indicators selected to cover social, economic, housing and environmental concerns. The index covers the districts and new unitary authorities in England. The most deprived area, at number one, was Liverpool. The lowest ranking is 310 at which a number of areas were equally ranked. Newcastle upon Tyne was the nineteenth most deprived area in England, Sunderland the twenty-first, and Middlesbrough the twenty-fourth. However, Castle Morpeth and Tynedale both had a rank of 299 (ONS 1999a). Again, there is considerable variation in ranking on income. Out of 459 local

authorities in Great Britain, Castle Morpeth was ranked at 95 and Tynedale at 161, in 1999. However, most of the North East's districts or unitary authorities fell below the UK average. For example, Newcastle had a rank of 379, Middlesbrough 418 and Sunderland 430. In fact, of the 100 poorest local government areas, fifteen were in the North East. This represents almost half of the thirty-two districts and unitary authorities in the region. The area with the lowest index of average household income, after Kingston upon Hull, was Easington, in County Durham, with a rank of 458 (CACI Information Solutions 1999). There is also a marked intra-regional geography of benefit. In 1999, for example, 5 per cent of households were in receipt of housing benefit in Castle Morpeth and 6 per cent in Tynedale, compared with 14 per cent in Sunderland, Middlesbrough and Newcastle (ONS 1999a).

The South East, too, contains considerable intra-regional variation in terms of its social and economic geography. There are some areas where multiple deprivation is a problem. For example, Brighton and Hove was ranked at 60 and Thanet, in Kent, at 64. However, and predictably, most areas in the South East had a relatively low ranking, with fifteen of the districts achieving the lowest rank of 310 (ONS 1999a). Not surprisingly, areas in the South East also figure among some of the wealthiest. The Elmbridge and Surrey Heath districts of Surrey are ranked third and fourth, respectively. Several other districts immediately to the south-east of London also ranked highly. Nevertheless, not all areas of the South East had a high average household income. Some of the coastal areas were amongst the least well-off. Thanet and Hastings, for example, were both ranked at 397 out of 459. Gosport, in Hampshire, ranked 270 and Shepway ranked 259 (CACI Information Solutions 1999). The Kent districts of Thanet and Shepway suffered badly as a result of a decline in the tourist industry during the 1970s and 1980s (Champion and Townsend 1990; Wood 1987) and have yet to recover. The percentage of households receiving income support in the South East ranged from 3 in Wokingham to 13 in Hastings and Thanet (ONS 1999a).

London, however, presents a less rosy picture, with no fewer than fourteen of its thirty-two boroughs, plus the City of London, amongst the thirty most deprived parts of England. The Inner London boroughs of Newham, Hackney and Tower Hamlets were ranked second, fourth and sixth, respectively. Only the Outer London boroughs of Sutton and Kingston upon Thames, in the south-west of the city, were ranked below 200, at 284 and 220, respectively (ONS 1999a). Clearly deprivation indices depend on the choice of variables used and the spatial scale employed. The DETR index measures material deprivation. One study of the 1991 Census of Population data measured both material and social deprivation separately. While there were fifteen London boroughs among the top twenty areas on the *material* deprivation index, there was no area from the North East. However, there were seven urban areas from the North East among the top twenty areas on the *social* deprivation index, and only seven London boroughs (cited in Goodwin 1995).

Despite the relative deprivation of many parts of the Metropolis, some of the highest ranking areas in the UK in terms of household income were London boroughs. In 1999, these included Kensington and Chelsea, which had the second highest household income in the UK, Richmond upon Thames, City of Westminster, Wandsworth, Kingston upon Thames, Barnet, Bromley, Camden and Harrow. The City of London had the highest household income at almost double the UK average. Only one London borough had a relative income index lower than the national average of 100 – Barking and Dagenham with an index of 90 (CACI Information Solutions 1999). The boroughs with the highest levels of income support were Hackney and Newham where 22 per cent of households had to have their incomes supplemented by benefit, compared with only 5 per cent in Richmond upon Thames (ONS 1999a).

Smaller scale variation is particularly evident in London, especially Inner London. At the level of ward or **enumeration district** there can be enormous variety in terms of social class, ethnicity, household composition, lifestyle, health and so on. This heterogeneity is the result of the crystallisation of historical developments, immigration, gentrification, economic expediency and public interventions in the built environment.

It is clear, then, that while there are variations in a range of social and economic phenomena between regions, there is also variation within regions. This highlights a drawback to a regional approach and indeed illustrates one of the biggest problems in geography – choice of scale. We have seen how there are some better-off districts in the North East and some less well-off districts in the South East. We have also seen the sharp contrasts within London. Bear in mind that within each London borough, each district and each unitary authority there will also be variation between one area and another. This is part of the rich mosaic which makes the study of geography so interesting.

14.5 SUMMARY

Over the last two decades socioeconomic change in the two case study regions of the South East and the North East has been radically different. The South East has remained an area of relative affluence – a 'social escalator' region – while the North East has experienced persistent economic problems. London has seen a strengthening of its global functions, such as in the economic functions of the City, and in strengthening international migration flows to the capital. The foregoing shows that in many social and economic indicators the North East is at a considerable disadvantage compared with the South East and, to a lesser extent, London. However, analysis at the regional level masks considerable internal variation. The long-standing and increasing socioeconomic and life chance differentials between the southern regions of England and the northern regions are likely to continue, with adverse consequences for both North and South. Only major structural changes can reverse the destructive and wasteful regional imbalance in the UK. It remains to be seen if political devolution to the English regions will be the answer to this problem.

R E V I S I O N   Q U E S T I O N S

- To what extent is the North East more homogeneous socioeconomically than the South East?
- To what extent has London's strengthening global functions impacted on demographic and social change in the city?
- Is London truly a part of the South East?
- Why have regional disparities been such a long-term feature of the social, economic and cultural geographies of the UK? What, if anything, do you think might stop such disparities?
- 'The past is a foreign country.' To what extent is this true of the economy of the North East?

KEY TEXTS

Allen, J., Massey, D. and Cochrane, A. (1998) *Rethinking the Region*, London: Routledge.

Claval, P. (1998) *An Introduction to Regional Geography*, trans. I. Thompson, Oxford: Blackwell.

European Commission (1995) *The Regions of the United Kingdom in the European Union*, London: HMSO.

Hudson, R. and Williams, A.M. (1995) *Divided Britain* (2nd edition), Chichester: John Wiley & Sons.

Morris, L. (1995) *Social Divisions: Economic Decline and Social Structural Change*, London: UCL Press.

Philo, C. (ed.) (1995) *Off the Map: The Social Geography of Poverty in the UK*, London: Child Poverty Action Group.

THE HAVES AND THE HAVE-NOTS

15.1 INTRODUCTION

As has been noted in earlier chapters, with the arrival of the Conservative government of Margaret Thatcher in 1979 government policy shifted from the Keynesian demand management of the economy, the principles of social justice, full employment and the eradication of poverty to deflationary macro-economic management and the advocacy of flexible labour markets (see Chapters 3, 4, 6 and 12). The net result has been that paid work and earnings have become more and more polarised in the UK. And as 'income is one of the most important influences over the patterns of life chances' (Hamnett 1999: 129), in this chapter we highlight the socioeconomic impacts of the 'fissuring' of the population into segments, while in Chapter 16 the spatial consequences are explored in detail.

An individual or household may have several different sources of income. However, it is useful to make a distinction between three main types. First, there is earned income, from employment or self-employment. Second, there is unearned income, accruing from property, investments, rent and the like. Third, there is transfer income, comprising benefits and pensions 'transferred' to the individual or household on the basis of entitlement. For some sub-groups of the population, for example the unemployed and pensioners, incomes are made up largely of transfer income, and for the very wealthy incomes from sources other than earnings (such as investment income and other types of unearned income) are important.

It is estimated that 70 per cent of total income in the UK is from earnings (Green 1999), and in a recent review of trends in pay in the 1980s and 1990s Blanchflower *et al.* (1996) identified four main developments in pay which have occurred since the 1970s. The first development was the rapid widening of the earnings distribution; second, a substantial growth in real earnings; third, a reversing of the public sector–private sector wage differential, and fourth, the widening of regional differentials.

In the remaining part of this chapter we explore economic and social polarisation at the household level, looking at the idea of 'work-poor'–'work-rich'

households, but also ethnic minority dimensions of polarisation, the feminisation of polarisation and life transitions, poverty and social exclusion (Chapters 3 and 4). These are dimensions of polarisation which to date have been explored relatively rarely (Modood 1998).

15.2 THE UK: THE EMERGENCE OF TWO NATIONS?

We have already seen (Chapter 14) how the UK is often viewed geographically as two nations – the North and the South; constitutionally, though, it is composed of four nations. The idea of the UK as two nations in terms of a division of rich and poor is not new. The nineteenth-century politician and novelist, Benjamin Disraeli wrote *Sybil: Or the Two Nations* in 1845. 'Two nations; between whom there is no intercourse and no sympathy; who are ignorant of each other's habits, thoughts and feelings, as if they were dwellers in different zones, or inhabitants of different planets . . . fed by different food . . . not governed by the same laws' (Disraeli 1969: 67). This work transformed ideas in the nineteenth century as it highlighted the social problems of the industrialisation. However, as was noted in the introduction to this chapter, since 1979 there has been a growing disparity between the rich and poor – the haves and have-nots – in this country, after what had seemed a period of social justice from the end of the Second World War to the late 1970s (Byrne 1999).

15.2.1 Social polarisation in the UK

Theories of social polarisation and widening disparities between the affluent and the poorer strata trace this development to the effects of economic restructuring. Some analyses focus on the exclusion of particular groups from participation in society, such as immigrants or the long-term unemployed, others concentrate on the geographical locations in which these processes manifest themselves most strongly, such as declining industrial areas or global cities.

During the 1980s a strong spatial dimension to polarisation was given through an analysis of an increasing North–South divide, which received much publicity shortly before the general election of 1987 (Champion *et al.* 1987; Woodward 1995). What all agreed on was that the country was divided, and increasingly so. For those on the right of the political spectrum, growth in London and the South East was proof that the new market philosophy could succeed: here was entrepreneurialism, new sunrise industries, few trade unions. In the 'north', by contrast, people were in trade unions and in sunset industries. The left saw it differently, with the South East and London a land of yuppies, greed and spiralling house prices where the triumph of individualism simply meant selfishness. The north for them, was where the old values and verities were hung on to (Allen *et al.* 1998).

Andre Gorz (1989) in his *Critique of Economic Reason* drew attention to the widening gulf between the haves and the have-nots in advanced capitalist economies like the UK. His divide was between a group of workers able to monopolise the well paid professional jobs and another group of workers 'servicing' the professional classes, in terms of their domestic and personal service needs. As was noted in Chapter 8, Will Hutton (1995: 14) presents a more complex picture of division in the UK, describing the UK as 'a 30–30–40 society'. Hutton's analysis focused on individuals, but a more holistic picture of division can be gleaned from adopting a household focus embracing wealth as well as income.

Rather than merely focusing on individuals there has also been some empirically based conceptual work on social and spatial divisions in the UK, and the idea of a polarising nation at the level of the household, which was first developed by Ray Pahl (1984). A process is identified based on distinctive work practices, creating a growing polarisation between households containing non-working ('work-poor') couples and those containing dual-earner ('work-rich') couples, particularly between two-earner and no-earner couples (Pinch 1993; see Chapter 8). This thesis is more concerned with the amount of wealth in each category, and on the ways in which the life chances of individuals are influenced by how economic restructuring affects other members of the households to which individuals belong (Pinch 1993). Thus, if one partner becomes disconnected from the labour market and begins to receive benefits, the probability of his/her partner also becoming disconnected is high because of the working of the benefit system, in what has been described as the unemployment trap (see Walker and Walker 1997).

Not all households today are made up of a heterosexual couple, married or cohabiting, with children or other relatives. Socioeconomic changes in the post-war period have been reflected in a move away from the stereotypical nuclear family towards a greater diversity of household forms. Today, about 40 per cent of marriages end in divorce, and for many female partners, divorce results in financial instability. The increase in single person and single parent households is widely documented in official statistics. Women head the majority of both of these types of households, and single parent households are among the most materially deprived in the UK today (see Chapter 7).

In the post-war period inequalities in income in the UK had been decreasing until the end of the 1970s, but since then the rate of increase in income inequality has risen more rapidly than in any other OECD country apart from New Zealand (Atkinson 1995). Atkinson (1995) argues that concern about poverty in the midst of a seeming redistribution of income from the 1960s onwards was due to the fact that the top 10 per cent of the population (in income terms) had reduced its share in favour of the next 40 per cent of the population; that is, the upper middle echelons. The bottom 50 per cent of the population had hardly changed their share of income. Since 1977 (the low point in inequality), the proportion of the population with less than half the average income has more than trebled; the wages of the lowest earners were lower in real terms in 1992 than in 1977.

Hills (1995) notes that inequalities in wealth are more pronounced than inequalities in income. And in the following section we explore the wealth, or lack of it, of the UK population, drawing on Hills' study on income and wealth for the Joseph Rowntree Foundation. Hills focused on the household, in all its diversity, and looked at income and wealth distribution by quintiles based on earned, unearned and transfer income.

15.2.2   The haves

The haves are composed of the 'seriously' wealthy; that is, those who derive their income from land or investments (property, shares, etc.) rather than undertaking paid work for a salary. But most of the haves are dependent on income from paid employment, such as managers and professionals. In 1911, about 10 per cent of the British workforce were managers and professionals, and by 1991 about 30 per cent of the workforce held such jobs (Mills 1995). The expansion in managerial and professional occupations occurred in the post-war period because of the development of employee status managers (often with professional qualifications) within the manufacturing sector as well as in the expanding service sector and the expansion of the professions in both the private and public sector. And these jobs have been feminised (see Chapter 8).

Using survey data for incomes and wealth for 1990, which although rather dated highlight the relative position of the haves and have-nots, the top fifth of households identified by Hills (1995) included two-earner households, in which at least one partner commanded a salary well above the average wage, and solo managers and professionals. These workers are not paid by the hour but by the task, and undertaking tasks may now demand working above contracted hours (see Chapter 8). Such things as company cars, private health insurance and performance-related pay can boost the salaries of managers and professionals, and the latter is certainly not the preserve of City yuppies! Clearly, in socioeconomic terms the haves form a privileged group, better able to compete economically, and exercise their influence in achieving priorities, than many other population groups. These households have more economic power than

other households in the UK which is derived in part from salaries but also from unearned income from rising house prices, especially in the South East (Allen *et al.* 1998; Hamnett 1999). But their economic power may be the result of credit. Their net incomes were above £437 per week for a couple; £267 for a single person and £629 for couples with two dependent children (aged 5 and 10 years). The second fifth of households commanded net incomes in the range of £150–£209 per week for single people and for a couple with children aged 5 and 10, the range would be £354–£492 per week.

15.2.3   The have-nots

The remaining three quintiles are composed of households who do not possess the economic power to make the lifestyle choices of the haves. Adults in the have-not households were in lower skilled jobs, both full-time and part-time, with income from paid work supplemented by transfer income; or some totally relied on transfer income (especially in the lower quintiles) because of labour market disconnection (through childbirth, single parenthood, redundancy, retirement, including forced early retirement). The low income of the have-nots is in part the result of economic processes such as deindustrialisation, combined with technological change and a restructuring of the divisions of labour (see Chapter 8; Philo 1995; Sibley 1995). In addition to these changes, other changes – wider social changes – in family and society, such as discrimination and alienation of members of ethnic minority communities and family breakdown are at play and social polarisation is accompanied by spatial polarisation. This means that the most disadvantaged end up in certain localities, spatially entrapped for example in the inner cities and on estates on the periphery of urban areas, sometimes termed 'problem estates' (see Chapters 4 and 16). Problem estates have been investigated by the Social Exclusion Unit (see Chapter 4).

The middle fifth in the income distribution had net incomes of £111–£150 per week for single people, and for a couple with children aged 5 and 10 years the range was £262–£354 per week. Many of these households are in fact struggling to acquire the material resources needed today to maintain or reach the lifestyle they aspire to. Today's 'consuming' lifestyle, with more and more material goods regarded as being necessities not luxuries, requires levels of weekly disposable income greater than the middle fifth command (see Chapter 9). Indeed, two professional/managerial salaries are not always sufficient to buy property in some desirable parts of London, where today house prices are spiralling (see Chapter 14).

The second poorest fifth had net incomes for single people in the range of £77–£111 per week and for a couple with children aged 5 and 10 years the range of £181–£262 per week. In the poorest fifth, for single people income was less than £77 per week, and for a couple with children aged 5 and 10 years income was less than £181 per week. Life for the poorest fifth is one of 'existence' and 'survival' through lack of choice. Disadvantage tends to be spatially concentrated and poverty is becoming entrenched (see Chapter 16). In some neighbourhoods networks of friends and families exist on benefit, perhaps supplemented with income from the informal sector, mixed with periodic participation on government schemes.

15.3 DIMENSIONS OF POLARISATION

In the following case studies three dimensions of polarisation worthy of further analysis are highlighted: polarisation within ethnic minority communities, the feminisation of social polarisation, as well as longitudinal aspects of polarisation.

15.3.1 Case studies of polarisation

Ethnic minority haves and have-nots

As we saw in Chapter 2, the post-war mass migrations, especially from the New Commonwealth, began because of labour shortages in the UK. The 1972 Nuffield Social Mobility Survey revealed that nearly a quarter of the non-white migrants had professional qualifications, which was twice the proportion of the host population, and over half had social origins in either the **petty bourgeoisie** or farming classes (compared with 16 per cent for the host population (Modood 1998). But they experienced downward occupational mobility as most secured manual work; and their material conditions upon arrival reflected their role as a replacement population, in terms of jobs and housing. They found work in low-status, poorly paid occupations, which held no attraction to white workers (D. Phillips 1998).

Around 6 per cent of the working population belong to a minority ethnic group and for them the labour market still demonstrates racial inequality, with discrimination and racism penalising some communities more than others. They experience much higher levels of unemployment than white workers, and when in paid employment they are more likely to be in occupations with low pay and low status (Bloch 1997). The unemployment rate is about double that for the white population, although there are large variations between the various ethnic minority communities (see below). The unemployment rates amongst African-Caribbean youth are three times that of white youth (Bloch 1997).

The differences between the minorities are as important as their position in relation to the white majority population. During the 1980s and 1990s the ethnic minority communities have witnessed a growing social polarisation, with polarisation in economic activity rates, educational attainment, professionalisation and entrepreneurial activity (Modood 1998; Phillips and Sarre 1995). Some groups, notably the Chinese and Indians, are re-establishing their premigration middle-class profile, while other minority groups (African-Caribbeans, Pakistanis and

Bangladeshis) are among the most marginal and disadvantaged people in the UK. The social class profile of the ethnic minority communities is now bimodal, with both a relatively large percentage in the upper echelons of white collar work (Social Classes I and II) and also a large percentage of people in semi-skilled manual work (Social Class IV). Unemployment levels in some communities (African-Caribbean, Pakistanis and Bangladeshis) are relatively high, along with problems of poor housing, low educational attainment, and health-related issues (Modood 1998; Owen 1996; see Chapters 12 and 16).

Feminisation and social polarisation

As was noted in Chapter 8, women have increased their share of employment in virtually all industries, occupations and areas in the UK in the last two decades. While women are more 'visible' in the workplace not all possess the same earning power and job security. Most women begin their working life in full-time employment, but many hold part-time jobs after childbirth or withdraw from the labour market because of the problems associated with access to adequate, affordable childcare and transport, combined with the shortage of 'quality' full-time jobs. One of the most disadvantaged groups are single mothers.

Patterns of childbirth have become more diverse as many mothers have children outside marriage, and at a variety of ages (see Chapter 7). The most significant feature has been the huge increase in the number of working mothers, with participation in the labour market being the norm even for mothers of children less than one year old. For a significant minority – 10 per cent – the birth of a child leads the family into poverty. Thus for women whose partners are 'workless', or single mothers, or if they have low levels of educational attainment, the transition into parenthood can lead them and their families into poverty.

But one striking feature of this period has been the feminisation of management and the professions. For example, in 1974 women comprised 2 per cent of managers; by 1998 the proportion was 18 per cent (ONS, EOC 1998). The proportion of women in some professional occupations has also increased, accounting in 1998 for 64 per cent of teaching professionals; 33 per cent of solicitors holding practising certificates in England and Wales (compared with 20 per cent in 1988); 24 per cent of barristers in 1998, compared with 14 per cent in 1987. But in 1998, 88 per cent of architects, town planners and surveyors were men; as were 71 per cent of business and financial professionals and 93 per cent of judges (ONS/EOC 1998). Compared to women from lower socioeconomic groups women managers and professionals, whether in solo or dual career households, are more likely to have the economic resources to make choices, such as access to a motor vehicle for journeys-to-work, or pay for childcare. But their ability to prioritise their career temporally and spatially is also dependent on the presence of dependent children and/or elderly relatives.

Life transitions, poverty and social exclusion

In the UK today some people encounter difficulty in managing the transitions from one life stage to another. The timing and nature of transitions from employment, parenthood and retirement have been profoundly reshaped by a combination of economic, social and cultural change. For some, greater choice, fulfilment and opportunity mark these transitions, but for others the story is one of increasing risk and uncertainty. For a significant minority, transitions from one life stage to another lead straight into poverty (Gurumurthy 1999). In this section we focus on the transition from education to paid work and the problems of sustaining paid work until the statutory retirement age.

Perhaps the most significant is the growing difficulty men in particular have in entering and sustaining paid work, especially at the beginning and end of their working lives (see Chapter 11). The transition from school to work has undergone a dramatic shift in the last twenty years. Many young people continue to make successful transitions into employment, usually following an extended period of further/higher education (see Chapter 8) and gaining the skills and credentials demanded by employers. But youth unemployment has been more than twice the average for the past two decades, and more than half a million or one in ten of under-24-year-olds are off the unemployment register (Gurumurthy 1999). This problem has been recognised by 'new' Labour with the New Deal for Welfare for Work, one key target group being the young who have failed to get connected to the labour market (see Chapter 12).

The transition into retirement has also undergone substantial change. People are experiencing extended retirement as many live longer or move out of paid work during their fifties (see Chapters 3 and 8). While for some retirement is a choice, characterised by the maintenance of high levels of purchasing power, thanks to occupational pensions and equity in housing, increasing numbers are being forced out of the labour market, and retirement comes too early, with too little preparation, support and savings. Indeed some move onto sickness or disability benefits (see Chapters 8 and 9). The scale of the problem today is illustrated by the fact that two-thirds of all people experiencing persistent poverty are pensioners (Gurumurthy 1999).

15.4 SUMMARY

Evidence presented in this chapter has highlighted social divisions within the UK, where today one in five households now live on less than half the national average income while the rich are getting richer. For most households income from paid work is critical in determining their level of economic well-being. Insecure paid work and disconnection from the labour market have resulted in more and more households being plunged into poverty (see Chapter 8). Moreover, the last two decades have seen income differentials grow in the UK. The haves possess the economic power to make lifestyle choices, such as where to live, the purchase of consumer goods, choice of school for children and

so on. But this is often achieved by consumer credit. For the have-nots the economic reality is much starker, their purchasing power ensures existence and survival, but not 'choices' as to where to live, which holiday to take, whether to have a meal out or to buy a car. One key strategy of the Labour government is to improve people's employability through the New Deal (see Chapter 12). But New Dealers tend to be clustered in localities with weak labour markets (Peck 1999; Turok and Webster 1998; Chapter 12). In the next chapter we will add a spatial dimension to the social polarisation debate.

Patterns of childbirth have become more diverse as many mothers have children outside marriage, and at a variety of ages (see Chapter 7). The most significant feature has been the huge increase in the number of working mothers, with participation in the labour market being the norm even for mothers of children less than one year old. For a significant minority – 10 per cent – the birth of a child leads the family into poverty. Thus for women whose partners are 'workless', or single mothers, or if they have low levels of educational attainment, the transition into parenthood can lead them and their families into poverty.

R E V I S I O N   Q U E S T I O N S

- What are the reasons for the increase in inequality in the UK?
- What approaches have been used to study social polarisation?
- Which groups are particularly vulnerable to social polarisation?

KEY TEXTS

Atkinson, A.B. (1995) *Incomes and the Welfare State: Essays on Britain and Europe*, Cambridge: Cambridge University Press.

Byrne, D. (1999) *Social Exclusion*, Milton Keynes: Open University Press.

Hills, J. (1995) *Inquiry into Income and Wealth, Volume 2*, York: Joseph Rowntree Foundation.

Modood, T. (1998) 'Ethnic diversity and racial disadvantage in employment',' in T. Butler and M. Savage (eds) *Social Change and the Middle Classes*, London: UCL Press, pp. 53–73.

Woodward, R. (1995) 'Approaches towards the study of social polarization', *Progress in Human Geography* 19(1): 75–89.

THE GEOGRAPHY OF POLARISATION AND DIVISION

- Introduction
- The forces of polarisation in urban and rural parts of the UK
- The impacts of urban and rural change
- Summary
- Revision questions
- Key texts

16.1 INTRODUCTION

In the previous chapter, the 'fissuring' of the UK population into segments, the haves and the have-nots, was described, and in Chapters 3 and 4 the forces of socioeconomic change in the UK were discussed. In this chapter, we add a spatial dimension to the analysis, highlighting the geography of polarisation and division at a number of spatial scales – regional/local and urban/rural.

Since the 1980s, a strong spatial dimension to polarisation has been present through an analysis of regional division – an increasing North–South divide (see Chapters 3, 4 and 13). The North–South divide received much publicity shortly before the general election of 1987, and more recently with the Blair administration (Champion *et al.* 1987; Woodward 1995). A variety of analyses of earnings at the regional level have highlighted a pattern of London and the South East versus the rest of the UK as the key feature of pay differentials in the UK (Champion *et al.* 1987). These differentials are the result of a clear regional differential in employment opportunities for men and women, with a location in London being especially favourable to the development of managerial and professional careers. Fielding (1992) has described London as a 'social escalator' region attracting many upwardly mobile young adults living in single person households and then encouraging their out-migration in nuclear family or 'empty-nest' households to other regions in later middle age or approaching retirement. A second reason for the London differential has been the house price booms of the 1980s and late 1990s–2000s in London and the South East, underlining the important interaction between housing markets and labour markets, which play an important part in determining the geography of division through patterns of earnings variations across regions (see Chapter 14).

In general, one could say that academics were showing that polarisation and social disparities were growing between those who had benefited from the measures of the successive Thatcher administrations and those who had lost out. However, the Thatcher regime at the time tried to deny the argument. The North–South divide debate re-emerged in autumn 1999 with concerns of the differential performance of the economy of London *vis-à-vis* the rest of the UK (Cabinet Office 1999a; see case study overleaf). Another way of looking at regional division is comparing the so-called Celtic fringe of the UK (Scotland, Wales and Northern Ireland) and England. Poverty and the have-nots are clustered in the Celtic lands (McKendrick 1995; see Chapter 7).

A regional focus does indeed disguise the extent of local variability, and detailed analyses of published data, including Census of Population data at ward level, have been used to map the way that poverty and disadvantage is concentrated. These studies have highlighted the fact that the haves and have-nots live in contiguous localities, in what have been termed 'two-speed' cities (Green 1994). Using Census of Population statistics for England and Wales, Dorling and Tomaney (1995) have produced cartograms for the main indicators of poverty (see Chapter 4). Figure 16.1 shows the geography of 'want' at ward level on a base map which converts population to area – hence the distortion. This shows children living in households where no one is officially in paid work. The darkest areas represent the 'poorest' third of households and the lightly shaded areas the 'richest'. Figure 16.2 employs the same methodology – demographic base map and ternary division – but demonstrates the geography of 'ignorance'. Here the darkest areas represent areas where less than 34 per cent of school leavers passed five or more GCSE examinations at grade C or above. The lightly shaded areas are where more than 48 per cent achieved at least this level of qualification. Again the unshaded areas show the middle range. In both maps the predictably high scores are to be seen in some of the conurbations, but pockets of rural deprivation can be seen as well as the detailed level of polarisation and division.

While urban poverty in inner city areas and on peripheral housing estates is very visible, affecting large numbers of people, the majority of the areas ranked the lowest in earnings terms are in fact rural in character (Green 1999; Hetherington 1999b). In the second part of this chapter we explore the nature of the processes shaping division in contemporary

**16.1.1
Case study**

Divided Britain

1 North West
The region's diversity manifests itself in many ways. For example, Merseyside has one of the most serious concentrations of unemployment and social exclusion in Europe, yet neighbouring Cheshire has the fifth highest GDP per capita in England.

2 North East
The region also has the lowest average house prices in Britain, but has some of the most severely deprived communities in the country. It has the highest unemployment rate and the highest percentage of benefit claimants.

3 Yorkshire and Humberside
At the regional level, the problem of two-speed cities and local economies needs to be tackled. This gap is more important to regional prosperity than perceptions of a North–South divide. The recent development of Leeds illustrates the point. Though one of Europe's most successful job generating cities, ten wards in the city – home for 225,000 people – are amongst the 10 per cent most deprived wards in England. Creating prosperity and achieving social inclusion must be tackled together.

4 West Midlands
The most striking feature of the West Midlands is its diversity. Although a relatively prosperous region, Birmingham and Sandwell both rank in the ten most deprived districts in England, with the country's most deprived ward being in Birmingham.

5 East Midlands
A region of relatively low unemployment. On the other hand, within areas of prosperity there are significant problems of deprivation and social exclusion. The New Deal for Communities programme, which targets the most deprived neighbourhoods in the country, includes inner city areas of Nottingham, Leicester and Derby.

6 East of England
Compared with the other regions, the East of England stands out as having a high rate of economic growth – averaging 3.5 per cent per annum from 1993–8, and lower unemployment and higher GDP per capita than most other regions. Although the region is relatively prosperous there are districts that suffer from significant deprivation. These are primarily in peripheral coastal locations, urban areas and the isolated rural areas around the Fens.

7 South West
Overall economic prospects are good, strengthened by excellent transport links with Wales, London and the South East, the Midlands and Europe (via motorway network and ports). But there is continued structural economic decline (inner cities, market towns, coastal resorts) – significant local pockets of deprivation, social exclusion and unemployment in urban and rural areas. Cornwall has the lowest average earnings and GDP of any county in England.

**16.1.1
Case study
continued**

8 South East
Economically the South East is strong. This general picture disguises considerable variation within the region. The Index of Local Deprivation shows that areas of Kent and the South coast have fared particularly poorly so far this decade. Overall unemployment is well below the national average, but varies from 0.8 per cent in Winchester and Woking to 8.6 per cent in Thanet. There are particular concentrations of unemployment in East Kent and towns on the South Coast.

9 London
London is both the capital city and a city region. With a population of over seven million it is the largest metropolis in the European Union. London's apparent prosperity disguises problems of severe poverty. The cost of living in London is higher than in the rest of the UK, significantly reducing the benefit of higher salaries. The 1998 Index of Deprivation shows that London has thirteen of the twenty most deprived districts in the country and unemployment rates have remained above the national average during the 1990s. In Inner London, almost one-third of all households received means-tested benefits in 1996/7.

10 Wales
The gap between unemployment rates in Wales and the UK rate is closing. However, less than three-quarters of people of working age in Wales are economically active compared with around four-fifths in the UK as a whole. There are also high levels of permanent sickness – about one in six of the population.

11 Scotland
Within Scotland there are significant variations in economic activity by local area. In October 1999, the claimant count rates by Local Authority area ranged from 2 per cent in the Shetland Isles to 9.7 per cent in West Dumbarton. Even in areas with average unemployment rates, there are concentrations of long-term unemployment (for example, in peripheral estates and areas of former heavy industry).

12 Northern Ireland
Long-term unemployment, in particular, accounts for over a third of total unemployment, a considerably higher proportion than for the United Kingdom. There is evidence of pockets of disadvantage (evidenced by higher unemployment, lower employment and lower qualification rates) within the urban areas of Belfast and Derry and in rural areas, particularly in the north and in the west and south of Northern Ireland.

Source: Cabinet Office (1999a)

Figure 16.1 The geography of 'want': England and Wales

Source: Dorling and Tomaney, 1995. Reproduced by permission of CPAG.

Britain. In the third section the impacts of these processes are examined including urban decline, crime and **rust belts**; social exclusion in Northern Ireland; and the threat to green belts and **NIMBYism**.

16.2 THE FORCES OF POLARISATION IN URBAN AND RURAL PARTS OF THE UK

Throughout the Industrial Revolution the country-side depopulated as people moved to urban areas,

largely in the industrial heartlands and London, in search of paid work. Since the nineteenth century, however, cities were seen to have significant disadvantages when compared to more idyllic 'rural' locations (Urry 1995). But for most people these disadvantages were balanced against the economic reality of having to live close to employment and city facilities.

During the twentieth century, urbanisation has been replaced by first suburbanisation and later by counterurbanisation as the dominant force shaping the UK's settlement patterns (see Chapters 7 and 12). In England especially, there has been a long history of anti-urban ideology (Glass 1964). The movement

Figure 16.2 The geography of 'ignorance': England and Wales

Source: Dorling and Tomaney, 1995. Reproduced by permission of CPAG.

has been of largely white households, serving to reinforce the pattern of racial segregation (D. Phillips 1998; see pp. 217–18). In England, the population of rural areas grew by 24 per cent between 1971 and 1996 compared with an increase of 6 per cent overall, and it has been estimated that between 1981 and 1991 there was an average of 77,000 in-migrants to England's rural districts per year (Cabinet Office 1999b). This reflects the attractiveness of rural environments but creates increasing demand for housing in rural areas and pressure for new roads, infrastructure and services. Indeed, in-migration to rural areas – which is undertaken at different stages in the life course (families with young children, older adults, and so on) – may be

conceptualised as part of a wider 'counter-urbanisation cascade' or the suburbanisation of the countryside (Champion *et al.* 1998).

Enhanced individual mobility, employment decentralisation, improved electronic communications (telecommunications, computers, facsimile machines, modems, cellular telephones and the like) and new working practices (notably moves towards more flexible working; see Chapter 8) have resulted in increasing numbers of people choosing to live in rural areas – whether or not they also work there (Barnett and Scruton 1999; Lewis 1998). Some 50 per cent of British people would like to live in the countryside (Countryside Agency 2000). The perceived quality of the rural environment is one of the most important factors in the appeal of rural areas as places to live, and surveys reveal that rural dwellers are more content than urban dwellers. Some 89 per cent of people living in rural areas said they were content with where they live, compared with 20 per cent in cities. Moreover, in another poll, 71 per cent of people believed that the quality of life is better in the countryside than elsewhere, and 66 per cent said that they would move there if there were no obstacles to doing so (Cabinet Office 1999b).

To some people, life in an urban area has become synonymous with the negative features of life in a postindustrial society. These negative features include insecurity arising from high crime rates and fear, so that people are afraid to walk or drive in certain urban areas (Cabinet Office 1999a; Valentine 1989), and an increasingly polluted atmosphere as commuting by car compounds traffic congestion (see Chapter 12). In Northern Ireland the geography of fear takes a different dimension, with division on the basis of religious tradition cutting across class boundaries (see pp. 231–3). Finally, some argued that urban life was becoming more socially isolating for individuals with the demise of the nuclear family and rising levels of social deprivation (see Chapter 15). Over the period 1981–91 there has been a remarkably low level of non-white ethnic minority spatial redistribution (see Chapter 7). Indeed, there has been a growing metropolitan concentration, with non-white minority ethnic communities in regions, such as Greater London (which now accounts for 45 per cent of Britain's non-white ethnic minority population), the West Midlands, Greater Manchester and West Yorkshire, increasing their share of the non-white ethnic minority population (D. Phillips 1998).

These social changes have been accompanied by profound economic changes. Today, the city is no longer the dominant focus of work or the location of basic services such as retailing in the high street or the city centre (see Chapters 3 and 8). As the UK economy has moved from being a manufacturing to a service-based one, there is now greater spatial flexibility with regard to workplace locations, embracing 'out-of-town' or **'edge city'** parks and

centres (Garreau 1991) and free-standing market towns in rural areas. One consequence has been the depletion of those services that were once the cornerstone of city living. At a time when Britain as a whole lost nearly 43 per cent of its manufacturing employment (1961–91) and the major conurbations lost in excess of 60 per cent of theirs, rural areas experienced a 45 per cent increase in manufacturing jobs (North 1998).

There is also evidence that in the late 1980s and early 1990s rural businesses were more dynamic than their urban counterparts. By 1997, some 42.5 per cent of VAT (value added tax) registered businesses in England were located in rural districts (Cabinet Office 1999b), and two-thirds of rural firms are set up by in-migrants, compared with one-third of new urban firms. Moreover, jobs in high technology industry are an important source of employment in rural southern England (see Chapter 8).

16.3 THE IMPACTS OF URBAN AND RURAL CHANGE

In the last few decades there have been considerable changes in both urban and rural areas as the result of the above processes, as well as a direct result of government policy, or lack of it (see Chapter 12). We now explore some of the impacts of change in urban and rural areas.

16.3.1 Rust belts, urban decline and crime

Disadvantage is spatially concentrated and poverty is becoming entrenched following the collapse of the UK's manufacturing base and as a result cities are an important locus of the have-nots. Spatial polarisation has its own impact on culture and life chances (see Chapters 3, 4 and 12), leading to residents of some areas becoming have-nots because of the collapse of industry, such as steel in Sheffield (Beattie 1986) and discussed in *The Full Monty* film, or in former mining communities (Beatty and

Fothergill 1996) as in the film *Brassed Off*. The old industrial heartlands in the north-east USA, which have undergone the same processes of restructuring, are called the Rust Belt (Bluestone and Harrison 1982). In the UK, the old industrial heartlands do not form a contiguous belt, they are more like 'rust pockets'. The decline in manufacturing discussed in Chapter 8 has resulted in levels of high, long-term unemployment in many of these 'rust pockets', particularly former coalfields and areas of heavy engineering. This has had its greatest impact on men, but the cycle of decline impacts on the whole community as average incomes and investment decline and poor health, crime, drug abuse, suicide and family breakdown increase – the pathologies of despair.

The story of urban decline is graphically illustrated with the case study below of the St Ann's area of inner city Nottingham. It describes the area as it was in the 1960s, the extracts being taken from a report on research undertaken at that time. This area was largely the result of late nineteenth-century railway development. It is a classic example of nineteenth-century over-development and subsequent 'landlordism'. Although some of the better property was owner occupied much was rented from landlords who had no interest in maintaining, let alone improving, the property. This resulted in the slum conditions described in the case study. There was thus considerable redevelopment in the late 1960s and early 1970s. Indeed, the St Ann's redevelopment scheme was one of the biggest and most controversial in Britain. Some 10,000 Victorian slums were removed and a new estate of 3,500 houses was built. Some of the older housing was retained (Figure 16.3) and there are areas of modern and older housing. The area has a population of 8,000 in 3,800 homes. St Ann's has the fourth highest level of unemployment in Nottingham at almost 20 per cent.

16.3.2 Case study

St Ann's, Nottingham – the 1960s situation

'St Ann's is a slum . . . 10,000 houses, crushed into a space of 340 acres . . . a tightly cramped neighbourhood; smallish houses are spilled together in terraces and blocks very close to one another, at 40 houses per acre' (p. 67).

'A large deteriorated district, geographically distinct, with a certain sense of identity; perhaps, it might be expected, even a sense of community . . . threatened with comprehensive demolition and reconstruction. It is an area of manifest environmental and social deprivation . . . where until recently there have been no play facilities for the children except the yards and streets . . . where the schools are old and decrepit; with dingy buildings and bleak factories and warehouses, functionally austere chapels, a host of second hand shops stacked with shabby, cast off goods; overhung in the winter with a damp pall of smoke' (pp. 66–7).

'[An] extraordinary variety of residents, the Poles and Ukrainians from war time days, the Italians shortly after, more recently the Asians and West Indians . . . the Scots and Irish, the Geordies and Liverpudlians, all drawn to the Midlands in the pursuit of work. Some stay [in St Ann's] for a few days or weeks . . . but they all live with, in and among the people born and bred in St Ann's, a key part of Nottingham's working class' (pp. 95–6).

Source: Coates and Silburn (1970)

**Figure 16.3
Traditional inner city
housing: St Ann's,
Nottingham**

Source: David T.
Graham.

**Figure 16.4
Recently
modernised
housing: St Ann's,
Nottingham**

Source: David T.
Graham.

Much of the area has been further enhanced since through City Challenge and Estate Action urban regeneration programmes (Figure 16.4). Physical conditions have certainly improved, but are residents of St Ann's relatively any better off? Data from the 1991 Census of Population and from more recent sources show that the area still has very high levels of disadvantage (Table 16.1). Despite economic recovery and investment of public funds, St Ann's remains an area of concentrated deprivation – one of the 300 most deprived electoral wards in the country (Glennerster *et al.* 1999).

In areas like St Ann's there is a high incidence of crime, notably joy riding and 'ram raiding', as well as violent disturbances. The illegal drug industry has also increased in St Ann's and other inner city districts in the city. Drug activity is part of the informal economy (Williams and Windebank 1999). Such is the scale of these activities that the use of firearms has increased dramatically as rival gangs compete for the markets. Armed police officers now routinely patrol the streets of these areas (Whitehead 2000). Nottingham is by no means exceptional in this – it is by a variety of measures a very typical English

Table 16.1 Socioeconomic indicators, St Ann's, Nottingham, 1991–9

	St Ann's (%)	Nottingham (%)	GB average (%)
Families which are lone parents with children 1991	26	13	8
Residents who are black or Asian 1991	23	11	5
Population aged over 16 economically inactive 1991	44	41	39
Residents in households without a car 1991	60	49	33
Rate of unemployment among economically active population[a] 1999	16	9	5
Pupils passing 5 GCSEs at grades A*–C 1999	11	29	48
Pupils passing no GCSEs 1999[b]	21	10	6
Estimated population with low or very low literacy 1999[c]	19	19	15
Standard mortality ratio (1991–5) for people under 65 1999	134[d]	109	100

Source: Glennerster *et al.* 1999; ONS 1999a, *Regional Trends 34*
Notes: [a] 'Official' ward level unemployment rates are not published. To enable a comparison, these rates are calculated using the claimant count unemployed in March 1999 divided by the economically active population at the 1991 Census. They do not correspond with 'official rates' which use a different denominator. [b] Because of pupil mobility, not all children attend the local school. [c] Average for schools in England. [d] Data relate to the Nottingham East Parliamentary Constituency.

city (Graham 1994c). An escalation of serious crime in cities and towns means that armed police are increasingly becoming a regular feature on the streets of certain parts of Liverpool, London, Manchester and elsewhere.

Not all inner city areas are pockets of deprivation, however. We have seen elsewhere (Chapters 8 and 9), and above, how parts of the former industrial areas of inner cities have been renovated, as in Nottingham's Lace Market. One criticism of such development is that it tends to be exclusive, attracting mainly the young career people with a certain lifestyle. There are also areas close to city centres that have traditionally been affluent and have managed to fend off many of the problems and stigma of inner citydom. The Park Estate in Nottingham is just fifteen minutes walk from St Ann's yet could not present a bigger contrast.

16.3.3 Case study

The Park Estate

The Park is a private estate and was originally part of the demesne of the famous Nottingham Castle. Building on the Park Estate began in the early nineteenth century, the layout being carefully planned. This was a prestigious area with large houses and mansions in small but secluded grounds. In its heyday the Park housed around 4,000 people in some 650 houses. These were people of stature – lace barons, professional people, manufacturers, merchants and industrialists. Among the better known of the Park's residents were Jesse Boot (chemist/retailer), A. J. Mundella (politician/ reformer) and John Player (tobacco baron). There has been much subsequent in-filling of grounds and more recently subdivision of the larger and smaller houses. There are a number of examples of badly designed and uncharacteristic buildings, especially those built in the 1960s. The Park still attempts to retain its exclusivity by closing the gates once a year. Access to the Estate from Castle Boulevard can only be gained by means of a residents' card (Figure 16.5). It has its own street cleaners and security personnel paid for by the residents through a special levy. New residents must sign and adhere to special regulations. There is a management committee, estate manager and office. There are, however, no shops, pubs, businesses or bus services. Some exclusive sporting facilities exist.

Figure 16.5
Exclusivity: The Park Estate, Nottingham

Source: David T. Graham.

16.3.4 Religious division, segregation and exclusion in Northern Ireland

Northern Ireland is by no means the only case of exclusion on the basis of 'tradition' rather than class, but represents the most extreme form of exclusion in the UK. There is no doubt that since the cease-fires many people are beginning, in a positive way, to reflect upon wider questions of cultural and political identity. However, and without doubt, most people still possess a dislike, fear and mistrust of what they denote as the 'other' community. As a result of conflict, violence and the perpetual residential segregation of Catholics and Protestants, it is clear that geography really matters in Northern Ireland. In Belfast, for example, 80 per cent of the population lives in wards, which are either 90 per cent Roman Catholic or Protestant. However, Northern Ireland has a plethora of geographies: geographies dedicated to political and cultural devotion; geographies of violence, fear and sacrifice; geographies of victimhood and at times remorse; geographies of class and very different experiences of conflict. In sum, Northern Ireland is probably the most culturally disunified place within the UK. A disunity which was, and is being, reproduced by links across space and time and resistance to the dilution of an identified ontological 'self' (Shirlow and McGovern 1997).

In particular, within urban areas the division and control of communities by either Ulster Loyalists or Irish Republicanism is an important aspect of everyday life. In particular, the emergence of paramilitary groups in the late 1960s and early 1970s was predicated upon the defence of Protestant and Catholic territory. Moreover, in recent times the disputes over Orange marches testify to the significance of territory. A central problem in Northern Ireland is that violence has meant that many communities, mostly working-class communities, are divided in terms of religious affiliation. In the early 1970s around 50,000 people fled their homes and moved into areas which were predominantly of one religion. Territorial segregation is so pronounced that in Northern Ireland around 90 per cent of children go to either a Catholic or Protestant school, and in addition 24 out of 25 children have parents of the same religion.

As such, in areas of violent conflict people tended to seek sanctuaries of ontological togetherness. During the conflict many Protestant and Catholic communities became separated by the euphemistically called 'peaceline'. In Belfast, a city which covers a mere 100 square miles, there are over fifty miles of walls, mostly in deprived areas, which are designed to keep Protestant and Catholic communities apart. These walls, usually around 30 feet high, weave their way through working-class areas of Belfast and in so doing aim to reduce the potential for violence. Even now, in a period of sustained peace, violence, such as attacks, beating, stone throwing and petrol bombing, continues unabated along many of the 'peacelines' in Belfast. In many ways Northern Ireland has merely shifted from being a violent to a less violent society. In such a situation, where contact between communities is minimal, it is not surprising that environments have been created within which myths, hatred and fears of the 'other' community are easily reproduced.

However, not everyone in Northern Ireland lives in communities ravaged by violence and upheld by political affiliation. For many middle-class people the conflict of the past thirty years has had a less immediate impact, as it was always possible to live in affluent areas and rarely be affected by violence, biased policing and cultural discrimination. This is not to say that middle-class people did not live in fear, as they were at times victims themselves. Moreover, it is not the case that sectarian asperity is the sole preserve of the working classes, as some of the most bigoted people live within affluent areas. But it is the case that the conflict impacts more upon those who tend to be economically powerless. Indeed, within mixed workplaces you will find more victims of sectarian violence on the shopfloor than you will in the middle-class environs of the office. A reason for this is that, as is the case with all middle-class communities, points of contestation and conflict are not mentioned within 'polite company'. Bigotry, racism and hatred are kept for behind closed doors.

In more general terms not only are deprived Catholic and Protestant communities divided by loyalty but also by aspiration and feelings of self-worth. In many instances Protestants in working-class areas are obsessed with a deeply felt sense they are losing out to political advances within Republicanism, whereas many Republicans feel that their political objectives will be fulfilled, especially given demographic trends. In many instances, such emotions create confidence for one community and a strong sociocultural sense of decline within the other. However, in deprived areas one common and shared perspective is that of fear.

A survey into fear and mobility was conducted in two deprived communities of North Belfast: Upper Ardoyne (Protestant) and Ardoyne (Roman Catholic) (Shirlow and Stewart 1999). The level of victimhood within each community can be explained in relation to the 'brutalising effect of violence'. To begin with a quarter of all fatalities in the contemporary conflict occurred within a mile radius of Alliance Avenue, which is the main street which divides these communities. Such high levels of violence had been inscribed upon the consciousness of these conflictual communities.

In Upper Ardoyne the decline of the population from about 3,000 in 1971 to around 1,500 today illustrates a process of decline. The slow decline of the Protestant primary school and the restocking of the community with fewer homes provide a sense of cultural dissipation and betrayal. In many ways this decline produces a resolve to remain and to try and keep possession of a Protestant/Unionist territory. In antithesis to this the continual growth in the Ardoyne population (about 6,000 in 1991), and the inability to accommodate population overspill in Upper Ardoyne, provides a sense that policy-makers are accommodating Protestant/Unionist demands before the needs of Ardoyne. As such, community relationships are conditioned by suspicion, mistrust and low levels of reciprocity. Both communities feel besieged but in different ways.

The two areas are also peculiar because of the facilities located within them. Ardoyne possesses a street of twenty shops, bars and other retail outlets. These shops are a mere 0.2 miles from the Protestant Upper Ardoyne area. Upper Ardoyne has only one small mobile shop but contains a large leisure centre. Given that the communities are so close to each other it would be assumed that facilities are shared, especially when it is observed that few people own cars. However, for the majority of people the fear of using services in areas inhabited by people of the opposite religion means that most people would, if the service they require is not within their area, travel to use services located in areas dominated by their own religious group (Shirlow and Stewart 1999).

As a result only 18 per cent of Protestants in Upper Ardoyne shop in Ardoyne. Of these the majority are pensioners who believe that because of their age they will not be attacked. Similarly, only 19 per cent of Catholics use the leisure centre located within the Protestant Ardoyne. The majority of these respondents were women who were involved in youth clubs. In overall terms Protestant respondents travelled six times further than their Catholic counterparts when shopping and eight times further when socialising. Furthermore, Catholics travelled eight times further for leisure centres and six times further when using job clubs and welfare services. But only 12 per cent of Catholics used public transport because of the fact that the services offered all travelled through predominantly Protestant areas (Shirlow and Stewart 1999).

What ultimately emerges between the two communities is a geography of sociocultural domination and/or resistance in which power relationships were spatialised and imagined in distinct and observable ways. In particular, reactive ideological forms were primarily concerned with the definition and defensive reaction to specific cultural and social forms, which are construed as alien, hostile and unacceptable. Given that the two areas are clearly demarcated it is obvious that modes and patterns of avoidance are played out through a reactive consciousness based upon fear, mistrust and threat.

Obviously the reality and perception of fear which is tied to the religious 'other' means that many individuals cling tenaciously to the values they have acquired and feel threatened when confronted with others who live according to different conceptions of what is desirable. Thus cultural and political identity are like a 'security blanket' which has 'great meaning to its owner'. The nature and durability of fear is an evident part of contemporary living in Northern Ireland's deprived communities. Territoriality may well provide a sense of security but it is when an individual leaves their community that their fears are most pronounced. In many ways it is difficult to determine how such fears and avoidance strategies will be removed. Undoubtedly, the hope is that political stability will alter attitudes although it should be stressed that this will take a great deal of

time. As with other issues within the UK it is evident that ethnicity, racism and violence are major constituents in many people's lives, and in terms of state responses little is being done to resolve these issues (Shirlow and Stewart 1999).

16.3.5 Gentrification and the inner city

As has been noted, not all the haves have deserted urban areas in search of their 'rural idyll' – far from it. One important aspect of the changing face of urban areas today in the UK is gentrification, a residential choice of growing numbers of single and two adult heterosexual largely childless households, as well as gay and lesbian households. When Ruth Glass (1964) coined the term 'gentrification' in the context of her study of London she placed emphasis on the frustrations of commuting to London, and implied that they had probably moved from the suburbs.

While the image of urban life today is viewed in a negative way by some, this is not a view held by all. Gentrification has become a major component of the revival of specific urban localities in cities throughout the UK, as popularised in the film *Notting Hill* or in recent television programmes such as *Queer as Folk*. Through the upgrading of districts such as Clapham or Islington in London, and the redevelopment of waterfront areas in Castlefields, Manchester, London Docklands, Liverpool, Swansea or Leeds, there has been a reversal of the social status gradient of the city.

Although the extent and impact of gentrification is disputed, in some parts of British cities and for some members of the population, city living is a choice. The process of gentrification has challenged urban theory and prompted adaptations to economic models. While some have felt that gentrification was a 'back to the city' movement of disillusioned suburbanites (Laska and Spain 1980), others have noted that gentrifiers were not by and large households moving from suburban residential locations but were people moving from other inner urban addresses, such as young professionals (Smith 1979; Zukin 1982).

Fielding (1992) focuses on the capital as a 'social escalator' region, and in a recent study of solo households the intertwining of labour markets and housing markets has been emphasised (Hall *et al.* 1999). They conclude that migration emerges as a key link between professionalisation and solo households, with large amounts of rented property in inner London providing the necessary flexibility of the housing market to permit high levels of in-migration. People living alone epitomise a professional, independent mobile lifestyle – postmodern individualism (Harvey 1989) – facilitated by the nature of the housing and labour markets in some inner cities like London. Another aspect of inner city living is that work and leisure spaces are becoming blurred for some young City professionals who regularly work in excess of 10–12 hours per day (McDowell 1997). Work and leisure can be combined in the gym, the restaurant or at a party (see Chapter 9).

In a study of the meanings of particular housing moves to two inner areas of Edinburgh, Liz Bondi (1999) found that some areas attracted a range of house purchasers, not all of whom were gentrifiers in the sense of contributing to changes in the class composition of an area. Of those that can be classified as gentrifiers, there is diversity – not all are in their twenties and thirties, and there are some nuclear families. She concludes that gentrification is less a lifetime alternative to suburban lifestyles and more a staging post on a journey likely to proceed towards parenthood and suburban or rural and semi-rural living.

16.3.6 In search of the rural idyll, NIMBYism and green belts

The process of outward movement from cities to the surrounding countryside has accelerated in the last two decades within a context of increasing affluence and a rising rate of private car ownership, but the rate of the rural population turnaround is not uniform throughout the UK (Champion *et al.* 1998; Lewis 1998). Counterurbanisation is a complex and multifaceted process, and the physical expansion of cities through suburban development and in freestanding towns and villages has been governed by planning controls since the late 1930s. Counterurbanisation accentuates polarisation in both urban and rural areas, as the haves leave urban areas (see below).

The reasons for migrating to rural areas has linked service class migration and rural gentrification, but the service class connection has also been critiqued (M. Phillips 1998). A study of twelve areas in England and Wales in the early 1990s highlighted the diversity of in-migrants, especially the fact that not all commanded high annual incomes (Cloke *et al.* 1995). Rural in-migrants include households with dependent children seeking what they perceive to be a better environment (lower crime levels and less pollution), an 'idyllic' alternative to the perceived problems of urban areas coupled with better schools (M. Phillips 1998), as well as younger 'retirement' migrants no longer constrained locationally by the demands of paid work or the needs of children still at home, who are regarded as moving for 'amenity' or 'lifestyle' reasons (Hardill *et al.* 1997). In-migrants also include those seeking to maximise their assets by moving to areas with lower house prices (working age and early retirees), as well as those 'opting out' of the 'rat race' (see Case study 8.6.1 on pp. 91–2).

As was noted above, it has been argued that in-migration to rural areas reflects the desire for a particular kind of living space. And this space extends beyond the house itself into the surrounding environment (M. Phillips 1998). This feature is related to the notion of NIMBYism, whereby recent rural in-migrants seek to prevent further developments in their backyards, even though they themselves may be living in new housing developments. Recent debates over future scales of housing provision have been particularly acute in the South East resulting in a number of cases of NIMBYism. They came to a head recently with the Deputy Prime Minister's announcement (Prescott 2000) that some 860,000 new homes will be needed within the South East, outside London, over a twenty-year period. This figure falls between a lower recommended scale of provision of 718,000 new homes, as set out in draft Regional Planning Guidance (SERPLAN 1998), and the report of a Panel led by Professor Stephen Crow (DETR 1999b), which oversaw the subsequent public examination of this draft Regional Planning Guidance and called for 1.1 million additional homes.

Rural in-migration has adversely impacted on locals in a number of ways. There is now a shortage of affordable housing, caused by the sale of rural council housing (which has always been in short supply) and by rising house prices due to demand from in-migrants (Cloke and Little 1997). Even discounting the needs of in-migrants, it has been estimated that in rural districts in England alone up to 10,000 additional affordable homes would be needed annually (Countryside Agency 2000). In addition to in-migration, second and holiday home purchases are fuelling the private housing market. For example, 58 per cent of England's second homes are in rural districts, even though the percentage of homes which are second homes remains low (0.9 per cent in rural areas compared to 0.4 per cent nationally). In some areas, however, the proportion is significant, such as 20 per cent in the Isles of Scilly and 16 per cent in Windermere (Cabinet Office 1999a). Holiday homes, rented to others, and occasionally used by the owners, account for a further 300,000 properties (Hetherington 1999c). In some localities, especially in North Wales and Scotland, second homes bought by English people have caused much resentment and even violence.

While some rural in-migrants have bought 'character' properties, most have bought houses in small estates in villages and small towns. The physical expansion of the housing stock in rural areas has been subject to planning control, especially in the post-war period. Concerns about the physical expansion of urban areas, such as London and other big cities, have been raised since the end of the nineteenth century (Johnson 1974). The merits of controlling developments on the urban fringe were debated as suburbanisation accelerated. In the urban fringe, scattered and piecemeal residential and commercial developments occurred and there were problems of reserving land for agriculture and ensuring that it remained economically viable, along with the reservation of recreation land. This concern led to the enactment of the Green Belt Acts in 1938 and 1947 (Cullingworth and Naden 1997).

In subsequent years, as a result of the general continuity of green belt policy by successive central administrations, green belts have become probably the best known tool of land use policy in both England and Scotland and have enjoyed widespread public support. Despite continued pressures for development, the amount of land designated as green belt has not fallen in recent years. Nevertheless, the use of green belts as a policy tool has not been without its critics. Green belts, along with restrictive planning policies in a more general sense, have come under fire, especially from free market supporters during the 1980s, for exerting a stranglehold around cities and preventing necessary economic growth and prosperity. Despite local perceptions that green belt policies were often ignored by central government and their planning inspectors during the Thatcher/Ridley era of the mid- to late 1980s – perhaps partly because of an understandable blurring in the public's mind of the distinction between formally designated green belt and other local designations (such as green wedges, landscape areas and the like) and wider countryside areas which were afforded less protection – green belt policy (and indeed the wider planning system) actually emerged fairly intact into the more environmentally sensitive 1990s (Thornley 1993).

Having survived the free marketers of the 1980s, ironically it is current concerns to promote the most sustainable forms of future development, linked to debates over housing pressures particularly in the South East, that are increasingly challenging the future role and purpose of green belts. Economic and demographic pressures from cities do not disappear with the creation of green belts, and it has long been recognised that the physical presence of green belts has thus shifted developments elsewhere. Ideally, this should help to encourage the reuse of derelict urban land, but it also increases development pressures beyond the green belt (Barker 1999). Indeed, there is little evidence that green belts have had much impact in encouraging the redevelopment of land within their core cities.

For example, a recent study of Nottingham (completely surrounded by green belt), Derby (green belt mainly to north and east) and Leicester (no green belt), revealed little fundamental difference between levels of urban regeneration in the cities with or without a green belt (Baker Associates 1999). Green belts have also been accused of artificially keeping house prices higher than they would be otherwise, thus exacerbating problems of social exclusion and affordability, although again firm research evidence either way is somewhat lacking and counter arguments have been put forward.

Recent statements of government housing policy (DETR 2000; Prescott 2000) have embraced the need to make efficient use of land and respect the countryside, leading to the introduction of a national target that 60 per cent of new homes should use recycled land or buildings. However, this is within the context of a projected increase of some 3.8 million households from 1996–2021 (DETR 1999c) and the recognition that much of the resulting demand for houses – perhaps up to two million homes – will still need to be accommodated on greenfield sites.

Irrespective of the eventual scale of new development, considerations of how to provide the necessary housing land allocations in the most environmentally sustainable way, particularly focusing on reductions in private transport, have led government policy to emphasise a 'sequential' search approach to land supply, starting with the reuse of previously developed land, followed by urban extensions that can be well connected into the existing urban areas and existing public transport networks, and finally by new development elsewhere around nodes in good transport corridors. Such an approach can, however, conflict with established green belts that do not generally allow for urban expansion. Although still considered as an exception to normal green belt policy, the government's latest policy guidance for housing (DETR 2000) thus acknowledges that there may be a case for reviewing some existing green belt boundaries, and planning for new development where an extension into the green belt may be preferable to new development elsewhere in a less sustainable location.

Nevertheless, development pressures driven by household growth and the continuing urban exodus of both people and jobs will remain at odds with the equally strong desire to protect the countryside from further encroachment. This is amply witnessed by a current high-profile *Sunday Times* newspaper campaign to save the countryside which stresses ideals of sustainability and environmental protection but, given the newspaper's generally middle-class readership, is no doubt also driven by elements of NIMBYism and a desire to protect a relatively affluent quality of life in the 'rural idyll'. In such circumstances, green belts as a long-standing, popular, easily understood and generally robust policy tool are not going to disappear overnight. Related to NIMBYism and countryside issues is the emergence of collective action against some developments (especially transport) by 'eco-warriors'. Two much publicised protests – the Newbury bypass and the extension to the runway at Manchester Airport – made one ecowarrior, 'Swampy', a national figure.

16.3.7 The 'other' rural

Many rural people, however, experience lives characterised by low income and poverty rather than prosperity and choice (Cloke *et al.* 1995). The majority of the ten areas with the lowest gross domestic product (GDP) per head in England in 1996 were rural – Cornwall and the Scilly Isles rank below Merseyside on GDP per head (Cabinet Office 1999a). Poverty in rural areas is a significant and persistent problem, though less prevalent than in urban areas. The proximity of affluent and deprived households in rural areas makes it harder to identify social exclusion in statistical data, the have-nots 'being out of sight and out of mind' (Cloke *et al.* 1995: 83). That said, socioeconomic polarisation is more acute in rural areas. As more wealthy people leave the towns for the rural areas this polarisation increases. A third of households in one Wiltshire village have incomes below £6,000, with a further third above £40,000; 27 per cent of Cotswolds District households have annual incomes of less than £7,000 (Countryside Agency 2000).

A recent survey, based on a sample of 5,000 household incomes, suggested that 30 per cent of individuals in rural Britain had experienced poverty in the last decade, compared with 40 per cent in urban areas (Cabinet Office 1999b). Hidden unemployment is higher in rural areas (16.3 per cent), and a higher proportion of rural people are dependent on part-time work (26.4 per cent) and seasonal jobs. Thus the lowest weekly wages in England are in Cornwall (£297), Northumberland (£315), Isle of Wight (£323) and Shropshire (£339), compared with an average of £405 in England (Countryside Agency 2000).

The underlying structural causes of rural poverty include the restructuring both of traditional rural industries (notably agriculture) and of those industries (particularly in the manufacturing sector) which have replaced them; the persistence of low pay and poor working conditions; the paucity and cost of 'affordable' housing to rent or buy; the withdrawal of available services and the collapse of public transport systems – the latter obliging poorer rural families to deplete their disposable incomes by buying and running a car (Cloke *et al.* 1995). According to the Rural Development Commission people retiring to rural areas and low wages are the main contributors to rural poverty (RDC 1996).

Housing is a major issue facing local rural areas and is of considerable importance to the viability of rural economies and rural communities. It is estimated that the number of households in rural districts is projected to increase by one million from 1991 to 2011 (Countryside Agency 1999). Yet the report notes that there was an estimated need for 80,000 additional affordable homes in rural areas between 1990 and 1995; but since 1990, less than 18,000 have been built. Thus, while homelessness is falling in urban areas it is increasing in the most rural areas. One in six young adults leave rural areas because they feel forced out by the lack of affordable housing (Countryside Agency 2000). Lack of services is also a major problem, particularly for the rural poor. Something like 70 per cent of rural parishes in England have no general store, 43 per cent have no post office, 83 per cent no general practitioner based in the parish, 91 per cent no dentist, 89 per cent no pharmacy and 75 per cent have no daily bus service (Countryside Agency 2000).

As previously noted, in many rural areas accessibility to public transport is severely limited outside the main market towns, and most rural settlements do not have a rail service (Countryside Agency 2000). Only 16 per cent of households in rural areas are within a six-minute walk of a bus stop which has a service at least every half an hour, compared to 77 per cent in Great Britain as a whole (Cabinet Office 1999b: 65). This makes rural dwellers more dependent upon a car for access to paid work, and for shopping and leisure activities (see Chapters 8 and 9). As a result 22 per cent of rural households do not have a car, compared with three in ten of all households in Great Britain (DETR 1998). Some rural areas still face barriers to economic dynamism associated with remoteness or an over-dependence on declining industries such as mineral extraction or agriculture. Social exclusion exists in rural areas and can be harder to tackle because individuals are geographically dispersed.

16.4 SUMMARY

As we have seen throughout this book, the patterns evident from any aspect of the human geography of the UK will depend on the lens employed. The spatial scale of analysis is thus important in any analysis. However, whatever scale is adopted, the UK can perhaps best be viewed as a state riven with divisions – many of them, such as the gap between rich and poor, healthy and unhealthy, British or 'other', excluded or included, growing wider. The geography of polarisation and division is fairly obvious at the macro level of the region, as we saw in Chapter 14. However, at finer resolution, as this chapter has shown, there are significant variations in the lived experience of people and communities who are geographical neighbours but might be in entirely different countries. They may be occupying the same geographical space but are operating in entirely separate social and economic spheres. As Hetherington (1999d: 24) summed up in an analysis of the micro-geographical divide in the UK, 'Poverty and plenty cheek by jowl. That is the real divide.'

REVISION QUESTIONS

- How important is the issue of geographical scale in understanding socioeconomic polarisation?
- Why have successive governments found it so difficult to tackle the problem of socioeconomic polarisation?
- What role does London play in the social polarisation debate?

KEY TEXTS

Barnett, A. and Scruton, R. (eds) (1999) *Town and Country*, London: Vintage.

Dorling, D. and Woodward, R. (1996) 'Social polarisation 1971–1991: a microgeographical analysis of Britain', *Progress in Planning* 45(2): 69–122.

Green, A.E. (1994) *The Geography of Poverty and Wealth*, Warwick: IER.

Philo, C. (ed.) (1995) *Off the Map: The Social Geography of Poverty in the UK*, London: CPAG.

Woodward, R. (1995) 'Approaches towards the study of social polarization', *Progress in Human Geography* 19(1): 75–89.

Glossary

(* terms appear in bold at first mention in text)

Aetiology: assignment of cause.

Ageism: discrimination on the basis of age, and embraces lack of opportunity, being disregarded and undervalued; a feeling of being 'over the hill'.

AIDS: acquired immune deficiency syndrome. A condition whereby the immune system is so compromised that it succumbs to opportunistic infections, particularly pneumonia and TB.

Bicameral: having two legislative chambers.

Brownfield: land previously used for industry or housing, often contaminated and in an urbanised area.

BSE: bovine spongiform encephalopathy (mad cow disease). One of many diseases that infect the brains of animals. Once infected the subject's brain rapidly degenerates as sponge-like holes develop in the organ.

Chronic disease: degenerative diseases, non-communicable diseases often associated with older populations – cancers, stroke, heart disease and so on. Sometimes called 'diseases of affluence'.

Citizenship: the various rights and obligations exercised by members of a community. In its modern form it is coupled with the nation-state, although since 1994 the European Union has implemented a European citizenship. The best-known exponent of the development from the eighteenth to the twentieth centuries of the civil, political and social rights associated with citizenship is the British sociologist T.H. Marshall (1950).

Consumption: the use of commodities for the satisfaction of needs and desires. A key, and increasingly more important, feature of capitalism.

Counterurbanisation: a process of population deconcentration. The movement of people from the larger towns to smaller towns and villages. This has qualitative connotations as well as quantitative.

Crude birth rate: number of births in a population over a given time period per 1,000 total population. Takes no account of age and sex structure.

Crude death rate: number of deaths in a population over a given time period per 1,000 total population. Takes no account of age and sex structure.

Culture: set of social customs and symbols in a given group.

Deindustrialisation: a term coined to describe the sustained decline in industrial (especially manufacturing) activity. It may involve the absolute and/or relative decline in industrial output, employment and the means of production. The causes of deindustrialisation are complex. In the case of the UK economy, which was the first advanced capitalist economy to exhibit signs of deindustrialisation, it was a combination of local circumstance and locational adjustment to global circumstances (for a fuller discussion see Martin and Rowthorn 1986).

Demographic transition theory: a paradigm that attempts to account for changes in population structure that occur during economic and social development. In the early stages, both fertility and mortality are high and fluctuate. With development, mortality declines, while fertility remains high. This results in population expansion. Through time, fertility also declines and both remain low and fluctuating.

Devolution: the passing down of power from a higher authority.

Diaspora: a dispersal of people, originally referred to the Jews. Now refers to any communities with common roots but living outside the homeland.

Disneyfication: theming or stylising a development around a brand image. A process of creating uniformity and banality at the expense of distinctiveness and personality. The triumph of style over substance.

Disposable income: that part of a person's or household's income remaining when the basic bills have been paid. Not all leisure will be met out of disposable income – a telephone and television, for example, are now generally regarded as necessities. Similarly, with consumption. Food, drink and clothes are necessities, but here the quantities and qualities of those that are consumed can cause the boundaries to blur. Are designer clothes a necessity? Are mangoes? Is spring water? Also, many consider cigarettes and alcohol necessities.

Ecowarrior: environmental protester, usually prepared to take militant action to prevent environmental threat – airport expansion, bypass road, quarry development and the like. Often associated with a certain lifestyle – new age traveller, 'drop out', and so on.

Edge city: term first coined in North America (Garreau 1991) for developments (housing, retail such as shopping malls and commercial, offices and so on). The car creates Edge City as much as ICTs do. They provide an escape from the more threatening aspects of metropolitan life (Zukin 1991).

Enumeration district (ED): the smallest geographical unit for the gathering of Census of Population data. An ED would usually include between 150 and 250 households.

Epidemiological transition theory: as with the demographic transition theory this is based on the idea of economic and social development causing a population shift. Here, development causes a shift from infectious diseases as the main cause of death to chronic diseases. In less developed societies infectious disease flourishes and kills mainly infants and children. As infrastructure, medical care, diet and education improve so infectious disease is less common and less deadly. Population expands and ages. Since death is inevitable more and more people die in old age of chronic degenerative diseases, such as cancer and heart disease.

Ethnic group: a group within a larger society that considers itself to be different because of a common ancestry, a shared past, and a cultural focus upon one or more symbolic elements which define the group's identity, such as kinship, religion, language, shared territory, nationality or physical appearance.

Ethnicity: set of attributes, such as language, lineage, religion, culture, physical features and the like, which characterise certain (usually minority) groups of people.

Faustian bargain: making a deal that has a high price, after the character who makes a deal for immortality with the devil in return for his soul.

Feminisation of the labour market: a process whereby women have increased their share of employment, with more women remaining in the labour market after childbirth/childrearing. But feminisation is also associated with low wages and insecure employment.

Fenian: secret Irish organisation founded in 1858 to overthrow British rule in Ireland, named after the *féne*, an ancient Irish people.

First-past-the-post: an electoral system whereby a party or individual contesting an election wins as a result of a simple majority of the votes.

Flexitime: a mosaic of people working on different, more individualised schedules.

Gàidhealtachd: the Gaelic-speaking part of the Highlands and Islands.

Gazumping: raising the price of a house after accepting an offer from a potential buyer.

Gentrification: term first coined by Ruth Glass (1964) in the context of a study of London. The term is used to describe the process of neighbourhood regeneration by relatively affluent incomers, who displace lower income groups and invest substantially in improvements to homes which have deteriorated. Such neighbourhoods are usually accessible to the city centre and comprise older dwellings (Johnston *et al.* 1983).

Global city: a city with a range of corporate and cultural functions, and the site of global trading and communications networks.

Governance: process of formation and implementation of public policy involving elected and non-elected organisations.

Green belts: part of the post-war planning response to the fears of urban sprawl. They comprise an area of open land surrounding urban settlements where further building is strictly controlled.

Greenfield: land previously in agricultural or horticultural use, usually in urban fringe or rural area.

Hegemon: dominant power, especially by one state within a confederacy.

HIV: human immunodeficiency virus. One of several viruses causing failure of the immune system.

Hot-desking: is where employees do not have a dedicated desk at their place of employment.

Hotelling: this is a variation on hot-desking, whereby business meetings are scheduled for coffee shops and hotel lounges.

Housing associations: private, non-profit-making bodies established to supply a 'third arm' of housing policy, providing a gap between local authority rental and owner occupation.

Identity: feeling of belonging characterised by a set of cultural determinants.

Infant mortality rate: deaths under age one per 1,000 live births.

Infectious disease: parasitic, communicable diseases often associated with younger populations – malaria, cholera, typhoid, plague and so on. Sometimes called 'diseases of poverty'.

Inner city: ill-defined area close to the city centre which is usually associated with physical decay and economic and social deprivation, but also with crime, drugs and, in some cities, with migrant communities such as Brick Lane in London.

Joined up government/thinking: looking more holistically at the likely impacts of policies, for example the environmental impact of transport policies.

Keynesian: economic focus named after Lord John Maynard Keynes working in the 1930s. Keynes believed that advanced capitalist economies worked best with strong government control of the economy and vigorous public spending programmes.

Labour market: a mechanism whereby labour power is exchanged for the material conditions of its reproduction. Labour markets are characterised by internal discontinuities, such as those induced by gender, race, age or skill.

Laddism: a male culture which may be seen as a reaction to the idea of the caring, sensitive 'new man' produced by the feminist movement. So laddism is characterised by a climate of rough behaviour, excessive drinking (lager louts) and all-male attendance at football matches (Storry and Childs 1997).

Leisure: an attitude, social activity or period of time which is generally regarded as distinct from work or required household tasks. Again the boundaries can blur – for example, shopping and DIY.

Life expectancy at birth: average number of years that a group of people born in a given year could expect to live assuming that the age-specific mortality rates of that year would be maintained throughout the life history of the group.

McDonaldisation: see Disneyfication.

Morbidity: in a state of disease or ill health.

Multicultural: consisting of a number of cultures whose influence impacts on the overall cultural milieu.

Multiethnic: consisting of a number of ethnic groups whose different identities help create a hybridised culture and identity.

Nationalism: an idea of belonging to a group linked by common heritage, language and religion, which manifests itself in a claim for territorial integrity.

Net migration: the difference between in-migration and out-migration.

NHS: National Health Service. The inclusive service set up in 1948 by the post-war Labour government.

NIMBYism: acronym for 'not in my backyard', associated with an attitude adopted towards new housing and other developments in rural and gentrified urban areas. They tend to campaign for developments to take place elsewhere, not in their village/town.

Non-standard employment: is the term that includes a whole range of job contracts such as part-time work, temporary job contracts, etc.

nvCJD: new variant Creutzfeldt–Jakob disease. Creutzfeldt–Jakob disease is the human spongiform encephalopathy. There are a number of modes of transmission but the new variant is thought to be the result of BSE crossing the species barrier from cattle to humans.

Paid work: for wages is also referred to as work for economic relations.

Perinatal mortality rate: stillbirths and deaths in the first week of life per 1,000 total births (both live and still).

Petty bourgeoisie: small scale entrepreneurs and shopkeepers.

Poverty trap: the result of efforts to relieve poverty by providing benefits. It arises because benefits are withdrawn the higher up the income scale a household moves. With higher income, therefore, a household faces a rising tax bill and the reduction or withdrawal of its social security benefits.

Proportional representation: an electoral system whereby a party contesting an election receives the same percentage of seats as it does of the votes.

Proto-industrial: refers to the phase that preceded and prepared for the industrialisation of capitalist economies.

Quangos: quasi-autonomous non-governmental organisations. Bodies, staffed by political acolytes, set up ostensibly as executive bodies that could deliberate quickly at the expense of local democracy.

Race: measure of categorisation based on certain secondary physical characteristics determined by genetics rather than environment.

Regional: study of geography through areal differentiation.

Restructuring: change in and/or between the constituent parts of the economy.

Rivers of blood: emotive and inflammatory speech on 'race' relations made by Conservative Shadow Defence Minister Enoch Powell to an audience in Birmingham in April 1968. Predicting large-scale ethnic conflict, he declared: 'As I look ahead I am filled with foreboding. Like the Roman, I seem to see the River Tiber foaming with much blood.' Although Powell was dismissed from the Shadow Cabinet he received over 100,000 letters of support for the speech and the sentiments contained therein.

Rust belt: this is a term that is more often used for the older manufacturing areas of the north-east USA than the UK.

Services: activities which are relatively detached from material production and which as a consequence do not directly involve the processing of physical materials. The main difference between manufacturing and service production seems to be that the expertise provided by services relies much more directly on workforce skills, experience, and knowledge than on physical techniques embodied in machinery or processes.

Social class: a sevenfold classification of the population based on occupation: I professional (e.g. doctors, lawyers); II managerial and technical (e.g. managers, teachers); IIIN skilled non-manual (e.g. office workers); IIIM skilled manual (e.g. bricklayers, coalminers); IV semi-skilled (e.g. postal workers); V unskilled (e.g. porters, labourers); OTHER (i.e. armed forces, social class not stated, unemployed whose previous occupation was more than eight years ago, or those who have never had a job).

Social escalator: a function of metropolitan areas, identified by Fielding (1992), allowing professional people to advance their career in their younger years by moving to a metropolitan area then get off the 'escalator' later in their careers and move out of the metropolitan area.

Standardised mortality ratio: a figure calculated by comparing the age- and sex-specific mortality rates in the study population with those in a reference population. The rate for the reference population will be 100, rates below this show favourable mortality while those above show unfavourable mortality. This is much more accurate than the crude death rate.

Suburbanisation: the development of the urban fringe, whereby new homes are constructed in an outlying district lying within the commuting zone of an area.

Suez: débâcle in 1956 when British and French forces attempted to force the Egyptians, who had recently nationalised the Suez Canal Company under Nasser, out of this strategic area. The British and French were forced to back down as a result of pressure from the USA, who were in turn under pressure from Nasser's sponsors, the USSR.

Systematic: the study of a particular element of geography such as agriculture or settlement, seeking to understand the processes that influence it and the spatial patterns that cause it.

TB: tuberculosis is a very infectious disease usually attacking the pulmonary system. It is a common cause of death among AIDS patients. New strains have developed that are resistant to standard chemotherapeutical measures.

Teleworking: this refers to someone who works (for economic relations) at a place other than where the results of the work are needed using Information and Communication Technologies (ICTs) (Stanworth 1998: 53).

Thatcherism: can be viewed in three different ways. First it is said to be a style of government, one of determination to ensure that a certain policy is carried, popularly called 'handbagging' as Mrs Thatcher always carried a handbag. A second way is through the imposition of 'new' right thinking on both the shape and content of state policy, through marketisation, and 'rolling back' the frontiers of the state (privatisation, the supremacy of the market and capital). A third way of viewing Thatcherism is as an electoral coalition, of an attempt to mobilise members of the working class to support the Tories, thereby detaching their traditional allegiance from the Labour Party.

Total period fertility rate: a very precise measure of fertility which indicates the number of children that would be born to a woman before age 50 assuming that none of the women die before this age and that age-specific birth rates remain the same as for the year of calculation. A figure of 2.1 children is considered to be enough for a population to replace itself. This is much more accurate than the crude birth rate.

Unemployment trap: similar to the poverty trap in that its existence is due to the availability of benefits. However, whereas the poverty trap affects those in employment, the unemployment trap affects those who are unemployed. In some cases the benefits available when unemployed are equal to or greater than the after-tax income that would be earned by accepting employment. The effect is to act as a disincentive for those currently unemployed to seek employment.

Unpaid work: is also called work for the tasks of social reproduction.

Ward: a geographical unit used for the purposes of local elections. Also used for the Census of Population. Each ward is divided into several enumeration districts.

Welfare state: set of socioeconomic policies that seek to alleviate poverty and equalise social and economic well-being through redistributive mechanisms controlled by central government.

Work-poor households: households containing two or more adults living as a couple, with or without other household members, in which both partners are not in paid employment.

Work-rich households: households containing two or more adults living as a couple, with or without other household members, in which both partners are in paid employment.

Bibliography

Abrams, F. (1999) '"Deprived" English call for equality', *The Independent*, 12 May, p. 2.

Adam, B. (2000) 'The media timescapes of BSE news', in S. Allan, B. Adam and C. Carter (eds) *Environmental Risks and the Media*, London: Routledge, pp. 117–29.

Adams, T. (1999) 'Francis thought it was all over. He still thinks so now . . .', *The Observer Review*, 6 June, pp. 2–3.

Adelstein, A.M. and Marmot, M.G. (1984) 'Migrant studies in Britain', *British Medical Bulletin* 40(4): 315–19.

Ahmed, K. (2000a) 'Blair's birthday blues', *The Observer*, 30 April, p. 17.

—— (2000b) 'Women lose patience with Labour', *The Observer*, 18 June, p. 2.

Ahrendt, D. and Young, K. (1994) 'Authoritarianism updated', in R. Jowell *et al.* (eds) *British Social Attitudes the 11th Report*, Aldershot: Dartmouth Publishing, pp. 75–93.

Allen, J., Massey, D. and Cochrane, A. (1998) *Rethinking the Region*, London: Routledge.

Amin, A. (ed.) (1994) *Post-Fordism: A Reader*, Oxford: Blackwell.

Anderson, A. (2000) 'Environmental pressure politics and the "risk society"', in S. Allan, B. Adam and C. Carter (eds) *Environmental Risks and the Media*, London: Routledge, pp. 93–104.

Anderson, B. (1991) *Imagined Communities: Reflections on the Origins and Spread of Nationalism*, London: Verso.

Anderson, P. (1992) *English Questions*, London: Verso.

Andorka, R. (1978) *Determinants of Fertility in Advanced Societies*, London: Methuen & Co. Ltd.

—— (1982) 'Lessons from studies on differential fertility in advanced societies', in C. Höhn and R. Mackensen (eds) *Determinants of Fertility Trends: Theories Re-examined*, Liège: Ordina Editions, pp. 21–33.

Anning, V. (1999) 'Time to talk about sex', *Evening Post*, 19 October, pp. 6–7.

Anon. (1999) 'Linguistics', *Private Eye*, 10 December, p. 11.

Anthony, A. (1999) 'Passport to Padstow', *The Observer Magazine*, 3 January, pp. 22–7.

Arlidge, J. (1999a) 'Revealed: Britain's filthiest air', *The Observer*, 21 March, p. 5.

—— (1999b) 'Sad culture kills 12 "lads" a week', *The Observer*, 17 October, p. 13.

—— (1999c) 'England expects a day to remember', *The Observer*, 18 April, p. 10.

—— (1999d) 'New lads to new sads', *The Observer*, 14 March, p. 14.

—— (1999e) 'Girl power gives boys a crisis of confidence', 14 March, p. 3.

Arlidge, J. and McVeigh, T. (1999) 'There's 14 and there's 14', *The Observer*, 14 November, pp. 10–11.

Armitage, B. (1997) 'Variation in fertility between different types of local area', *Population Trends* 87: 20–8.

Armstrong, G. (1999) 'Kicking off with the wannabe warriors', in M. Perryman (ed.) *The Ingerland Factor: Home Truths from Football*, Edinburgh: Mainstream, pp. 45–58.

Arthur, M.B., Inkson, K. and Pringle, J.K. (1999) *The New Careers: Individual Action and Economic Change*, London: Sage.

Ascherson, N. (1999) 'England's Lords prepare for

another thousand years in fear of the mob', *The Observer*, 18 July, p. 33.

Atkinson, A.B. (1995) *Incomes and the Welfare State: Essays on Britain and Europe*, Cambridge: Cambridge University Press.

Atkinson, M. (2000) 'Factories fight back but strong pound takes its toll', *The Guardian*, 10 May, p. 23.

Bagehot (1999) 'An independence beyond their ken', *The Economist*, 20 November, p. 47.

Baggott, R. (1998) *Health and Health Care in Britain* (2nd edition), Basingstoke: Macmillan.

Baines, S. (1999) 'Servicing the media: freelancing, tele-working and "enterprising" careers', *New Technology, Work and Employment* 14(1): 18–31.

Bakari, I. (1989) 'Calypso and reggae', in R. Samuel (ed.) *Patriotism: The Making and Unmaking of British National Identity, Volume III, National Fictions*, London: Routledge, pp. 99–122.

Baker Associates (1999) *Strategic Sustainability Assessment of the Nottingham–Derby Green Belt in the East Midlands Region*, London: DETR.

Baker, R. (1999) 'The death of the nuclear family', *The Independent*, 19 April, p. 6.

Baldwin, S. (1937) *On England*, Harmondsworth: Penguin.

Bale, J. (1982) *Sport and Place: A Geography of Sport in England, Scotland and Wales*, London: C. Hurst and Company.

—— (1989) *Sports Geography*, London: E. & F.N. Spon.

Balsom, D. (1990) 'Wales', in M. Watson (ed.) *Contemporary Minority Nationalism*, London: Routledge, pp. 8–23.

Bannister, C. and Gallent, N. (1998) 'Trends in commuting in England and Wales – becoming less sustainable?', *Area* 30(4): 331–42.

Bardsley, M. and Morgan, D. (1996) *Deprivation and Health in London: An Overview of Health Variations within the Capital*, London: The Health of Londoners Project.

Barker, P. (1999) 'Edge city', in A. Barnett and R. Scruton (eds) *Town and Country*, London: Vintage, pp. 206–16.

Barnes, J. (1997) *Federal Britain? No Longer Unthinkable*, London: Centre for Policy Studies.

—— (1998) *England, England*, London: Cape.

Barnett, A. (1999) 'Girl of 12 makes her mother a granny at 26', *The Observer*, 29 August, p. 1.

Barnett, A. and Scruton, R. (1999) 'Introduction', in A. Barnett and R. Scruton (eds) *Town and Country*, London: Vintage, pp. xi–xx.

Bartley, M., Montgomery, S., Cook, D. and Wadsworth, M. (1996) 'Health and work insecurity in young men', in D. Blane, E. Brunner and R. Wilkinson (eds) *Health and Social Organization: Towards a Health Policy for the 21st Century*, London: Routledge, pp. 255–71.

Baruch, Y. (2000) 'Teleworking benefits and pitfalls as perceived by professionals and managers', *New Technology, Work and Employment* 15(1): 34–49.

BBC1 (2000a) *Panorama: Speed*, 14 February.

—— (2000b) *Panorama: England's Shame*, 20 June.

—— (2000c) *Panorama: Losing Control*, 28 February.

Beattie, G. (1986) *Survivors of Steel City: A Portrait of Sheffield*, London: Chatto & Windus.

Beatty, C. and Fothergill, S. (1996) 'Labour market adjustment in areas of chronic industrial decline: the case of the UK coalfields', *Regional Studies* 30(7): 627–40.

Beaumont, A. (1998) 'Oh baby! Times are a changin'', *Evening Post*, 8 July, p. 5.

Beck, U. (1992) *Risk Society: Towards a New Modernity*, London: Sage.

Bell, A., Bradley, M. and Kemp, A. (1999) 'New nation, old bigotry', *The Observer*, 15 August, p. 17.

Bell, D. and Valentine G. (eds) (1997) *Consuming Geographies*, London: Routledge.

Bell, I. (1999) 'Wake up England', *Red Pepper*, April, http://www.redpepper.org.uk/xscot.html

Benjamin, B. (1965) *Social and Economic Factors in Mortality*, Paris: Mouton.

—— (1989) *Population Statistics: A Review of UK Sources*, Aldershot: Gower.

Benneworth, P. (2000) 'Reaching out, regional development agencies and evolving regional governance', *Regions* 227: 22–40.

Bennie, D. (1995) *Not Playing for Celtic*, Edinburgh: Mainstream.

Beynon, H. (1999) 'The end of the industrial worker?', in J. Bryson, N. Henry, D. Keeble and R.

Martin (eds) *The Economic Geography Reader*, Chichester: Wiley, pp. 357–60.

Beynon, H., Hudson, R. and Sadler, D. (1991) *A Tale of Two Industries: The Contraction of Coal and Steel in the North East of England*, Milton Keynes: Open University Press.

—— (1994) *A Place Called Teesside: A Locality in a Global Economy*, Edinburgh: Edinburgh University Press.

Bibby, A. (1999) 'Show me the way to go home . . .', *The Observer, Cash*, 11 July, p. 2.

Birley, D. (1993) *Sport and the Making of Britain*, Manchester: Manchester University Press.

—— (1995) *Land of Sport and Glory: Sport and British Society 1887–1910*, Manchester: Manchester University Press.

Bishop, P. (1991) 'Constable country: diet, landscape and national identity', *Landscape Research* 16(2): 31–6.

Black, J. (2000) *Modern British History since 1900*, Basingstoke: Macmillan.

Blackburn, C. (1999) 'Poor health, poor health care', in M. Purdy and D. Banks (eds) *Health and Exclusion*, London: Routledge, pp. 26–44.

Blair, T. (2000) 'Britain and Britishness', speech, 28 March.

Blake, A. (1999) 'Chants would be a fine thing', in M. Perryman (ed.) *The Ingerland Factor: Home Truths from Football*, Edinburgh: Mainstream, pp. 109–18.

Blake, J. (1968) 'Are babies consumer durables?', *Population Studies* 22: 5–25.

Blanchflower, D.G., Oswald, A.J., Baker, B. and Sandbach, J. (1996) 'The area labour cost adjustment: empirical analysis and evidence of a new approach, project C', Report prepared for the Department of the Environment, National Economic Research Association.

Blane, D., White, I. and Morris, J. (1996) 'Education, social circumstances and mortality', in D. Blane, E. Brunner and R. Wilkinson (eds) *Health and Social Organization: Towards a Health Policy for the 21st Century*, London: Routledge, pp. 171–87.

Bloch, A. (1997) 'Ethnic inequality and social security policy', in A. Walker and C. Walker (eds) *Britain Divided: The Growth of Social Exclusion in the 1980s and 1990s*, London: CPAG, pp. 111–22.

Bluestone, B. and Harrison, B. (1982) *The De-industrialisation of America*, New York: Basic Books.

Blunkett, D. (1999) 'Press release', 3 November, http://www.dfee.gov.uk/news/news.cfm?PR_ID=48)

Blunt, A. and Wills, J. (2000) *Dissident Geographies: An Introduction to Radical Ideas and Practice*, Harlow: Prentice-Hall.

Bocock, R. (1999) 'Consumption and lifestyles', in J. Bryson, N. Henry, D. Keeble and R. Martin (eds) *The Economic Geography Reader*, Chichester: Wiley, pp. 279–90.

Bogdanor, V. (1999) *Devolution in the United Kingdom*, Oxford: OPUS.

Bondi, L. (1999) 'On the journeys of the gentrifiers: exploring gender, gentrification and migration', in P. Boyle and K. Halfacree (eds) *Gender and Migration in the Developed World*, London: Routledge, pp. 204–22.

Bonefeld, W. (1999) 'British experience: monetarism hiding behind Europe', *Journal of European Area Studies* 7(1): 55–71.

Boniface, B.G. and Cooper, C. (1994) *The Geography of Travel and Tourism*, Oxford: Butterworth Heinemann.

Booth, C. (ed.) (1889–1902) *Life and Labour of the People of London*, 17 volumes, London: Macmillan.

Boseley, S. (1999a) 'Giving birth by numbers', *The Guardian*, 26 January, p. 13.

—— (1999b) 'Glasgow and Belfast top heart attack list', *The Guardian*, 7 May, p. 16.

—— (1999c) 'Car exhaust fumes linked to worsening asthma in children', *The Guardian*, 12 March, p. 10.

—— (1999d) 'British teenagers have the worst sexual health in Europe', *The Guardian*, 14 May, p. 1.

Botting, B. (1998) 'Teenage mothers and the health of their children', *Population Trends* 93: 19–28.

Bower, H. and Boseley, S. (1999) 'Cancer: a culture that kills', *The Guardian*, 16 March, p. 15.

Bowie, F. (1993) 'Wales from within: conflicting interpretations of Welsh identity', in S. Macdonald (ed.) *Inside European Identities*, Oxford: Berg, pp. 167–93.

Boyle, K. (1991) 'Northern Ireland: allegiances and

identities', in B. Crick (ed.) *National Identities: The Constitution of the United Kingdom*, Oxford: Blackwell, pp. 68–78.

Bradbury, J. (1997) 'Conservative governments, Scotland and Wales: a perspective on territorial management', in J. Bradbury and J. Mawson (eds) *British Regionalism and Devolution: The Challenges of State Reform and European Integration*, London: Jessica Kingsley, pp. 74–98.

Bradley, A. and Baker, O. (1999) 'Drugs in the United Kingdom – a jigsaw with missing pieces', *Social Trends* 29: 15–28.

Bragg, B. (1999) 'Two World Wars and one world cup', in M. Perryman (ed.) *The Ingerland Factor: Home Truths from Football*, Edinburgh: Mainstream, pp. 37–43.

Brand, J. (1990) 'Scotland', in M. Watson (ed.) *Contemporary Minority Nationalism*, London: Routledge, pp. 24–37.

Brand, J. and Mitchell, J. (1997) 'Home rule in Scotland: the politics and bases of a movement', in J. Bradbury and J. Mawson (eds) *British Regionalism and Devolution: The Challenges of State Reform and European Integration*, London: Jessica Kingsley, pp. 35–54.

Brandwood, G. (1999) 'Slap in the face for city theme bars', *What's Brewing?*, September, p. 10.

Brazier, C. (1999) 'Gender canyon', *New Internationalist* 315: 12–16.

Breheny, M. (ed.) (1999) *The People: Where Will They Work?*, London: Town and Country Planning Association.

Brett, D.U. (1976) *Aspects of Nationalism and Regionalism: The Cases of Scotland and the North East*, North-East Area Study Working Paper 28, University of Durham.

Bright, M. (1998) 'Boys performing badly', *The Observer*, 4 January, p. 13.

—— (1999) 'UK eyes Dutch sex lessons', *The Observer*, 21 February, p. 12.

—— (2000) 'Top teacher attacks "Posh-and-Becks" culture', *The Observer*, 25 June, p. 1.

Brimson, D. and Brimson, E. (1996) *England, My England: The Trouble with the National Team*, London: Headline.

Britton, M. (1990) 'Geographic variation in mortality since 1920 for selected causes', in M. Britton (ed.) *Mortality and Geography: A Review in the Mid-1980's*, Registrar-General's Decennial Supplement for England and Wales Series DS No. 9, London: HMSO, pp. 27–45.

Brockliss, L. and Eastwood, D. (1997) 'A union of multiple identities', in L. Brockliss and D. Eastwood (eds) *A Union of Multiple Identities: The British Isles c.1750–c.1850*, Manchester: Manchester University Press, pp. 1–8.

Brodie, A. (1999) 'The years of living dangerously . . .', *The Observer, Cash*, 31 January, pp. 2–3.

Brook, S. (1997) 'Arise, Lord Oik', *The Observer*, 26 October, p. 6.

Brooks, R. (1999) 'Game boys just want to play in the street', *The Observer*, 7 March, p. 4.

Brown, A. (1997) 'Scotland: paving the way for devolution?, *Parliamentary Affairs* 50(4): 658–71.

Brown, A., McCrone, D. and Paterson, L. (1998) *Politics and Society in Scotland* (2nd edition), Basingstoke: Macmillan.

Brown, D. (2000) 'Apathy wins the day', *The Guardian*, 5 May, p. 5.

Brown, P. (1998) 'Alarm at killer traffic fumes', *The Guardian*, 14 January, p. 2.

Browne, A. (1998) 'To have and have not – by postcode', *The Observer*, 25 October, p. 10.

—— (1999a) 'Your children can expect to live to 100', *The Observer*, 26 December, p. 1.

—— (1999b) 'Killer that shames Britain', *The Observer*, 12 December, p. 13.

—— (1999c) 'Aids patients threatened by cutback on "miracle drugs"', *The Observer*, 28 November, p. 5.

—— (1999d) 'In Seattle you're 15 times more likely to survive cardiac arrest than in London. Why?', *The Observer*, 19 December, p. 9.

—— (1999e) 'Homes: boom or bust?', *The Observer, Business*, 11 July, p. 2.

—— (2000a) 'The last days of a white world', *The Observer*, 3 September, p. 17.

—— (2000b) 'Pint of bitter . . . and a tonic, please', *The Observer*, 27 February, p. 16.

Browne, A. and Reeves, R. (1999) 'Just 500 Japanese. And empty Europe. The world dying out', *The Observer*, 8 August, p. 18.

Bryson, B. (1997) *Notes from a Small Island*, London: Black Swan Books.

Bryson, J., Henry, N., Keeble D. and Martin R. (eds) (1999) *The Economic Geography Reader*, Chichester: Wiley.

Buchanan, C.D. (1963) *Transport in Towns (Buchanan Report)*, London: HMSO.

Bulmer, M. (1996) 'The ethnic group question in the 1991 Census of Population', in D. Coleman and J. Salt (eds) *Ethnicity in the 1991 Census*, London: OPCS, pp. 33–62.

Buswell, R.J., Champion, A.J. and Townsend, A.R. (1987) 'The Northern Region', in P. Damesick and P. Wood (eds) *Regional Problems, Problem Regions, and Public Policy in the United Kingdom*, Oxford: Clarendon Press, pp. 167–90.

Butler, E. (2000) 'Whose side are you on?', *The Observer*, 26 March, p. 19.

Byrne, D. (1999) *Social Exclusion*, Milton Keynes: Open University Press.

Cabinet Office (1999a) *Sharing the Nation's Prosperity: Variations in Economic and Social Conditions Across the UK*, London: Cabinet Office.

—— (1999b) *Rural Economies: A Performance and Innovation Unit Report*, London: Cabinet Office.

CACI Information Solutions (1999) *The Geodemographic Pocket Book*, Henley-on-Thames: NTC Publications.

Cameron, S. and Crompton, P. (1988) 'Housing', in F. Robinson (ed.) *Post-Industrial Tyneside: An Economic and Social Survey of Tyneside in the 1980s*, Newcastle: Newcastle upon Tyne City Libraries and Arts, pp. 120–55.

Campaign for the English Regions (2000) http://www.cfer.org.uk

Campbell, D. (1999) 'How soccer mafia sank Wembley', *The Observer*, 5 December, pp. 16–17.

—— (2000a) 'Schools rear crop of couch potatoes', *The Observer*, 27 February, p. 6.

—— (2000b) 'Scandal of playing fields that Labour didn't save', *The Observer*, 13 February, p. 13.

—— (2000c) 'National anthem faces red card', *The Observer*, 5 March, p. 6.

Campbell, J. (1984) *Invisible Country: A Journey through Scotland*, London: Weidenfeld & Nicolson.

Cannadine, D. (1995) 'British history as a "new subject": politics, perspectives and prospects', in A. Grant and K.J. Stringer (eds) *Uniting the Kingdom? The Making of British History*, London: Routledge, pp. 12–28.

—— (1998) *The Rise and Fall of Class in Britain*, New York: Columbia University Press.

Carrington, B. (1999) 'Too many St George Crosses to bear', in M. Perryman (ed.) *The Ingerland Factor: Home Truths from Football*, Edinburgh: Mainstream, pp. 71–86.

Carter, C.F. (ed.) (1962) *Manchester and its Region*, Manchester: British Association.

Carter, E., Donald, J. and Squires, J. (1993) *Space and Place: Theories of Identity and Location*, London: Lawrence & Wishart.

Casey, A.M. (1977) 'Cornish nationalism', in G. Ashworth (ed.) *World Minorities, Volume One*, Sunbury: Quartermaine House, pp. 53–5.

Castells, M. (1989) *The Informational City*, Oxford: Blackwell.

—— (1996) *The Network Society*, Oxford: Blackwell.

CDP Information and Intelligence Unit (1974) *Inter-Project Report*, London: CDP Information and Intelligence Unit.

Cecil, R. (1993) 'The marching season in Northern Ireland: an expression of politico-religious identity', in S. Macdonald (ed.) *Inside European Identities*, Oxford: Berg, pp. 146–66.

Cerny, P. (1990) *The Changing Architecture of Politics: Structure, Agency and the Future of the State*, London: Sage.

Chadwick, E. (1842) *Report on the Sanitary Condition of the Labouring Population of Great Britain* (reprinted 1965), Edinburgh: Edinburgh University Press.

Champion, A., Green, A., Owen, D., Ellin, D. and Coombes, M. (1987) *Changing Places: Britain's Demographic, Economic and Social Complexion*, London: Edward Arnold.

Champion, A.G. and Townsend, A.R. (1990) *Contemporary Britain: A Geographical Perspective*, London: Edward Arnold.

Champion, T. (2000) 'Demography', in V. Gardiner and H. Matthews (eds) *The Changing Geography of the United Kingdom* (3rd edition), London: Routledge, pp. 169–89.

Champion, T. and Ford, T. (1999) *Attempts at Isolating the Main Components of the Distinctive Social Composition of London's Migration Exchanges*, Working Paper 2 of ESRC Cities Programme Project on Migration, Residential Preferences and the Changing Environment of Cities, University of Newcastle.

Champion, T., Fotheringham, S., Rees, P., Boyle, P. and Stilwell, J. (1998) *The Determinants of Migration Flows in England: A Review of Existing Data and Evidence*, Newcastle upon Tyne: University of Newcastle and University of Leeds.

Champion, T., Wong, C., Rooke, A., Dorling, D., Coombes, M. and Brunsdon, C. (1996) *The Population of Britain in the 1990s: A Social and Economic Atlas*, Oxford: Clarendon Press.

Channel 4 (1999) *Green and Pleasant Land*, 12 December.

—— (2000) *The Real Queen Mother*, 10 July.

Chaudhary, V. (1999) 'Warning shots fired in battle for TV rights', *The Guardian*, Sport, 13 January, p. 22.

Chen, S. and Wright, T. (eds) (2000) *The English Question*, London: The Fabian Society.

Childs, P. (1997) 'Place and environment: nation and region', in M. Storry and P. Childs (eds) *British Cultural Identities*, London: Routledge, pp. 41–82.

Chisholm, M. (1995) 'Some lessons from the review of local government in England', *Regional Studies* 29(6): 563–9.

Clammer, J. (1999) 'Aesthetics of the self: shopping and social being in contemporary urban Japan', in J. Bryson, N. Henry, D. Keeble and R. Martin (eds) *The Economic Geography Reader*, Chichester: Wiley, pp. 327–32.

Clarke, J. (1999) 'Smoke screening', *The Observer Magazine*, 29 August, p. 45.

Clarke, J., Cricher, J. and Johnson, R. (eds) (1979) *Working Class Culture: Studies in History and Theory*, London: Hutchinson.

Claval, P. (1998) *An Introduction to Regional Geography*, trans. I. Thompson, Oxford: Blackwell.

Cliff, A. and Haggett, P. (1990) 'Epidemic control and critical community size: spatial aspects of eliminating communicable diseases in human populations', in R.W. Thomas (ed.) *Spatial Epidemiology*, London: Pion, pp. 93–110.

Cloke, P. and Little, J. (1997) *Contested Countryside Cultures: Otherness, Marginalisation and Rurality*, London: Routledge.

Cloke, P., Milbourne, P. and Thomas, C. (1995) 'Poverty in the countryside: out of sight and out of mind', in C. Philo (ed.) *Off the Map: The Social Geography of Poverty in the UK*, London: Child Poverty Action Group, pp. 83–102.

Close, A. (1992) 'A woman's place', *Scotland on Sunday*, 1 March, pp. 27–8.

Coates, B.E. and Rawstron, E.M. (1971) *Regional Variations in Britain: Studies in Economic and Social Geography*, London: B.T. Batsford.

Coates, K. and Silburn, R. (1970) *Poverty: The Forgotten Englishmen*, Harmondsworth: Penguin.

Cohen, A.P. (1982) 'Belonging: the experience of culture', in A.P. Cohen (ed.) *Belonging: Identity and Social Organisation in British Rural Cultures*, Manchester: Manchester University Press, pp. 1–20.

—— (1986) 'Of symbols and boundaries, or, does Ertie's greatcoat hold the key?', in A.P. Cohen (ed.) *Symbolising Boundaries: Identity and Diversity in British Cultures*, Manchester: Manchester University Press, pp. 1–19.

Cohen, N. (1999) *Cruel Britannia: Reports on the Sinister and Preposterous*, London: Verso.

Cohen, R. (1994) *Frontiers of Identity: The British and Others*, Harlow: Longman.

—— (1995) 'Fuzzy frontiers of identity: the British case', *Social Identities* 1(1): 35–62.

Cohn, N. (1999) 'Republic of England', *The Observer Review*, 28 March, p. 3.

Cole, G.D.H. (1947) *Local and Regional Government*, London: Cassell.

Coleman, D. and Salt, J. (1992) *The British Population: Patterns, Trends and Processes*, Oxford: Oxford University Press.

—— (1996) 'The ethnic group question in the 1991 Census: a new landmark in British social statistics', in D. Coleman and J. Salt (eds) *Ethnicity in the 1991 Census*, London: OPCS, pp. 1–32.

Colley, L. (1992) *Britons: Forging the Nation 1707–1837*, New Haven: Yale University Press.

Collis, C. and Mallier, A. (1996) 'Third age male activity rates in Britain and its regions', *Regional Studies* 30: 803–11.

Colls, R. (1992) 'Born-again Geordies', in R. Colls and B. Lancaster (eds) *Geordies: Roots of Regionalism*, Edinburgh: Edinburgh University Press, 1–34.

Compton, P.A. (1991) 'The changing population', in R.J. Johnston and V. Gardiner (eds) *The Changing Geography of the United Kingdom* (2nd edition), London: Routledge, pp. 35–82.

—— (1996) 'Indigenous and older minorities', in D. Coleman and J. Salt (eds) *Ethnicity in the 1991 Census*, London: OPCS, pp. 243–82.

Congdon, P. (1996) 'The epidemiology of suicide in London', *Journal of the Royal Statistical Society* 159(3): 515–33.

Conner, W. (1986) 'The impact of homelands upon diasporas', in G. Sheffer (ed.) *Modern Diasporas in International Politics*, London: Croom Helm, pp. 16–46.

Cooper, J. (1991) 'Births outside marriage: recent trends and associated demographic and social changes', *Population Trends* 63: 8–18.

Cooper, J. and Botting, B. (1992) 'Analysing fertility and infant mortality by mother's social class as defined by occupation', *Population Trends* 70: 15–21.

Cope, R. (2000) 'UK outbound', *Travel and Tourism Analyst* 1: 19–39.

Corden, A. and Eardley, T. (1999) 'Sexing the enterprise: gender, work and resource allocation in self employed households', in L. McKie, S. Bowlby and S. Gregory (eds) *Gender, Power and the Household*, Basingstoke: Macmillan, pp. 207–25.

Cornelius, W.A., Martin, P.L. and Hollifield, J.F. (1994) 'Introduction: the ambivalent quest for immigration control', in W.A. Cornelius, P.L. Martin and J.F. Hollifield (eds) *Controlling Immigration: A Global Perspective*, Stanford: Stanford University Press, pp. 3–41.

Cosgrave, S. (1998) 'Scotland's away days', *The Observer Guide to the World Cup*, 7 June, pp. 10–12.

Cosgrove, D. (1985) *Social Formation and Symbolic Landscapes*, London: Croom Helm.

Cosgrove, D. and Jackson, P. (1987) 'New directions in cultural geography', *Area* 19: 95–101.

Countryside Agency (1999) *Planning for Quality in Rural England*, Wetherby: Countryside Agency.

—— (2000) *State of the Countryside 2000*, Wetherby: Countryside Agency.

Coward, J. (1986) 'The analysis of regional fertility patterns', in R. Woods, and P. Rees (eds) *Population Structures and Models*, London: Allen & Unwin, pp. 45–67.

Cox, B. (2000) 'In defence of political apathy', *The Observer*, 16 July, p. 28.

Crang, P. (1998) *Cultural Geography*, London: Routledge.

Creton, D. (1991) 'Fertility changes and the Irish family', *Geography* 76: 154–7.

Crewe, C. (1992) 'The lean and hungry', *The Guardian*, 10 July, p. 21.

Crewe, L. and Beaverstock, J.V. (1998) 'Fashioning the city: cultures of consumption in contemporary urban spaces', *Geoforum* 29(3): 287–308.

Crick, B. (1991) 'The English and the British', in B. Crick (ed.) *National Identities: The Constitution of the United Kingdom*, Oxford: Blackwell, pp. 90–104.

—— (1995) 'The sense of identity of the indigenous British', *New Community* 21(2): 167–82.

Critchley, J. (1986) 'Introduction', in J. Critchley (ed.) *Britain: A View from Westminster*, Poole: Blandford, pp. 7–8.

Croft, J. (1997) 'Youth culture and age', in M. Storry and P. Childs (eds) *British Cultural Identities*, London: Routledge, pp. 163–200.

Crolley, L. (1999) 'Lads will be lads', in M. Perryman (ed.) *The Ingerland Factor: Home Truths from Football*, Edinburgh: Mainstream, pp. 59–70.

Crompton, R. (1997) *Women and Work in Modern Britain*, Oxford: Oxford University Press.

Crompton, R. and Mann, M. (1986) *Gender and Social Stratification*, Cambridge: Polity.

Crouch, D. (2000) 'Leisure and consumption', in V. Gardiner and H. Matthews (eds) *The Changing Geography of the United Kingdom* (3rd edition), London: Routledge, pp. 261–75.

CSO (1961) *Annual Abstract of Statistics*, London: HMSO.

Cullingworth, J.B. and Naden, V. (1997) *Town and Country Planning in the UK*, London: Routledge.

Curtice, J. and Park, A. (1999) 'Region: New Labour, new geography', in G. Evans and P. Norris (eds) *Critical Elections: British Parties and Voters in Long-Term Perspective*, London: Sage, pp. 124–47.

Curtis, S. (1995) 'Geographical perspectives on poverty, health and health policy in different parts of the UK', in C. Philo (ed.) *Off the Map: The Social Geography of Poverty in the UK*, London: Child Poverty Action Group, pp. 153–74.

Curtis, S. and Taket, A. (1996) *Health and Societies: Changing Perspectives*, London: Arnold.

Cusick, E. (1997) 'Religion and heritage', in M. Storry and P. Childs (eds) *British Cultural Identities*, London: Routledge, pp. 277–314.

Damesick, P. and Wood, P. (eds) (1987) *Regional Problems, Problem Regions, and Public Policy in the United Kingdom*, Oxford: Clarendon Press.

Daniels, P. (1999) 'Services in a shrinking world', in J. Bryson, N. Henry, D. Keeble and R. Martin (eds) *The Economic Geography Reader*, Chichester: Wiley, pp. 156–66.

Danson, M. and Lloyd, G. (2000) 'Land matters: the land reform debate in Scotland', *Local Economy* 15(2): 214–25.

Davey, K. (1999) *English Imaginaries*, London: Lawrence & Wishart.

David, P. (1999) 'Undoing Britain', *The Economist*, 6 November, pp. 3–18.

Davie, G. (1994) *Religion in Britain since 1945*, Oxford: Blackwell.

Davies, N. (1999) *The Isles: A History*, Basingstoke: Macmillan.

Davison, J. (1998) 'Misery of going to bed without food', *The Independent*, 17 October, p. 10.

Day, L.H. (1995) *The Future of Low-Birthrate Populations*, London: Routledge.

Dean, H. (1998) 'Popular paradigms and welfare regimes', *Critical Social Policy* 18(2): 131–56.

Demangeon, A. (1927) *Les Isles Britanniques*, Paris: Librairie Armand Colin.

Denver, D. (1998) 'The British electorate in the 1990s', *West European Politics* 21(1): 197–217.

Department of Energy (various years) Digest of United Kingdom Energy Statistics, London: The Stationery Office.

Department of Health (1998) *Our Healthier Nation: A Contract for Health*, London: The Stationery Office.

DETR (1998) *A New Deal for Transport: Better for Everyone*, London: The Stationery Office.

—— (1999a) *Urban Renaissance: Sharing the Vision*, http://www.detr.gov.uk/regeneration/urbanren/index.htm

—— (1999b) *Regional Planning Guidance for the South East: Public Examination May–June 1999: Report of the Panel of the London and South East Regional Planning Conference*, Guildford: GOSE.

—— (1999c) *Projections of Households in England to 2021*, London: The Stationery Office.

—— (2000) *PPG3: Housing (revised)*, London: The Stationery Office.

DETR, MAFF (1999) *Rural England: A Discussion Document*, London: The Stationery Office.

Devine, T. (1999a) *The Scottish Nation 1700–2000*, London: Allen Lane.

—— (1999b) 'Spot the Catholic', *The Times Higher*, 10 September, p. 19.

Dex, S. and McCulloch, A. (1997) *Flexible Employment: The Future of Britain's Jobs*, Basingstoke: Macmillan.

DfEE (1996) *Labour Market and Skills Trends 1997/1998*, London: Department for Education and Employment.

—— (1998) *The Learning Age: A Renaissance for a New Britain*, London: The Stationery Office.

Dicken, P., Peck, J. and Tickell, A. (1997) 'Unpacking the global', in R. Lee and J. Wills (eds) *Geographies of Economies*, London: Edward Arnold, pp. 158–66.

Dickson, J. (1999) 'What is "Scottish"?', *The Scots Magazine* 151(1): 9–11.

Dillner, L. (1999) 'Why Britain is lagging behind in cancer care', *The Observer*, 14 March, p. 15.

Disraeli, B. (1969) *Sybil: Or The Two Nations*, London: Oxford University Press.

Dodd. P. (1986) 'Englishness and the national culture', in R. Colls and P. Dodd (eds) *Englishness: Politics and Culture 1880–1920*, Beckenham: Croom Helm, pp. 1–28.

DoE (1995) *PPG2: Green Belts (revised)*, London: HMSO.

Dorling, D. (1995) *A New Social Atlas of Britain*, Chichester: John Wiley.

Dorling, D. and Shaw, M. (2000) 'Life chances and lifestyles', in V. Gardiner and H. Matthews (eds) *The Changing Geography of the United Kingdom*, London: Routledge, pp. 230–60.

Dorling, D. and Tomaney, J. (1995) 'Poverty in the old industrial regions: a comparative view', in C. Philo (ed.) *Off the Map: The Social Geography of Poverty in the UK*, London: Child Poverty Action Group, pp. 103–22.

Dorling, D. and Woodward, R. (1996) 'Social polarisation 1971–1991: a microgeographical analysis of Britain', *Progress in Planning* 45(2): 69–122.

Driver, F. (1999) 'Imaginative geographies', in P. Cloke, P. Crang and M. Goodwin (eds) *Introducing Human Geographies*, London: Arnold, pp. 209–16.

Dungey, J. (1997) 'Committed to the regions', *Local Government Chronicle*, 16 May, p. 13.

Dwyer, C. (1999) 'Migrations and diasporas', in P. Cloke, P. Crang and M. Goodwin (eds) *Introducing Human Geographies*, London: Arnold, pp. 287–95.

East Midlands Regional Development Agency (1999) *The Agenda*, January, Nottingham.

Edwards, K.C. (ed.) (1966) *Nottingham and its Region*, Nottingham: British Association.

Election UK (1997) *Election UK 1997*, http://www.election.co.uk

Ellen, B. (2000) 'Boys are the new girls', *The Observer*, 30 January, p. 31.

Elton, G. (1992) *The English*, Oxford: Blackwell.

Emmett, I. (1982) 'Place, community and bilingualism in Blaenau Ffestiniog', in A.P. Cohen (ed.) *Belonging: Identity and Social Organisation in British Rural Cultures*, Manchester: Manchester University Press, pp. 201–21.

Employment Service Research Division (1999) Reports ESR33 and ESR34. Tel. 0114 259 6217 or email erd.es.rh@gtnet.gov.uk

Endean, R. and Harris, T. (1998) 'A picture of poverty in the UK', *Social Trends Quarterly*, Winter: 9–13.

Engel, M. (2000) 'Chucking them out of London', *The Guardian*, 4 July, p. 18.

EOC (1994) *Black and Ethnic Minority Women and Men in Britain 1994*, Manchester: EOC.

—— (1995) *The Lifecycle of Inequality: Women and Men in Britain 1995*, Manchester: EOC.

Ermisch, J. (1989) 'Divorce: economic antecedents and aftermath', in H. Joshi (ed.) *The Changing Population of Britain*, Oxford: Basil Blackwell, pp. 42–55.

Esler, G. (1999) 'Britain's fly over people get restive', *The Scotsman*, 29 March, p. 13.

European Commission (1995) *The Regions of the United Kingdom in the European Union*, London: HMSO.

Evans, E. (1995) 'Englishness and Britishness: national identities, c.1790–c.1870', in A. Grant and K.J. Stringer (eds) *Uniting the Kingdom? The Making of British History*, London: Routledge, pp. 223–43.

Evans, G. (1999) 'Europe: a new electoral cleavage?', in G. Evans and P. Norris (eds) *Critical Elections: British Parties and Voters in Long-Term Perspective*, London: Sage, pp. 207–22.

Evans, G., Heath, A. and Payne, C. (1999) 'Class: Labour as a catch-all party', in G. Evans and P. Norris (eds) *Critical Elections: British Parties and Voters in Long-Term Perspective*, London: Sage, pp. 87–101.

Farmer, R.D.T. and Miller, D.L. (1983) *Lecture Notes on Epidemiology and Community Medicine* (2nd edition), Oxford: Blackwell Scientific.

Fawcett, C.B. (1960) *Provinces of England: A Study of Some Geographical Aspects of Devolution*, London: Hutchinson (original published in 1919).

Fay, M.-T. and Meehan, E. (2000) 'British decline and European integration', in R. English and M. Kenny (eds) *Rethinking British Decline*, Basingstoke: Macmillan, pp. 210–30.

Featherstone, M. (1987) 'Lifestyle and consumer culture', *Theory, Culture and Society* 4(1): 55–70.

Ferguson, W. (1978) *Scotland: 1689 to the Present*, Edinburgh: Oliver & Boyd.

Fielding, A.J. (1992) 'Migration and social mobility: South East England as an "escalator" region', *Regional Studies* 26: 1–15.

—— (1993) 'Migration and the metropolis: recent research on the causes of migration to southeast England', *Progress in Human Geography* 17(2): 195–212.

Fieldman, S. (1999) 'Once, twice, three times the money . . .', *The Observer, Cash*, 4 July, p. 2.

Findlay, A. (1994) 'An economic audit of contemporary immigration', in S. Spencer (ed.) *Strangers and Citizens: A Positive Approach to Migrants and Refugees*, London: IPPR/Oram Press, pp. 159–201.

Foley, P. (1998) 'The impact of the Regional Development Agency and Regional Chamber in the East Midlands', *Regional Studies* 32(8): 777–82.

Ford, A. (1999) 'Why shouldn't we be proud to say we are English', *The Mail on Sunday*, 18 April, p. 32.

Foster, H.D. (1992) *Health, Disease and the Environment*, London: Belhaven Press.

Fothergill, S. and Guy, N. (1990) *Retreat from the Regions: Corporate Change and the Closure of Factories*, London: Regional Studies Association.

Freely, M. (1999) 'Nice work if you can get it', *The Observer Review*, 4 July, pp. 1–2.

Friedman, J. (1994) *Cultural Identity and Global Processes*, London: Sage.

Gardiner, V. and Matthews, H. (eds) (2000) *The Changing Geography of the United Kingdom*, London: Routledge.

Garreau, J. (1991) *Edge City: Life on the New Frontier*, New York: Doubleday.

Garrett, R. (1997) 'Gender, sex, and the family', in M. Storry and P. Childs (eds) *British Cultural Identities*, London: Routledge, pp. 129–62.

George, S. (1996) 'The European Union, 1992 and the fear of "Fortress Europe"', in A. Gamble and A. Payne (eds) *Regionalism and World Order*, Basingstoke: Macmillan, pp. 21–54.

Gerrard, N. (1999a) 'Will you be lonely this Christmas?', *The Observer Review*, 12 December, p. 4.

—— (1999b) 'Goodbye to all that', *The Observer Review*, 31 January, pp. 1–2.

Gibbon, L.G. (1934) 'The antique scene', in L.G. Gibbon and H. MacDiarmid (eds) *Scottish Scene: The Intelligent Man's Guide to Albyn*, London: Jarrolds, pp. 19–36.

Gibbs, D. (1998) 'Regional Development Agencies: an opportunity for sustainable development', *The Regional Review* 8(2): 17–18.

Giddens, A. (1998) *The Third Way: The Renewal of Social Democracy*, Cambridge: Polity.

Gilbert, E.W. (1939) 'Practical regionalism in England and Wales', *The Geographical Journal* 94(1): 29–44.

—— (1948) 'The boundaries of local government areas', *The Geographical Journal* 111(4–6): 172–206.

—— (1960) 'Geography and regionalism', in G. Taylor (ed.) *Geography in the Twentieth Century* (3rd edition), London: Methuen, pp. 345–71.

Gill, D., Illsley, R. and Koplik, L.H. (1970) 'Pregnancy in teenage girls', *Social Science and Medicine* 3: 549–54.

Gillespie, A. (1999) 'The changing employment geography of Britain', in M. Breheny (ed.) *The People: Where will they Work?*, London: Town and Country Planning Association, pp. 9–28.

Gillespie, M. (1998) '"Being cool and classy": style hierarchies in a London–Punjabi peer culture', *International Journal of Punjabi Studies* 5(2): 159–78.

Gillespie, R. (1995) 'Health behaviour and the individual', in G. Moon and R. Gillespie (eds) *Society and Health: An Introduction to Social Science for Health Professionals*, London: Routledge, pp. 97–110.

Gillies, N. (1999) 'Smell a rat? It's probably the dead pigeon', *The Observer*, 28 November, p. 22.

Gilroy, P. (1993) *Black Atlantic: Modernity and Double Consciousness*, London: Verso.

Gilroy, S. (1999) 'Intra-household power relations and their impact on women's leisure', in L. McKie, S. Bowlby and S. Gregory (eds) *Gender, Power and the Household*, Basingstoke: Macmillan, pp. 155–72.

Ginn, J. and Sandell, J. (1997) 'Balancing home and employment: stress reported by social services staff', *Work, Employment and Society* 11(3): 413–34.

Glancey, J. (1999) 'In deep Bluewater', *The Guardian*, 27 September, pp. 12–13.

Glass, R. (1964) *London: Aspects of Change*, London:

Centre for Urban Studies and MacGibbon & Kee.

Glennerster, H., Lupton, R., Noden, P. and Power, A. (1999) *Poverty, Social Exclusion and Neighbourhood: Studying the Area Bases of Social Exclusion*, CASE Working Paper 22, London: Centre for Analysis of Social Exclusion, LSE.

Glucksmann, M. (1995) 'Why "work"? Gender and the "total social organisation of labour"', *Gender, Work and Organisation* 2(2): 63–75.

Gold, K. (2000) 'It's a family affair . . . but what is a family?', *The Times Higher*, 18 February, pp. 18–19.

Goldthorpe J., Llewellyn, C. and Payne, C. (1980) *Social Mobility and Class Structure in Modern Britain*, Oxford: Clarendon Press.

Goodwin, M. (1995) 'Poverty in the city: "you can raise your voice but who is listening?"', in C. Philo (ed.) *Off the Map: The Social Geography of Poverty in the UK*, London: Child Poverty Action Group, pp. 65–82.

—— (1999) 'Citizenship and governance', in P. Cloke, P. Crang and M. Goodwin (eds) *Introducing Human Geographies*, London: Arnold, pp. 189–98.

Gorman, T. (1999) 'A parliament for England?', *This England* 32(2): 20–1.

Gorz, A. (1989) *Critique of Economic Reason*, London: Verso.

Goss, J. (1999) 'The "magic of the mall": an analysis of form, function, and meaning in the contemporary retail built environment', in J. Bryson, N. Henry, D. Keeble and R. Martin (eds) *The Economic Geography Reader*, Chichester: Wiley, pp. 327–32.

Graham, B. (1997) *In Search of Ireland: A Cultural Geography*, London: Routledge.

Graham, C. (1972) *Portrait of Aberdeen and Deeside*, London: Robert Hale.

Graham, D.T. (1988) 'Female employment patterns and urban sociodemographic structure in Great Britain, 1881–1981', Unpublished PhD thesis, University of Dundee.

—— (1993) 'Twentieth-century demographic trends in the Scottish cities', in A.H. Dawson, H.R. Jones, A. Small and A. Soulsby (eds) *Scottish Geographical Studies*, Dundee: Departments of Geography, Universities of Dundee and St Andrews, pp. 268–89.

—— (1994a) 'Socio-demographic trends in Northern Ireland', *Scottish Geographical Magazine* 110: 168–76.

—— (1994b) 'Female employment and infant mortality: some evidence from British towns, 1911, 1931 and 1951', *Continuity and Change* 9, 313–46.

—— (1994c) 'Spatial variations in socioeconomic and demographic indicators in the East Midlands, 1981 and 1991: a multivariate approach', *Trent Geographer* 15: 38–55.

—— (1995) 'The new local political geography of Scotland', *Geography* 80(4): 407–14.

—— (1996) 'The reddening of Middle England: the 1995 local political geography of the East Midlands', *Trent Geographer* 16: 4–13.

—— (1998a) 'Patterns and trends in human fertility in the East Midlands, 1981 to 1991', *Trent Geographer* 17: 13–30.

—— (1998b) 'The geography of extra-marital fertility in the East Midlands, 1981–1991', *East Midland Geographer* 20: 3–19.

—— (2000) 'The people paradox: human movements and human security in a globalising world', in D.T. Graham and N.K. Poku (eds) *Migration, Globalisation and Human Security*, London: Routledge, pp. 186–216.

Graham, D.T. and Poku, N. (1998) 'Population movements, health and security', in N. Poku and D.T. Graham (eds) *Redefining Security: Population Movements and National Security*, Westport, Conn.: Praeger, pp. 203–34.

Grant, A. and Stringer, K.J. (eds) (1995) *Uniting the Kingdom? The Making of British History*, London: Routledge.

Grant, G.V.R. (2000) 'Heads you . . .', *The Scotsman*, 18 July, p. 15.

Grant, W. (1989) *Pressure Groups, Politics and Democracy in Britain*, London: Philip Allan.

Green, A.E. (1994) *The Geography of Poverty and Wealth*, Warwick: IER.

—— (1997) 'A question of compromise? Case study evidence on the location and mobility strategies of dual career households', *Regional Studies* 31(1): 643–59.

—— (1999) 'Employment opportunities and constraints facing in-migrants to rural areas in England', *Geography* 84(1): 34–44.

Green, A.E., Hogarth, T. and Shackleton, R.E. (1999) *Long Distance Living: Dual Location Households*, Bristol: Policy Press.

Green, E., Hebron, S. and Woodward, D. (1990) *Women's Leisure, What Leisure?*, Basingstoke: Macmillan.

Greenhalgh, P. (1988) *Ephemeral Vista: The Expositions Universelles, Great Exhibitions and World's Fairs 1851–1939*, Manchester: Manchester University Press.

Greer, G. (1970) *The Female Eunuch*, London: MacGibbon and Kee.

Gregoriadis, L. (1999) 'Football urged to fund grass roots', *The Independent*, 12 January, p. 7.

Gregson, N. and Crewe, L. (1994) 'Beyond the high street and the mall: car boot fairs and the new geographies of consumption in the 1990s', *Area* 26(3): 261–7.

Gregson, N. and Lowe, M. (1994) *Servicing the Middle Classes: Class, Gender and Waged Domestic Labour in Contemporary Britain*, London: Routledge.

Gruffudd, P. (1999) 'Nationalism', in P. Cloke, P. Crang and M. Goodwin (eds) *Introducing Human Geographies*, London: Arnold, pp. 199–206.

Gurumurthy, R. (1999) 'Tackling poverty and extending opportunity', *Political Quarterly* 70(3): 341–5.

Hague, E., Thomas, C. and Williams, S. (1999) 'Left out? Observations of the RGS-IBG conference on social exclusion and the city', *Area* 31(3): 293–6.

Hague, W. (1999) 'Identity and the British Way', speech to the Centre for Policy Studies, 24 January.

Hall, P. (1998) 'London's a unique world city', *The Observer*, 15 November, p. 28.

Hall, R., Ogden, P. and Hill, C. (1999) 'Gender variations in the characteristics of migrants living alone in England and Wales 1991', in P. Boyle and K. Halfacree (eds) *Migration and Gender in the Developed World*, London: Routledge, pp. 186–203.

Hall, S. (1995) 'New cultures for old', in D. Massey and P. Jess (eds) *A Place in the World? Places, Cultures and Globalization*, Oxford: Oxford University Press, pp. 175–214.

—— (2000) 'Reflections on British decline', in R. English and M. Kenny (eds) *Rethinking British Decline*, Basingstoke: Macmillan, pp. 104–16.

Hall, S. and Jefferson, T. (1976) *Resistance through Ritual: Youth Sub-Cultures in Post-war Britain*, London: Hutchinson.

Hamilton, W. (1975) *My Queen and I*, London: Quartet.

Hamnett, C. (1999) *Winners and Losers: Home Ownership in Modern Britain*, London: UCL Press.

Hannay, D. (1980) 'The iceberg of illness and trivial consultations', *Journal of the Royal College of General Practitioners* 30: 551–4.

Hardie, A. (2000) 'Gaelic plans to help seal sectarian wounds', *The Scotsman*, 2 March, p. 10.

Hardill, I. (1990) 'The recent restructuring of the British wool textiles industry', *Geography* 75(3): 203–10.

—— (1998) 'Gender perspectives on British expatriate households', *Geoforum* 29: 257–68.

—— (1999) 'Diasporic business connections: an examination of the role of female entrepreneurs in a South Asian business district', Economic Geography Research Group Working Paper 3/1999.

Hardill, I., Green, A.E. and Dudleston, A.C. (1997) 'The "blurring of boundaries" between "work" and "home": perspectives from case studies in the East Midlands', *Area* 29(3): 335–43.

Hardill, I. and Raghuram, P. (1998) 'Diasporic connections: case studies of Asian women in business', *Area* 30(3): 255–61.

Harding, S., Rosato, M. and Brown, J. (1998) 'Who becomes a lone mother?', *Population Trends* 91: 2–3.

Harrabin, R. (2000) 'Prosecute all drivers who kill', *The Independent*, 12 January, p. 12.

Harrison, M. (2000) 'Ford will limit jobs loss at Dagenham', *The Independent*, 12 May, p. 4.

Harvey, D. (1973) *Social Justice and the City*, London: Edward Arnold.

—— (1989) *The Condition of Postmodernity: An*

Enquiry into the Origins of Cultural Change, Oxford: Blackwell.

Harvie, C. (1982) *Against Metropolis, Fabian Tract 484*, London: Fabian Society.

—— (1989) 'Scott and the image of Scotland', in R. Samuel (ed.) *Patriotism: The Making and Unmaking of British National Identity, Volume II, Minorities and Outsiders*, London: Routledge, pp. 173–92.

—— (1991) 'English regionalism: the dog that never barked', in B. Crick (ed.) *National Identities: The Constitution of the United Kingdom*, Oxford: Blackwell, pp. 105–18.

—— (1998) *Scotland and Nationalism, Scottish Society and Politics 1707 to the Present* (3rd edition), London: Routledge.

Haskey, J. (1995) 'Trends in marriage and cohabitation: the decline in marriage and the changing pattern of living in partnerships', *Population Trends* 80: 5–15.

—— (1999) 'Cohabitational and marital histories of adults in Great Britain', *Population Trends* 96: 13–23.

Hassan, G. (1999) 'There will be no turf war with Westminster', *Parliamentary Brief* 6(1): 45, 47.

Hawthorn, G. (1982) 'The paradox of the modern: determinants of fertility in northern and western Europe since 1950', in C. Höhn and R. Mackensen (eds) *Determinants of Fertility Trends: Theories Re-examined*, Liège: Ordina Editions, pp. 283–96.

Health of Londoners Project (1996) *Transport in London and the Implications for Health*, London: The Health of Londoners Project.

Hebdige, D. (1979) *Subculture: The Meaning of Style*, London: Routledge.

Hechter, M. (1975) *Internal Colonialism: The Celtic Fringe in British National Development 1536–1966*, London: Routledge & Kegan Paul.

Heffer, S. (1999) *Nor Shall My Sword: The Reinvention of England*, London: Weidenfeld & Nicolson.

Hencke, D., Watt, N. and Wintour, P. (2000) 'Labour panic over poll meltdown', *The Guardian*, 31 March, p. 1.

Henry, N. and Massey, D. (1995) 'Competitive time-space in high technology', *Geoforum* 26: 49–64.

Hetherington, P. (1999a) 'Champions of England', *The Guardian*, 24 March, p. 4.

—— (1999b) 'Life's tougher up north, insist Labour MPs', *The Guardian*, 7 December, p. 10.

—— (1999c) 'Battle of second home front', *The Guardian*, 2 October, p. 3.

—— (1999d) 'Dirty old towns', *The Guardian*, 4 December, p. 24.

Hetherington, P. and Robinson, F. (1988) 'Tyneside life', in F. Robinson (ed.) *Post-Industrial Tyneside: An Economic and Social Survey of Tyneside in the 1980s*, Newcastle: Newcastle upon Tyne City Libraries and Arts, pp. 189–210.

Hills, J. (1995) *Inquiry into Income and Wealth, Volume 2*, York: Joseph Rowntree Foundation.

Hilpern, K. (1999) 'Oi, you! Read this', *The Guardian*, 29 September, p. 6.

Hirst, P. and Thompson, G. (1995) 'Globalisation and the future of the nation state', *Economy and Society* 24: 408–42.

—— (1996) *Globalisation in Question*, Cambridge: Polity Press.

Hitchens, P. (1999) *The Abolition of Britain: The British Cultural Revolution from Lady Chatterley to Tony Blair*, London: Quartet.

HMSO (1995) *Rural England: A Nation Committed to a Living Countryside*, London: HMSO.

Hogarth, T. (1986) 'Long distance weekly commuting', *Policy Studies* 8: 27–43.

Hogarth, T. and Daniel, W.W. (1988) *Britain's New Industrial Gypsies*, London: Policy Studies Institute.

Hoggart, R. (1957) *Uses of Literacy: Aspects of Working Class Life with Special Reference to Publications and Entertainments*, London: Chatto & Windus.

Hogwood, B.W. (1982) 'Introduction', in B.W. Hogwood and M. Keating (eds) *Regional Government in England*, Oxford: Clarendon Press, pp. 1–20.

—— (1996) *Mapping the Regions*, Bristol: Policy Press.

Hollifield, J.F. (1997) 'Immigration and integration in Western Europe: a comparative analysis', in E.M. Uçarer and D.J. Puchala (eds) *Immigration into Western Societies: Problems and Policies*, London: Pinter, pp. 29–69.

Holohan, A., Pledger, G. and Wilson, D. (1988) 'Health and health provision', in F. Robinson (ed.) *Post-Industrial Tyneside: An Economic and Social Survey of Tyneside in the 1980s*, Newcastle: Newcastle upon Tyne City Libraries and Arts, pp. 156–88.

Holt, R. (1989) *Sport and the British: A Modern History*, Oxford: Clarendon Press.

Horrell, S. and Humphries, J. (1995) 'Women's labour force participation and the transition to the male breadwinner family, 1790–1865', *Economic History Review* 48(1): 89–117.

House, J.W. (ed.) (1978) *The U.K. Space: Resources, Environment and the Future* (2nd edition), London: Weidenfeld & Nicolson.

Hudson, R. (1997) 'The end of mass production and the mass production worker? Experimenting with production and employment', in R. Lee and J. Wills (eds) *Geographies of Economies*, London: Arnold, pp. 302–21.

—— (2000) *Production, Places and Environment: Changing Perspectives in Economic Geography*, Harlow: Pearson.

Hudson, R. and Williams, A.M. (1986) *The United Kingdom*, London: Paul Chapman.

—— (1995) *Divided Britain* (2nd edition), Chichester: John Wiley & Sons.

Hugill, B. (1998) 'Minded out of their minds', *The Observer*, 29 March, p. 7.

Hunter, M. (1999) 'Extra planning to cut abortion rate', *Evening Post*, 24 December, p. 15.

Huq, R. (1996) 'Asian kool? Bhangra and beyond', in S. Sharma, J. Hutnyk and A. Sharma (eds) *Dis-Orienting Rhythms: The Politics of the New Asian Dance Music*, London: Zed Books, pp. 61–80.

Husband, C. (1982) 'Introduction: "race", the continuity of a concept', in C. Husband (ed.) *'Race' in Britain. Continuity and Change*, London: Hutchinson.

Hutchinson, R. (1996) *Empire Games: The British Invention of Twentieth-Century Sport*, Edinburgh: Mainstream.

Hutnyk, J. (1997) 'Adorno at Womad: South Asian crossovers and the limits of hybridity-talk', in P. Werbner and T. Modood (eds) *Debating Cultural Hybridity: Multi-Cultural Identities and the Politics of Anti-Racism*, London: Zed Books, pp. 106–36.

Hutton, W. (1995) *The State We're In*, London: Vintage.

Ignatieff, M. (1994) *Blood and Belonging: Journeys into the New Nationalism*, London: Vintage.

Ilbery, B., Bowler, I., Clark, G., Crockett, A. and Shaw, A. (1998) 'Farm-based tourism as an alternative farm enterprise: a case study from the Northern Pennines', *Regional Studies* 32(4): 355–64.

Illsley, R. and Gill, D. (1968) 'Changing trends in illegitimacy', *Social Science and Medicine* 2: 415–33.

Insley, J. (1999) 'Shopping 'til you drop', *The Observer*, *Cash*, 8 August, pp. 2–4.

Jackson, P. (1994) *Maps of Meaning*, London: Unwin Hyman.

—— (2000) 'Cultures of difference', in V. Gardiner and H. Matthews (eds) *The Changing Geography of the United Kingdom*, London: Routledge, pp. 276–96.

Jackson, S. (1998) *Britain's Population: Demographic Issues in Contemporary Society*, London: Routledge.

Jacques, M. (1997) 'The melting pot that is born-again Britannia', *The Observer*, 28 December, p. 15.

James, A. (1995) 'How British is food?', in P. Caplan (ed.) *Food, Health and Identity*, London: Routledge, pp. 71–86.

James, O. (1998) *Britain on the Couch: Treating a Low Serotonin Society*, London: Arrow.

Jarvie, G. (1999) 'Sport in the making of Celtic cultures', in G. Jarvie (ed.) *Sport in the Making of Celtic Cultures*', London: Leicester University Press, pp. 1–11.

Johnson, J.H. (1974) *Suburban Growth: Geographical Processes at the Edge of the Western City*, London: Wiley.

Johnston, R.J., Gregory, D., Haggett, P., Smith, D. and Stoddart, D.R. (eds) (1983) *The Dictionary of Human Geography*, Oxford: Blackwell.

Johnston, R.J. and Pattie, C.J. (1998) 'Composition and context: region and voting in Britain revisited during Labour's 1990s revival', *Geoforum* 29(3): 309–29.

Jones, B. (1997) 'Welsh politics and changing British and European contexts', in J. Bradbury and J.

Mawson (eds) *British Regionalism and Devolution: The Challenges of State Reform and European Integration*, London: Jessica Kingsley, pp. 55–73.

Jones, E. (1998) *The English Nation: The Great Myth*, Stroud: Sutton Publishing.

Jones, H. (1994) *Health and Society in Twentieth-Century Britain*, Harlow: Longman.

Jones, H. and Kandiah, M. (eds) (1996) *The Myth of Consensus: New Views on British History, 1945–64*, Basingstoke: Macmillan.

Jones, H.R. (1990) *Population Geography*, London: Paul Chapman.

Jones, K. and Moon, G. (1992) *Health, Disease and Society: An Introduction to Medical Geography*, London: Routledge.

Jones, S.J. (ed.) (1968) *Dundee and District*, Dundee: British Association.

Juss, S.S. (1994) *Immigration, Nationality and Citizenship*, London: Mansell.

Kay, B. (1986) *Scots: The Mither Tongue*, Edinburgh: Mainstream Publishing.

Kearney, H. (1991) 'Four nations or one?', in B. Crick (ed.) *National Identities: The Constitution of the United Kingdom*, Oxford: Blackwell, pp. 1–6.

Kearns, G. (1993) 'Prologue: fin de siècle geopolitics: Mackinder, Hobson and theories of global closure', in P. Taylor (ed.) *The Political Geography of the Twentieth Century*, London: Belhaven, pp. 9–30.

Keating, M. (1982) 'The debate on regional reform', in B.W. Hogwood and M. Keating (eds) *Regional Government in England*, Oxford: Clarendon Press, pp. 235–53.

Keeble, D. (1980a) 'The South East: II Greater London', in G. Manners, D. Keeble, B. Rodgers and K. Warren (eds) *Regional Development in Britain* (2nd edition), Chichester: John Wiley & Sons, pp. 123–54.

—— (1980b) 'The South East: III Outside London', in G. Manners, D. Keeble, B. Rodgers and K. Warren (eds) *Regional Development in Britain* (2nd edition), Chichester: John Wiley & Sons, pp. 155–75.

Kendall, I. (1995) 'The founding of the NHS', in G. Moon and R. Gillespie (eds) *Society and Health: An Introduction to Social Science for Health Professionals*, London: Routledge, pp. 143–61.

King, A., Crewe, I., Denver, D., Newton, K., Norton, P., Sanders, D. and Seyd, P. (1993) *Britain and the Polls 1992*, Chatham: Chatham House.

Knowsley, J. (1999) 'High-fliers quit the City rat race for a stress-free career in garden design', *Sunday Telegraph*, 4 July, p. 19.

Knox, P. and Pinch, S. (2000) *Urban Social Geography: An Introduction* (4th edition), Harlow: Prentice-Hall.

Kofman, E. and Youngs, G. (eds) (1996) *Globalization: Theory and Practice*, London: Pinter.

Kuper, S. (1999) 'Big soccer transfers to be axed', *The Observer*, 3 October, p. 1.

Lakeman, E. (1970) *How Democracies Vote: A Study of Majority and Proportional Electoral Systems*, London: Faber & Faber.

Lakha, S. (1999) 'The new international division of labour and the Indian computer software industry', in J. Bryson, N. Henry, D. Keeble and R. Martin (eds) *The Economic Geography Reader*, Chichester: Wiley, pp. 148–55.

Lane, H. and Ryle, S. (1999) 'Gazumpers win as house prices peak', *The Observer*, 20 June, p. 22.

Larsen, S.S. (1982) 'Two sides of the house: identity and social organisation in Kilbroney, Northern Ireland', in A.P. Cohen (ed.) *Belonging: Identity and Social Organisation in British Rural Cultures*, Manchester: Manchester University Press, pp. 131–64.

Laska, S. and Spain, D. (1980) *Back to the City: Issues in Neighbourhood Renovation*, New York: Pergamon.

Laurance, J. (1999) 'It's OK mum, I'm not pregnant: Why does contraception for teenagers cause such widespread alarm?', *The Independent*, 4 February, p. 8.

Laurie, N., Dwyer, C., Holloway, C. and Smith, F. (1999) *Geographies of New Femininities*, Harlow: Pearson.

Layton-Henry, Z. (1994) 'Britain: the would-be zero-immigration country', in W.A Cornelius, P.L. Martin and J.F. Hollifield (eds) *Controlling Immigration: A Global Perspective*, Stanford: Stanford University Press, pp. 273–300.

Lee, P. (1999) 'Where are the socially excluded? Continuing debates in the identification of poor neighbourhoods', *Regional Studies* 33(5): 483–6.

Lee, R. and Wills, J. (eds) (1997) *Geographies of Economies*, London: Arnold.

Le Fanu, J. (1999) *The Rise and Fall of Modern Medicine*, London: Little, Brown & Co.

Leith, S. (2000) 'Designs of the times', *The Observer Review*, 23 January, p. 13.

Lenman, B. (1980) *The Jacobite Risings in Britain 1689–1750*, London: Methuen.

Lewis, G. (1998) 'Rural migration and demographic change', in B. Ilbery (ed.) *The Geography of Rural Change*, Harlow: Longman, pp. 131–60.

Lewis, J. and Townsend, A. (eds) (1989) *The North–South Divide: Regional Change in Britain in the 1980s*, London: Paul Chapman Publishing.

Lewis, S. and Cooper, C. (1988) 'Stress in dual earner families', *Women and Work* 3: 139–68.

Linklater, M. (2000) 'We are a nation once again', *The Times*, 8 May, pp. 3–4.

Lowe, P. and Goyder, J. (1982) *Environmental Groups in Politics*, London: George Allen & Unwin.

Lowe, P. and Ward, N. (1998) 'Regional Policy, CAP reform and rural development in Britain: the challenge for New Labour', *Regional Studies* 32(5): 469–74.

Lynch, P. (1999) 'New Labour and the English Regional Development Agencies: devolution as evolution', *Regional Studies* 33(1): 73–89.

McAllister, I. (1997) 'Regional voting', *Parliamentary Affairs* 50(4): 641–57.

McCarthy, J. and Newlands, D. (1999) 'Introduction', in J. McCarthy and J. Newlands (eds) *Governing Scotland: Problems and Prospects – The Economic Impact of the Scottish Parliament*, Aldershot: Ashgate, pp. 1–7.

McCreadie, R. (1991) 'Scottish identity and the constitution', in B. Crick (ed.) *National Identities: The Constitution of the United Kingdom*, Oxford: Blackwell, pp. 38–56.

McCrone, D. (1992) *Understanding Scotland: The Sociology of a Stateless Nation*, London: Routledge.

McCrone, G. (1969) *Regional Policy in Britain*, London: George Allen & Unwin.

Macdonald, S. (1993) 'Identity complexes in Western Europe: social anthropological perspectives', in S. Macdonald (ed.) *Inside European Identities*, Oxford: Berg, pp. 1–26.

McDowell, L. (1991) 'Life without father and Ford: the new gender order of post-Fordism', *Transactions of the Institute of British Geographers* 16(4): 400–19.

—— (1997) *Capital Culture: Gender at Work in the City*, Oxford: Blackwell.

—— (2000) 'Making sense of masculinity: gender identities, school, locality and economic change', *Gender, Place and Culture* 7(4): 389–416.

MacErlean, N. (1999a) 'How the young will rule the world', *The Observer*, 7 February, p. 9.

—— (1999b) 'Who's left holding the babies?', *The Observer*, 1 August, p. 16.

MacFarlane, E.R. (1998) 'What – or who – is rural Britain', *Town and Country Planning* 67(5): 184–8.

McGhie, C. (1999) 'House and home', *The Sunday Telegraph Review*, 4 July, pp. 18–19.

McKay, G. (ed.) (1998) *DiY Culture: Party and Protest in Nineties Britain*, London: Verso.

Mackay, J. and Thane, P. (1986) 'The English-woman', in R. Colls and P. Dodd (eds) *Englishness: Politics and Culture 1880–1920*, Beckenham: Croom Helm, pp. 191–229.

McKendrick, J. (1995) 'Poverty in the United Kingdom: the Celtic divide', in C. Philo (ed.) *Off the Map: The Social Geography of Poverty in the UK*, London: Child Poverty Action Group, pp. 45–64.

Mackenzie, W.J.M. (1978) *Political Identity*, Harmondsworth: Penguin.

McKie, R. (2000) 'Gene bank will show what makes every one of us tick', *The Observer*, 13 February, pp. 12–13.

Mackinder, H.J. (1902) *Britain and the British Seas*, London: Heinemann.

McLean, G. (2000) 'Too busy to seek help from the WA', *The Scotsman*, 10 February, p. 5.

McRobbie, A. (1989) *Zoot Suits and Second Hand Dresses*, London: Macmillan.

McSmith, A. (1999a) 'They call it delay. But it's death', *The Observer*, 17 October, p. 8.

—— (1999b) 'Tories torn over English flag-waving', *The Observer*, 11 July, p. 8.

McVeigh, T. (1999) 'Girl power suffers major breakdown', *The Observer*, 7 November, p. 5.

—— (2000) 'Blackpool becoming the drink capital of Europe', *The Observer*, 20 February, p. 7.

Mallier, A. and Shafto, T. (1992) *The Economics of Flexible Retirement*, London: Academic Press.

Marmot, M. and Feeney, A. (1996) 'Work and health: implications for individuals and society', in D. Blane, E. Brunner and R. Wilkinson (eds) *Health and Social Organization: Towards a Health Policy for the 21st Century*, London: Routledge, pp. 235–54.

Marmot, M.G. (1984) 'Geography of blood pressure and hypertension', *British Medical Bulletin* 40(4): 380–6.

—— (1996) 'The social pattern of health and disease', in D. Blane, E. Brunner and R. Wilkinson (eds) *Health and Social Organization: Towards a Health Policy for the 21st Century*, London: Routledge, pp. 42–67.

Marquand, D. (1995a) 'How united is the modern United Kingdom', in A. Grant and K.J. Stringer (eds) *Uniting the Kingdom? The Making of British History*, London: Routledge, pp. 277–91.

—— (1995b) 'After Whig imperialism: can there be a new British identity?', *New Community* 21(2): 183–93.

—— (2000) 'Reflections on British decline', in R. English and M. Kenny (eds) *Rethinking British Decline*, Basingstoke: Macmillan, pp. 117–36.

Marqusee, M. (1994) *Anyone but England: Cricket and the National Malaise*, London: Verso.

Marr, A. (1995) *The Battle for Scotland*, Harmondsworth: Penguin.

—— (1996) *Ruling Britannia: The Failure and Future of British Democracy*, London: Penguin.

—— (1999a) 'So what kind of England do we really stand for?', *The Observer*, 31 October, pp. 22–3.

—— (1999b) 'Perils of ethnic purity', *The Observer*, 4 July, p. 24.

—— (1999c) 'Complacent. Ignorant. Uninterested. Thank goodness for the English', *The Observer*, 18 April, p. 26.

—— (2000) *The Day Britain Died*, London: Profile.

Marshall, T.H. (1950) *Citizenship and Social Class and Other Essays*, Cambridge: Cambridge University Press.

Martin, D.-C. (1995) 'The choices of identity', *Social Identities* 1(1): 5–20.

Martin, R. (1988) 'Industrial capitalism in transition: the contemporary reorganisation of the British space economy', in D. Massey and J. Allen (eds) *Uneven Redevelopment: Cities and Regions in Transition*, London: Hodder, pp. 202–31.

—— (1995) 'Income and poverty inequalities across regional Britain: the North–South divide lingers on', in C. Philo (ed.) *Off the Map: The Social Geography of Poverty in the UK*, London: Child Poverty Action Group, pp. 23–44.

Martin, R. and Rowthorn B. (eds) (1986) *The Geography of Deindustrialisation*, London: Macmillan.

Marwick, A. (1991) *Culture in Britain since 1945*, Oxford: Basil Blackwell.

—— (1996) *British Society since 1945*, Harmondsworth: Penguin.

Massey, D. (1984) *Spatial Divisions of Labour: Social Structures and the Geography of Production*, London: Macmillan.

—— (1994) *Space, Place and Gender*, Cambridge: Polity Press.

Massey, D. and McDowell, L. (1994) 'A woman's place', in D. Massey (ed.) *Space, Place and Gender*, Cambridge: Polity Press, pp. 191–211.

Mawson, J. (1997) 'The English regional debate: towards regional governance or government?', in J. Bradbury and J. Mawson (eds) *British Regionalism and Devolution: The Challenges of State Reform and European Integration*, London: Jessica Kingsley, pp. 180–211.

Mawson, J. and Spencer, K. (1997) 'The origins and operation of the Government Offices for the English Regions', in J. Bradbury and J. Mawson (eds) *British Regionalism and Devolution: The Challenges of State Reform and European Integration*, London: Jessica Kingsley, pp. 158–79.

Mayes, I. (1999) 'Sense of place', *The Guardian*, 25 September, p. 7.

Mayhew, H. (1851–62) *London Labour and the London Poor*, 4 volumes (reprinted 1967), London: Cass.

Mbanje, C. and Koch, R. (1999) 'HIV/AIDS and teenage girls – the new epidemic', *ActionAid*, News release, 15 September.

Meacher, M. (1997) 'Street of shame', *The Guardian*, 19 November, pp. 4–5.

Meikle, J. (1999) 'Tourists threaten to swamp historic cities', *The Guardian*, 19 July, p. 9.

Mess, H.A. (1928) *Industrial Tyneside: A Social Survey*, London: Ernest Benn Limited.

Miles, R. (1989) *Racism*, London: Routledge.

Miller, R. and Tivy, J. (eds) (1958) *The Glasgow Region: A General Survey*, Glasgow: British Association.

Miller, W.L. (1998) 'The periphery and its paradoxes', *West European Politics* 21(1): 167–96.

Mills, C. (1995) 'Managerial and professional work histories', in T. Butler and M. Savage (eds) *Social Change and the Middle Classes*, London: UCL Press, pp. 95–116.

MIND (1992) *The MIND Survey: Stress at Work*, London: National Association for Mental Health.

Minority Rights Group International (1997) *World Directory of Minorities*, London: MRGI.

Mitchell, J. (ed.) (1962) *Great Britain: Geographical Essays*, London: Cambridge University Press.

Modood, T. (ed.) (1997) *Ethnic Diversity and Disadvantage*, London: Institute of Policy Studies.

—— (1998) 'Ethnic diversity and racial disadvantage in employment', in T. Butler and M. Savage (eds) *Social Change and the Middle Classes*, London: UCL Press, pp. 53–73.

Moeller, S.D. (1999) *Compassion Fatigue: How the Media Sell Disease, Famine, War and Death*, New York: Routledge.

Mohan, J. (1995) 'Missing the boat: poverty, debt and unemployment in the South-East', in C. Philo (ed.) *Off the Map: The Social Geography of Poverty in the UK*, London: Child Poverty Action Group, pp. 133–51.

—— (1999) *A United Kingdom? Economic, Social and Political Geographies*, London: Arnold.

Montgomery, D. (1999) 'Scots face lowest crime risk in UK', *The Scotsman*, 4 June, p. 9.

Morris, L. (1995) *Social Divisions: Economic Decline and Social Structural Change*, London: UCL Press.

Morton, H.V. (1927) *In Search of England*, London: Methuen.

Mowat, C.L. (1968) *Britain between the Wars: 1918–1940*, London: Methuen.

Moxon-Browne, E. (1991) 'National identity in Northern Ireland', in P. Stringer and G. Robinson (eds) *Social Attitudes in Northern Ireland*, Belfast: Blackstaff Press, pp. 23–30.

Murray, I. (2000) 'Ten-year plan launched to cut heart disease', *The Times*, 7 March, p. 4.

Nairn, T. (1977) *The Break-Up of Britain*, London: New Left Books.

—— (1988) *The Enchanted Glass: Britain and its Monarchy*, London: Picador.

—— (1997) *Faces of Nationalism: Janus Revisited*, London: Verso.

—— (2000) *After Britain: New Labour and the Return of Scotland*, London: Granta.

Narayan, U. (1995) 'Eating cultures: incorporation, identity and Indian food', *Social Identities* 1(1): 63–86.

Nash, D. and Reeder, D. (1993) *Leicester in the Twentieth Century*, Stroud: Alan Sutton Publishing in association with Leicester City Council.

Newman, P. (1999) 'Exit from auto hell', *New Internationalist* 313: 24–5.

Nicolson, S. (2000) 'Motorists "ignored" over road charges', *The Scotsman*, 2 February, p. 4.

Norris, P. (1997) 'Anatomy of a Labour landslide', *Parliamentary Affairs* 50(4): 509–32.

—— (1999) 'Gender: a gender-generation gap?', in G. Evans and P. Norris (eds) *Critical Elections: British Parties and Voters in Long-Term Perspective*, London: Sage, pp. 148–63.

Norris, P. and Evans, G. (1999a) 'Conclusion: was 1997 a critical election?', in G. Evans and P. Norris (eds) *Critical Elections: British Parties and Voters in Long-Term Perspective*, London: Sage, pp. 259–71.

—— (1999b) 'Introduction: understanding electoral change', in G. Evans and P. Norris (eds) *Critical Elections: British Parties and Voters in Long-Term Perspective*, London: Sage, pp. xix–xl.

North, D. (1998) 'Rural industrialisation', in B. Ilbery (ed.) *The Geography of Rural Change*, Harlow: Longman, pp. 161–88.

NTC (1999a) *Lifestyle Pocket Book*, Henley-on-Thames: NTC Publications.

NTC (1999b) *Regional Marketing Pocket Book*, Henley-on-Thames: NTC Publications.

O'Brien, R. (1992) *Global Financial Integration: The End of Geography*, London: Royal Institute of International Affairs.

Ogilvie, A.G. (ed.) (1928) *Great Britain: Essays in Regional Geography*, Cambridge: Cambridge University Press.

Ohmae, K. (1990) *The Borderless World: Power and Strategy in the Interlocked Economy*, London: Harper Business.

Omran, A.R. (1971) 'The epidemiological transition: a theory of the epidemiology of population change', *Milbank Memorial Fund Quarterly* 49: 509–38.

ONS (1996) *Social Focus on Ethnic Minorities*, London: HMSO.

—— (1997) *Regional Trends 32*, London: The Stationery Office.

—— (1998a) *Regional Trends 33*, London: The Stationery Office.

—— (1998b) *How Exactly is Unemployment Measured*, London: The Stationery Office.

—— (1998c) *How Exactly is Employment Measured*, London: The Stationery Office.

—— (1999a) *Regional Trends 34*, London: The Stationery Office.

—— (1999b) *International Migration 1997*, Series MN no. 24, London: The Stationery Office.

—— (1999c) *Social Trends 29*, London: The Stationery Office.

—— (1999d) *Annual Abstract of Statistics*, London: The Stationery Office.

—— (1999e) *Social Focus on Older People*, London: The Stationery Office.

—— (2000) *Labour Market Statistics Bulletin 166*, 17 May, p. 1.

ONS, EOC (1998) *Social Focus on Women and Men*, London: The Stationery Office.

Osmond, J. (1988) *The Divided Kingdom*, London: Constable.

O'Sullivan, J. (1999) 'Welcome to Aberdeen, Britain's saddest city', *The Independent Review*, 29 October, p. 1.

Overbeek, H. (2000) 'Globalisation and British decline', in R. English and M. Kenny (eds) *Rethinking British Decline*, Basingstoke: Macmillan, pp. 231–56.

Owen, D.W. (1996) 'The Other-Asians : the salad bowl', in C. Peach (ed.) *Ethnicity in the 1991 Census*, London: HMSO, pp. 181–205.

Paddison, R. (1997) 'The restructuring of local government in Scotland', in J. Bradbury and J. Mawson (eds) *British Regionalism and Devolution: The Challenges of State Reform and European Integration*, London: Jessica Kingsley, pp. 99–117.

Pahl, J. (1989) *Money and Marriage*, London: Macmillan.

Pahl, R. (1984) *Divisions of Labour*, Oxford: Blackwell.

Painter, J. (2000) 'Local government and governance', in V. Gardiner and H. Matthews (eds) *The Changing Geography of the United Kingdom*, London: Routledge, pp. 296–314.

Palast, G. (1999) 'Sickness at the heart of private medicine', *The Observer, Business*, 25 April, p. 10.

Pandga, N. (1999) 'Jobs and money', *The Guardian*, 10 July, p. 19.

Paterson, L. (2000) 'Time to take the high road', *Red Pepper*, February, http://www.redpepper.org.uk/xscots.html

Paton Walsh, N. (1999) 'Slim boy fat', *The Observer Review*, 5 December, p. 2.

Pattie, C. (2000) 'A (dis)United Kingdom?', in V. Gardiner and H. Matthews (eds) *The Changing Geography of the United Kingdom*, London: Routledge, pp. 315–35.

Pattie, C. and Johnston, R. (2000) '"People who talk together vote together": an exploration of contextual effects in Great Britain', *Annals of the Association of American Geographers* 90(1): 41–66.

Pattie, C., Johnston, R., Dorling, D., Rossiter, D., Tunstall, H. and MacAllister, I. (1997) 'New Labour, new geography? The electoral geography of the 1997 British general election', *Area* 29(3): 253–9.

Pattie, C., Russell, A. and Johnston, R. (1991) 'Going Green in Britain? Votes for the Green Party and attitudes to Green issues in the late 1980s', *Journal of Rural Studies* 7: 285–97.

Paxman, J. (1998) *The English: A Portrait of a People*, London: Michael Joseph.

Peach, C. (ed.) (1996) *Ethnicity in the 1991 Census, Volume 2*, London: HMSO.

Pearce, N. (1998) *Britain 1999: The Official Yearbook of the United Kingdom*, London: The Stationery Office.

Peck, J. (1999) 'New Labourers? Making a New Deal for the workless class', *Environment and Planning C* 17(3): 345–72.

Peel, J. (1999) 'The Ingerland playlist: can you hear the English sing?', in M. Perryman (ed.) *The Ingerland Factor: Home Truths from Football*, Edinburgh: Mainstream, pp. 33–6.

Pennington, S. and Westover, B. (1989) *A Hidden Workforce: Homeworkers in England, 1850–1985*, Basingstoke: Macmillan.

Perryman, M. (1999) 'The Ingerland factor', in M. Perryman (ed.) *The Ingerland Factor: Home Truths from Football*, Edinburgh: Mainstream, pp. 15–32.

Phillimore, P. and Moffat, S. (2000) '"Industry causes lung cancer": would you be happy with that headline? Environmental health and local politics', in S. Allan, B. Adam and C. Carter (eds) *Environmental Risks and the Media*, London: Routledge, pp. 105–16.

Phillips, A. (2000) 'The rising tide of "sod them" politics', *The Observer*, 7 May, p. 21.

Phillips, D. (1998) 'Black minority ethnic concentration, segregation and dispersal in Britain', *Urban Studies* 35(10): 1681–702.

Phillips, D. and Sarre, P. (1995) 'Black middle-class formation in contemporary Britain', in T. Butler, and M. Savage (eds) *Social Change and the Middle Classes*, London: UCL Press, pp. 76–94.

Phillips, M. (1998) 'Social perspective', in B. Ilbery (ed.) *The Geography of Rural Change*, Harlow: Longman, pp. 31–55.

Philo, C. (ed.) (1995) *Off the Map: The Social Geography of Poverty in the UK*, London: Child Poverty Action Group.

Phizacklea, A. and Ram, M. (1996) 'Being your own boss: ethnic minority entrepreneurs in comparative perspective', *Work, Employment and Society* 10(3): 319–39.

Phizacklea, A. and Wolkowitz, C. (1990) *Homeworking Women: Gender, Racism and Class at Work*, London: Sage.

PHLS (1999a) Press release, 23 November.

—— (1999b) 'Infectious diseases in the news', 6 August.

—— (1999c) Press release, 14 December.

—— (2000) Press release, 27 January.

Pile, H. and O'Donnell, C. (1997) 'Earnings, taxation and wealth', in A. Walker and C. Walker (eds) *Britain Divided: The Growth of Social Exclusion in the 1980s and 1990s*, London: Child Poverty Action Group, pp. 32–47.

Pimlott, B. (1985) 'Unemployment and "the unemployed" in north-east England', *The Political Quarterly* 56(4): 346–60.

Pinch, S. (1993) 'Social polarisation: a comparison of evidence from Britain and the United States', *Environment and Planning A* 25: 779–95.

PIU (1999) *Rural Economies*, Performance and Innovation Unit Report, London: The Stationery Office, http://www.cabinetoffice.gov.uk/innovation/1999/rural/index/html

—— (2000) *Reaching Out: The Role of Central Government at the Regional and the Local Level*, London: The Stationery Office.

Pollert, A. (1999) 'Dismantling flexibility', in J. Bryson, N. Henry, D. Keeble and R. Martin (eds) *The Economic Geography Reader*, Chichester: Wiley, pp. 349–56.

Population Trends (1997) 'Our health – better or worse? Findings from "The Health of Adult Britain 1841–1994"', *Population Trends* 88: 43–7.

—— (1998) 'Subnational population projections for Scotland', *Population Trends* 91: 2.

—— (1999) 'Tables', *Population Trends* 96: 57–86.

Porter, D. (1999) *Health, Civilization and the State*, London: Routledge.

Poulton, E. (1999) 'Fighting talk from the press corps', in M. Perryman (ed.) *The Ingerland Factor: Home Truths from Football*, Edinburgh: Mainstream, pp. 119–35.

Pratt, G. (1994) 'Feminist geographies', in R. Johnston, D. Gregory, and D. Smith (eds) *Dictionary of Human Geography* (3rd edition), Oxford: Blackwell, pp. 192–6.

Prescott, J. (2000) *Statement by the Deputy Prime Minister to the House of Commons*, 7 March, London: DETR.

Priestley, J.B. (1994) *English Journey*, London: Mandarin (first published in 1934).

Purdy, M. (1999) 'Health of which nation? Health, social regulation and the new consensus', in M.

Purdy and D. Banks (eds) *Health and Exclusion*, London: Routledge, 62–77.

Ram, M., Sanghera, B., Abbas, T., Barlow, G. and Jones, T. (2000) 'Ethnic minority business in comparative perspective: the case of the independent restaurant sector', *Journal of Ethnic and Migration Studies* 26(3): 495–510.

Rawnsley, A. (1999) 'My moral manifesto for the 21st century', *The Observer*, 5 September, pp. 8–9.

Rayner, J. (2000) 'We want to be alone', *The Observer Review*, 16 January, p. 1.

RDC (1996) *Disadvantage in Rural Areas*, Salisbury: RDC.

Redhead, S. (1999) 'Labouring for a vin-da-loo nation', in M. Perryman (ed.) *The Ingerland Factor: Home Truths from Football*, Edinburgh: Mainstream, pp. 199–206.

Redman, T., Wilkinson, A. and Snape, E. (1997) 'Stuck in the middle? Managers in building societies', *Work, Employment and Society* 11(1): 101–14.

Rees, T. (1993) 'United Kingdom I: inheriting empire's people', in D. Kubat (ed.) *The Politics of Migration Policies* (2nd edition), New York: Center for Migration Studies, pp. 87–107.

Reeves, R. (1999a) 'The new baby boom', *The Observer*, 19 December, p. 18.

—— (1999b) 'Seeking southern comfort', *The Observer*, 4 July, p. 14.

—— (1999c) 'Ditch your man and be happy', *The Observer*, 17 October, p. 13.

—— (1999d) 'The mad rush to save time', *The Observer*, 3 October, p. 17.

—— (1999e) 'Fat! So? The flabby fight back in battle of the bulge', *The Observer*, 29 August, p. 8.

—— (1999f) 'Boozing women behaving badly', *The Observer*, 12 December, p. 17.

—— (2000a) 'Your country needs US', *The Observer*, 23 January, p. 18.

—— (2000b) 'Can't men live without a shoulder to cry on?', *The Observer*, 9 January, p. 14.

Reeves, R. and Zinn, C. (1999) 'Charles "ready to face British voters"', *The Observer*, 7 November, p. 1.

Rennell, T. (1999) 'Who says life begins at 50?', *The Observer*, 18 July, p. 23.

Report of the Archbishop of Canterbury's Commission on Urban Priority Areas (1985) *Faith in the City, A Call for Action by Church and Nation, Volume 15, Part 1*, London, Church House.

Roberts, P. (2000) *The New Territorial Governance: Planning, Developing and Managing the United Kingdom in an Era of Devolution*, London: Town and Country Planning Association.

Robertson, R. (1995) 'Glocalization: time–space and homogeneity–heterogeneity', in M. Featherstone, S. Lash and R. Robertson (eds) *Global Modernities*, London: Sage, pp. 25–44.

Robins, K. (1999) 'Europe', in P. Cloke, P. Crang and M. Goodwin (eds) *Introducing Human Geographies*, London: Arnold, pp. 268–76.

Robinson, F. (1988a) 'The labour market', in F. Robinson (ed.) *Post-Industrial Tyneside: An Economic and Social Survey of Tyneside in the 1980s*, Newcastle: Newcastle upon Tyne City Libraries and Arts, pp. 62–85.

—— (1988b) 'Introduction', in F. Robinson (ed.) *Post-Industrial Tyneside: An Economic and Social Survey of Tyneside in the 1980s*, Newcastle: Newcastle upon Tyne City Libraries and Arts, pp. 1–11.

Robinson, F. and Gregson, N. (1992) 'The underclass: a class apart?', *Critical Social Policy* 34: 38–51.

Room, G. (1995) 'Poverty and social exclusion: the new European agenda for policy and research', in G. Room (ed.) *Beyond the Threshold: The Measurement and Analysis of Social Exclusion*, Bristol: Policy Press, pp. 1–9.

Rowlat, J. (1999) 'Kids' teeth decay as NHS dentistry dies', *The Observer*, 24 January, p. 11.

Rowntree, B.S. (1937) *The Human Needs of Labour*, London: Longmans, Green & Co.

—— (1941) *Poverty and Progress: A Second Social Survey of York*, London: Longmans, Green & Co.

Rowntree, B.S. and Lavers, G.R. (1952) *English Life and Leisure: A Social Study*, London: Longmans, Green & Co.

Ryle, S. (1998) 'Away the lads and back to the dole', *The Observer, Business*, 16 August, p. 3.

—— (1999) 'To Lady Emma and Sir Michael "private" meant quality. But he dies and she wished they had gone to the NHS', *The Observer*, 19 September, p. 6.

Saggar, S. (1997) 'Racial politics', *Parliamentary Affairs* 50(4): 693–707.

Saggar, S. and Heath, A. (1999) 'Race: towards a multicultural electorate', in G. Evans and P. Norris (eds) *Critical Elections: British Parties and Voters in Long-Term Perspective*, London: Sage, pp. 102–23.

Samers, M. (1998) 'Immigration, "ethnic minorities", and "social exclusion" in the European Union: a critical perspective', *Geoforum* 29(2): 123–44.

Sassen, S. (1991) *The Global City. New York, London, Tokyo*, Princeton: Princeton University Press.

Savage, M. (1988) 'The missing link? The relationship between spatial mobility and social mobility', *British Journal of Sociology* 39: 554–77.

—— (1995) 'Class analysis and social research', in T. Butler and M. Savage (eds) *Social Change and the Middle Classes*, London: UCL Press, pp. 15–25.

Scase, R., Scales, J. and Smith, C. (1998) *Work Now Pay Later: The Impact of Working Long Hours on Health and Family Life*, Working Paper of the ESRC Research Centre on Micro-Social Change, University of Essex.

Schuman, J. (1999) 'The ethnic minority populations of Great Britain – latest estimates', *Population Trends* 96: 33–43.

Schuster, L. and Solomos, J. (1999) 'The politics of refugee and asylum policies in Britain: historical patterns and contemporary realities', in A. Bloch and C. Levy (eds) *Refugees, Citizenship and Social Policy in Europe*, Basingstoke: Macmillan, pp. 51–75.

Scottish Parliament Information Centre (1999), *Scottish Parliament Election Results 6 May 1999*, Research Paper 99/1, Edinburgh: The Scottish Parliament.

Scott, C. (1999) 'Tackling teen pregnancies', *Evening Post*, 14 August, p. 12.

Seenan, G. (1999) 'Composer denounces bigotry of fellow Scots', *The Guardian*, 9 August, pp. 1, 3.

Segal, L. (1999) 'Man made crisis', *Red Pepper*, March, http://www.redpepper.org.uk/xmen.html

Segall, A. (1999) 'Property rage is here as buyers battle for homes', *The Daily Telegraph*, 13 July, p. 5.

Self, P. (1967) 'Planning and housing in London and South-East England', *Town and Country Planning* 35(3): 134–42.

Sennett, R. (1998) *The Corrosion of Character: The Personal Consequences of Work in the New Capitalism*, New York: Norton.

SERPLAN (1998) *A Sustainable Development Strategy for the South East*, London: London and South East Regional Planning Conference SERP500.

Seton-Watson, H. (1977) *Nations and States*, London: Methuen.

Sharma, S. (1996) 'Noisy Asians or "Asian noise"?', in S. Sharma, J. Hutnyk and A. Sharma (eds) *Dis-Orienting Rhythms: The Politics of the New Asian Dance Music*, London: Zed Books, pp. 32–57.

Sharpe, L.J. (1997) 'British regionalism and the link with regional planning: a perspective of England', in J. Bradbury and J. Mawson (eds) *British Regionalism and Devolution: The Challenges of State Reform and European Integration*, London: Jessica Kingsley, pp. 121–36.

Shepherd, M. (1984) 'Urban factors in mental disorders – an epidemiological approach', *British Medical Bulletin* 40(4): 401–4.

Shirlow, P. and McGovern, M. (eds) (1997) *Who are the People? Unionism, Protestantism and Loyalism in Northern Ireland*, London: Pluto Press.

Shirlow, P. and Stewart, P. (1999) 'Northern Ireland: between peace and war?', *Capital and Class* 69: vi–xiv.

Sibley, D. (1995) *Geographies of Exclusion: Society and Difference in the West*, London: Routledge.

Simpson, W.D. (1963) 'The region before 1700', in A.C. O'Dell and J. Mackintosh (eds) *The North-East of Scotland*, Aberdeen: British Association, pp. 67–86.

Singh, A. (1977) 'UK industry and the world economy: a case of deindustrialisation', *Cambridge Journal of Economics* 1(2): 113–362.

Sked, A. and Cook, C. (1990) *Post-War Britain: A Political History* (3rd edition), London: Penguin.

Skelton, T. and Valentine, G. (eds) (1998) *Cool Places: Geographies of Youth Culture*, London: Routledge.

Smith, A. (2000a) 'Where did you find that voice?', *The Observer Review*, 12 March, p. 2.

—— (2000b) 'Three lions', *The Observer Magazine*, 21 May, pp. 20–4.

Smith, A.D. (1991) *National Identity*, Harmondsworth: Penguin.

Smith, B.C. (1965) *Regionalism in England: Its Nature and Purpose*, London: The Acton Society Trust.

Smith, C. (1999) 'Coalfield's joy at £24m cash boost', *Evening Post*, 15 July, p. 5.

Smith, N. (1979) 'The theory of gentrification: a back to the city movement by capital not people', *Journal of the American Planners Association* 45: 538–48.

Smyth, G. (1997) 'Ethnicity and language', in M. Storry and P. Childs (eds) *British Cultural Identities*, London: Routledge, pp. 241–76.

Social Exclusion Unit (1998) *Bringing Britain Together: A National Strategy for Neighbourhood Renewal*, Cm 4045, London: HMSO.

Spence, J., Walton, W.S., Miller, F.J.W. and Court, S.D. (1954) *A Thousand Families in Newcastle upon Tyne*, London: Oxford University Press.

Spencer, N. (1999) 'The trouble with boys', *The Observer Magazine*, pp. 14–18.

Stacey, J. (1996) *In the Name of the Family: Rethinking Family Values in the Postmodern Age*, Boston: Beacon Press.

—— (1998) *Brave New Families: Stories of Domestic Upheaval in Late Twentieth Century America*, Basic Books: New York.

Stalker, P. (1994) *The Work of Strangers*, Geneva: ILO.

Stamp, D. and Beaver, S.H. (1933) *The British Isles: A Geographic and Economic Survey*, London: Longman.

Stanworth, C. (1998) 'Telework and the information age', *New Technology, Work and Employment* 13(1): 51–62.

Stevens, R. and Legge, A. (1987) 'Illegitimacy obscured but not obliterated: an analysis of the family law reform act 1987', *Family Law* 17: 409–13.

Stokowski, P.A. (1994) *Leisure in Society: A Network Structural Perspective*, London: Mansell.

Storry, M. and Childs, P. (1997) 'Introduction: the ghost of Britain past', in M. Storry and P. Childs (eds) *British Cultural Identities*, London: Routledge, pp. 1–40.

Sugden, J. and Bairner, A. (1993) *Sport, Sectarianism and Society in a Divided Ireland*, Leicester: Leicester University Press.

Summerskill, B. (2000a) 'Teen sex, tough love', *The Observer*, 27 August, p. 22.

—— (2000b) 'Sorry no children', *The Observer Review*, 30 July, pp. 1–2.

—— (2000c) 'Playtime as kidults grow up at last', *The Observer*, 23 July, p. 20.

Surridge, P., Brown, A., McCrone, D. and Paterson, L. (1999) 'Scotland: constitutional preferences and voting behaviour', in G. Evans and P. Norris (eds) *Critical Elections: British Parties and Voters in Long-Term Perspective*, London: Sage, pp. 223–39.

Sutherland, V. and Cooper, C. (1990) *Understanding Stress: A Psychological Perspective for Health Professionals*, London: Chapman and Hall.

Sylvester, R. (1999) 'Forget the happy family – we're all just domestic units now', *The Independent*, 14 March, p. 8.

Taylor, A.H. (1973) 'The electoral geography of Welsh and Scottish nationalism', *Scottish Geographical Magazine* 89(1): 44–52.

Taylor, P. (1990) *Britain and the Cold War: 1945 as Geopolitical Transition*, London: Pinter.

Taylor, P.J. (1991) 'The English and their Englishness: "a curiously mysterious, elusive and little understood people"', *Scottish Geographical Magazine* 107(3): 146–61.

Taylor, P. and Flint, C. (2000) *Political Geography* (4th edition), Harlow: Pearson Education.

Taylor, S. (1999) 'It's never Father's day', *The Observer Review*, 8 August, p. 4.

Thane, P. (1989) 'Old age: benefit or burden', in H. Joshi (ed.) *The Changing Population of Britain*, Oxford: Basil Blackwell, pp. 56–71.

Thatcher, R. (1999) 'The demography of centenarians in England and Wales', *Population Trends* 96: 5–12.

The Economist (1998) 'Little countries: small but perfectly formed', 3 January, pp. 63–5.

—— (1999a) 'The new politics', 15 May, pp. 31–2.

—— (1999b) 'Les français sont arrivés', 23 October, p. 37.

—— (1999c) 'A nation once again', 1 May, pp. 21–7.

—— (1999d) 'A Geordie nation?', 27 March, pp. 26–7.

—— (1999e) 'Hague's breather', 15 May, pp. 38–9.

—— (2000a) 'Teenage pregnancy: baby dolls', 5 August, pp. 37–8.

—— (2000b) 'Demography: legions of centenarians', 5 August, p. 37.

—— (2000c) 'Europe's migrants: riding the tide', 5 August, pp. 45–6.

—— (2000d) 'Britain and NAFTA: dream on?', 15 April, pp. 30, 33.

The Guardian (2000a) 'Alliance wish list', *Society*, 23 August, p. 6.

—— (2000b) 'Official: Livingstone wins', 5 May, p. 3.

The Times (2000) 'The other referendum: an alliance for the Alternative Vote that should trouble the Tories', 30 June, p. 13.

Thomas, D.E. (1991) 'The constitution of Wales', in B. Crick (ed.) *National Identities: The Constitution of the United Kingdom*, Oxford: Blackwell, pp. 57–67.

Thomas, R. (1998) 'Smoking: the new apartheid, the fag end of society', *The Observer*, 6 December, p. 18.

—— (1999) 'Suburban bliss, seaside hell', *The Observer*, 2 May, p. 15.

Thompson, T. (1999) 'The rise of Crime plc', *The Observer*, 5 September, p. 17.

—— (2000) 'Obesity among children on the increase', *The Scotsman*, 28 February, p. 15.

Thornley, A. (1993) *Urban Planning under Thatcherism*, London: Routledge.

Thrift, N. and Leyshon, A. (1999) 'In the wake of money: the City of London and the accumulation of value', in J. Bryson, N. Henry, D. Keeble and R. Martin (eds) *The Economic Geography Reader*, Chichester: Wiley, pp. 333–40.

Tickell, A. and Peck, J. (1999) 'Social regulation after Fordism: regulation theory, neo-liberalism and the global–local nexus', in J. Bryson, N. Henry, D. Keeble and R. Martin (eds) *The Economic Geography Reader*, Chichester: Wiley, pp. 121–30.

Tisdall, S. (2000) 'Do they mean us?', *The Guardian*, 28 March, p. 15.

Tolson, A.R. and Johnstone, M.E. (1970) *A Geography of Britain*, Oxford: Oxford University Press.

Tomaney, J. (2000) 'Democratically elected regional government in England: the work of the North East Constitutional Convention', *Regional Studies* 34(4): 383–8.

Townsend, A.R. (1983) *The Impact of Recession: On Industry, Employment and the Regions, 1976–1981*, London: Croom Helm.

—— (1997) *Making a Living in Europe: Human Geographies of Economic Change*, London: Routledge.

Townsend, A.R. and Taylor, C.C. (1974) *Sense of Place and Local Identity in North-East England*, North-East Area Study Working Paper 4, University of Durham.

—— (1975) 'Regional culture and identity in industrialized societies: the case of north-east England', *Regional Studies* 9(4): 379–93.

Townsend, P. (1957) *The Family Life of Older People*, London: Routledge & Kegan Paul.

—— (1979) *Poverty in the United Kingdom: A Survey of Household Resources and Standards of Living*, Harmondsworth: Penguin.

Tresidder, R. (1999) 'Brand new approach', *Evening Post*, 17 April, p. 23.

Trueland, J. (2000) 'Stress leads to 20% rise in women's lung cancer', *The Scotsman*, 2 March, p. 7.

Tudor Hart, J. (1971) 'The inverse care law', *Lancet*, 27 February, pp. 405–12.

Tunstall, H., Rossiter, D.J., Pattie, C.J., MacAllister, I., Johnston, R.J. and Dorling, D.F.L. (2000) 'Geographical scale, the "feel-good factor" and voting at the 1997 general election in England and Wales', *Transactions of the Institute of British Geographers* 25(1): 51–64.

Turok, I. and Webster, D. (1998) 'The New Deal: jeopardised by the geography of unemployment', *Local Economy* 12(4): 309–28.

Unwin, T. (1992) *The Place of Geography*, Harlow: Longman.

Urry, J. (1995) *Consuming Places*, London: Routledge.

Utley, A. (1998) 'Women lose out on leisure', *Times Higher Education Supplement*, 7 August, p. 5.

Valentine, A. (2000) 'Hysterical claim', *The Scotsman*, 18 July, p. 15.

Valentine, G. (1989) 'The geography of women's fear', *Area* 21(4): 385–90.

Veash, N. (1999) 'Women: the new men', *The Observer Review*, 18 July, p. 12.

Veblen, T. (1953) *The Theory of the Leisure Class*, New York: Mentor Books.

Vickers, L. (1998) 'Trends in migration in the UK', *Population Trends* 94: 25–34.

Vidal de la Blache, P. (1922) *Principes de Géographie Humain*, Paris: Librairie Armand Colin.

Viner, K. (1994) 'Girls just wanna have fun', *The Guardian*, 15 December, p. 14.

Vulliamy, E. and Thompson, T. (1999) 'They're not afraid to kill or to die. Now Yardies are doing both here', *The Observer*, 18 July, p. 10.

Wainwright, M. (1999) 'Revolutionary Yorkshire-men edge towards devolution', *The Guardian*, 18 March, p. 11.

Walby, S. (1999) 'The new regulatory state: the social power of the European Union', *The British Journal of Sociology* 50(1): 118–40.

Walker, A. (1998) 'Kirk urged to sanction cohabiting', *The Scotsman*, 27 July, p. 6.

Walker, A. and Walker, C. (1997) *Britain Divided: The Growth of Social Exclusion in the 1980s and 1990s*, London: CPAG.

Walker, D. (1999) 'It's coming home, maybe', *The Guardian*, 13 April, p. 15.

Wallerstein, I. (1983) *Historical Capitalism*, London: Verso.

Walter, N. (1999) 'Don't change us, Change the world', *The Independent*, 30 June, p. 4.

Walters, J. (1999) 'Transport in chaos: fatal congestion at the heart of Britain', *The Observer*, 20 June, p. 18.

Ward, L. (2000) 'Typical woman earns £250,000 less than a man', *The Guardian*, 21 February, p. 5.

Warren, K. (1980) 'North East England', in G. Manners, D. Keeble, B. Rodgers and K. Warren (eds) *Regional Development in Britain* (2nd edition), Chichester: John Wiley & Sons, pp. 361–81.

Watson, W. and Sissons, J.B. (1964) *The British Isles: A Systematic Geography*, London: Nelson.

Wazir, B. (1999) 'Young, bored and pregnant', *The Observer*, 5 September, p. 18.

Webb, P. and Farrell, D.M. (1999) 'Party members and ideological change', in G. Evans and P. Norris (eds) *Critical Elections: British Parties and Voters in Long-Term Perspective*, London: Sage, pp. 44–63.

Weeks, J. (1985) *Sexuality and its Discontents*, London: Routledge.

Weiss, L. (1997) 'Globalization and the myth of the powerless state', *New Left Review* 225: 3–27.

Werther, C. (1997) 'Imaginary communities? The discourse of community in English local government restructuring', *Journal of Area Studies* 10: 24–34.

White, M. (2000) 'Blair admits case for regional governments', *The Guardian*, 29 March, p. 13.

Whitehead, T. (2000) 'Pledge to rid the city streets of menace', *Evening Post*, 25 May, p. 10.

Wilkinson, R.G. (1996) *Unhealthy Societies: The Afflictions of Inequality*, London: Routledge.

Willey, S. (2000) 'Booze blamed for violence in Notts', *Evening Post*, 5 January, p. 13.

Williams, C.C. and Windebank, J. (1999) 'The formalisation of work thesis: a critical evaluation', *Futures* 31: 547–58.

Williams, R. (1958) *Culture and Society 1780–1950*, London: Chatto & Windus.

——— (1981) *Culture*, London: Fontana.

Williams, T. (1989) 'The Anglicisation of South Wales', in R. Samuel (ed.) *Patriotism: The Making and Unmaking of British National Identity. Volume II, Minorities and Outsiders*, London: Routledge, pp. 193–203.

Wilson, M.G.A. (1978) 'A spatial analysis of human fertility in Scotland: reappraisal and extension', *Scottish Geographical Magazine* 94: 130–43.

Winter, M. (2000) 'Strong policy or weak policy? The environmental impact of the 1992 reforms to the CAP arable regime in Great Britain', *Journal of Rural Studies* 16: 47–59.

Wintour, P. (1998) 'Blueprint for a new Britain', *The Observer*, 1 November, p. 21.

Wintour, P. and Bright, M. (1999) 'Underage girls to get Pill at school', *The Observer*, 9 May, p. 1.

Women and Geography Study Group (1984) *Geography and Gender: An Introduction to Feminist Geography*, London: Hutchinson.

Wood, P.A. (1987) 'The South East', in P. Damesick and P. Wood (eds) *Regional Problems, Problem Regions, and Public Policy in the United Kingdom*, Oxford: Clarendon Press, pp. 64–94.

Woodward, R. (1995) 'Approaches towards the study of social polarization', *Progress in Human Geography* 19(1): 75–89.

Wright, T. (2000) 'Introduction: England, whose England?', in S. Chen and T. Wright (eds) *The English Question*, London: The Fabian Society, pp. 7–17.

Wrightson, K.E. (1989) 'Kindred adjoining kingdoms: an English perspective on the social and economic history of early modern Scotland', in R.A. Houston and I.D. Whyte (eds) *Scottish Society, 1500–1600*, Cambridge: Cambridge University Press, pp. 245–60.

Wrigley, N. and Lowe, M. (1997) *Retailing, Consumption and Capital: Towards the New Retail Geography*, Harlow: Longman.

—— (1999) 'New landscapes of consumption', in J. Bryson, N. Henry, D. Keeble and R. Martin (eds) *The Economic Geography Reader*, Chichester: Wiley, pp. 311–14.

Wyness, F. (1971) *Aberdeen: Century of Change*, Aberdeen: Impulse Books.

Yorkshire Forward (2000) http://www.yhrda.com/home.html

Young, H. (2000) 'What is Britishness? Tories dream while Labour defines', *The Guardian*, 28 March, p. 22.

Younge, G. (2000a) 'On race and Englishness', in S. Chen and T. Wright (eds) *The English Question*, London: The Fabian Society, pp. 111–16.

—— (2000b) 'Fractured Britain hits Blair's electoral plan', *Red Pepper*, February, http://www.redpepper.org.uk/keynote.html

Zukin, S. (1982) *Loft Living: Culture and Capital in Urban Change*, London: Century Hutchinson.

—— (1991) *The Landscapes of Power: From Detroit to Disneyworld*, Berkeley: University of California Press.

Index

Page numbers in *italic* refer to *figures* and *tables*